AF839487

Green Energy and Technology

For further volumes:
http://www.springer.com/series/8059

Maria Carmela Di Piazza · Gianpaolo Vitale

Photovoltaic Sources

Modeling and Emulation

Springer

Maria Carmela Di Piazza
National Research Council
Institute of Intelligent Systems
for Automation
Palermo
Italy

Gianpaolo Vitale
National Research Council
Institute of Intelligent Systems
for Automation
Palermo
Italy

ISSN 1865-3529 ISSN 1865-3537 (electronic)
ISBN 978-1-4471-4377-2 ISBN 978-1-4471-4378-9 (eBook)
DOI 10.1007/978-1-4471-4378-9
Springer London Heidelberg New York Dordrecht

Library of Congress Control Number: 2012943366

© Springer-Verlag London 2013

MATLAB® and Simulink® are registered trademarks of The MathWorks, Inc. See www.mathworks.com/trademarks for a list of additional trademarks
PLECS is a registered trademark of Plexim GmbH

This work is subject to copyright. All rights are reserved by the Publisher, whether the whole or part of the material is concerned, specifically the rights of translation, reprinting, reuse of illustrations, recitation, broadcasting, reproduction on microfilms or in any other physical way, and transmission or information storage and retrieval, electronic adaptation, computer software, or by similar or dissimilar methodology now known or hereafter developed. Exempted from this legal reservation are brief excerpts in connection with reviews or scholarly analysis or material supplied specifically for the purpose of being entered and executed on a computer system, for exclusive use by the purchaser of the work. Duplication of this publication or parts thereof is permitted only under the provisions of the Copyright Law of the Publisher's location, in its current version, and permission for use must always be obtained from Springer. Permissions for use may be obtained through RightsLink at the Copyright Clearance Center. Violations are liable to prosecution under the respective Copyright Law.
The use of general descriptive names, registered names, trademarks, service marks, etc. in this publication does not imply, even in the absence of a specific statement, that such names are exempt from the relevant protective laws and regulations and therefore free for general use.
While the advice and information in this book are believed to be true and accurate at the date of publication, neither the authors nor the editors nor the publisher can accept any legal responsibility for any errors or omissions that may be made. The publisher makes no warranty, express or implied, with respect to the material contained herein.

Printed on acid-free paper

Springer is part of Springer Science+Business Media (www.springer.com)

To my parents whose example teaches me to do anything as best as I can
To my children whose smile gives meaning to my every effort

Maria Carmela Di Piazza

To my child Gianluigi

Gianpaolo Vitale

Preface

Photovoltaic (PV) industry is growing exponentially and the availability of specialized equipment for the laboratory tests is becoming crucial especially for manufacturers and laboratories working in the field of power electronics related to PV plants.

All people, concerned with renewable sources' applications, experience the need of laboratory equipment to carry out measurements and tests. In particular, in the field of PV generators, the difficulties are tied to the use of an actual outdoor plant for performing tests involving the PV source as well as various kinds of loads, such as resistive loads, DC motors, storage batteries, and inverter-connected loads with their maximum power point trackers (MPPT).

A real PV plant needs a wide outer surface and high costs. Moreover, its produced energy is strongly dependent on uncontrollable weather conditions. Finally, the necessary investment to set up an experimental plant for testing purpose can be made more severe by the continuous evolution of the involved technologies. Therefore, an alternative solution to the use of an actual outdoor plant has to be taken into consideration.

The alternative is the setup of a PV emulator, i.e., a suitable laboratory equipment in which voltage and current are the same as in a real source. Such an experimental facility allows measurements and tests to be carried out without the constraints of the environmental conditions and, above all, more cheaply, since the use of an actual PV array is avoided.

Considering that the trend in PV industry moves toward more efficient PV plants, the setup of effective and reliable PV emulators is becoming more and more a challenging issue.

The emulation of a PV generator is performed by two main tasks: the first one is the knowledge of the electrical characteristics of the generator; the latter consists in their realization by a power amplifier.

On the score of such considerations, this book provides an extensive introduction to the modeling of PV generators and to their emulation by means of power electronic converters.

The authors have first focused on the definition of accurate PV source models, including the parameter extraction techniques. Then, they have devised a system to reproduce the behavior of a PV source, by using the characteristic voltage–current curves, obtained by the modeling, as a reference to suitably control a switching DC/DC converter.

The book collects the results of many years of research and development of the authors at the CNR (National Research Council)—ISSIA (Institute of Intelligent System for Automation), Renewable Sources Laboratory (RES).

The primary aim of the book is to guide the reader to design and set up a PV emulator based on controlled power converters, starting from an appropriate modeling of the PV source. The reader is thus allowed to manage a virtual plant in which the dependence of the PV source electrical behavior on weather conditions, partial shadow, and dynamics is accounted for.

On the other hand, the book gives an in-depth tool and an extensive reference in modeling PV generators and the more common topologies of DC/DC converters.

It is worth considering that the overall emulation concept, adopted in this book with reference to PV sources, is general; therefore, it can be easily extended to the other kind of electrical sources, such as fuel cells, batteries, or wind turbines with permanent magnet motor in which the DC output voltage is obtained by a rectifier.

Outline of the Book

The book is divided into two parts.

Part I embraces the chapters from 1 to 5; it is dedicated to the PV source modeling.

Part II includes the chapters from 6 to 8; it refers to PV emulation issues, including design and realization.

Part I

Chapter 1. This chapter deals with the energy generated inside the Sun that reaches the Earth. It is a fantastic trip in which the reader is involved with the two main concepts of very big (the scale of the universe and the produced energy) and of very small (inside semiconductor).

Chapter 2. After its trip as electromagnetic wave, energy reaches the Earth. Now, it is transformed into electric energy. This chapter deals with the behavior of matter, in particular of semiconductor materials, and explains how it can interact with the Sun's irradiance to obtain voltage and current at the terminals of a photovoltaic cell.

Chapter 3. In this chapter, the fundamentals on the derivation of PV source circuit models, including parasitic resistance effects and junction non-idealities, are given. Then, the model is extended and generalized to the case of single PV modules and their connection to form a PV field, considering the effects of partial shading as well.

Chapter 4. This chapter is devoted to parameters extraction methods for PV static models. The considered identification methods are mainly based on analytical or numerical solutions and use either the rated data given by the manufacturer or the experimental voltage-current (V-I) curves, directly measured at the PV source. A simplified method based on experimental measurement of the remarkable points is presented too. The possibility to simplify the parameter identification by using linear regression methods is explored and some hints on the PV characteristic determination by mapping techniques are given. The PV models, suitably identified, are finally implemented in Matlab/Simulink® environment and simulation results are shown.

Chapter 5. This chapter treats the dynamic modeling of PV sources and methods for the identification of its parameters on the basis of experimental measurements.

First, a dynamic model, including capacitive and inductive effects, is analyzed and an identification method based on least square regression (LSR) is proposed. Finally, the nonlinear junction capacitance effects are accounted for.

The corresponding models are implemented both in Simulink® and PLECS® software. Some practical examples show how transient phenomena can be investigated.

Part II

Chapter 6. This chapter gives an overview of PV emulators proposed in the technical literature in the last decades. The most critical and interesting aspects related to their design and operation are emphasized, including hardware structures, control features, dynamics, power ratings, and modularity. A brief survey of commercial solutions is included.

Chapter 7. This chapter deals with three main topologies of DC/DC converters: buck, boost, and buck-boost. After a short presentation of linear conversion, fundamentals on switching conversion are explained. The DC/DC converters are analyzed in continuous and discontinuous operating mode and main parameters as gain, critical inductance, and load voltage ripple are calculated. Then, the state space averaging approach is used to achieve the transfer functions in terms of output voltage versus duty cycle and versus input voltage.

Chapter 8. This chapter treats the design and realization of a DC/DC converter for emulating purpose. The chapter contains the fundamentals of feedback control and an example of emulator design, realized by a buck converter. In particular, both the PV model and the converter are at first simulated in Matlab-PLECS® environment. Then, the practical implementation of the control algorithm on a DSP board and the power circuit are illustrated and finally experimental results are presented.

Originality of the Approach

The book has the specificity to propose a novel method for the analysis of several issues related to PV plants operation, doing without the use of a real installation. This approach, leading to the design and setup of a virtual indoor photovoltaic generator, i.e., the PV emulator, implies the transversal investigation of several disciplines going from solar cell physics to power electronics and control. In this sense, the book gives a new perspective on this subject. As a matter of fact, these disciplines are presented in a synergic way to solve the problem of PV source emulation.

The reader can find, joined in a unique text, all information related to the emulation of a PV source (model choice, parameters identification, software implementation, design and control of the power amplifier) without the necessity of deducing them from different specific books.

Moreover, the PV source emulation approach can be extended to other kind of sources as fuel cells, batteries, or wind turbines with permanent magnet motor connected to a rectifier circuit, simply by using the corresponding model and maintaining the same hardware.

To Whom this Book is Addressed

This book is intended primarily to meet the demands of postgraduate level students but it should also prove useful to the professional engineers and researchers dealing with the problem of creating an indoor virtual PV plant for testing MPPT techniques, non-ideal operating conditions, control for the grid or load interfacing, etc. Moreover, the book can be useful to:

- Researchers and Engineers who work in the field of renewable sources;
- Engineers and Physicists who deal with PV sources modeling;
- Mathematicians and Statistics who work with parameters identification techniques.

How to Use the Book

First of all, this book represents a tool to obtain a PV source model and to reproduce it by a power converter.

Chapter 1 gives an overview of the Energy conversion from the Sun to the Earth.

The readers interested on modeling issues can read the first part, in particular Chap. 2 and 3; furthermore, they can utilize Chap. 4 to identify the related parameters.

The readers interested on emulation of the source can read the Chap. 6 if they want to have the scenario on existing solutions, or directly Chaps. 7 and 8 to design the DC/DC converter for emulation of the source. It should be noted that the approach used for emulating a PV source can be extended to other sources having their characteristics lying on I–V plane.

Prerequisites

Readers should be familiar with basic semiconductor physics, circuit analysis, and with the fundamentals of DC/DC converters circuits and control. The book has been conceived as an advanced textbook, designed primarily to meet the demands of a course taught at postgraduate (M.Sc., Ph.D. etc.) level. However, the authors have made an effort to present each subject to be understandable by MSc students. For this reason, the book can be also used as supporting test in some Engineering or Physics courses.

Acknowledgments

The authors are very grateful to Mr. Antonio Sauro and Mr. Giuseppe Scordato, technicians at CNR-ISSIA, for their irreplaceable contribution in building the emulator prototype.

The authors are very grateful to Dr. Massimiliano Luna, who has co-authored Chap. 5 and performed simulations included in the same chapter.

The authors would like to express their gratitude to Prof. Manuel Pèrez Donsiòn (Electrical Engineering Department of Vigo University, Spain) for the help and for the yearly organization of International Conference on Renewable Energy and Power Quality (ICREPQ) that has been an extraordinary occasion to discuss and to improve many issues treated in this book.

The authors would like to thank Prof. Jon C. Clare (Faculty of Engineering, University of Nottingham, UK) for the encouragement to undertake the writing of this book.

The authors are grateful to CNR-ISSIA for institutional support.

The authors would like to acknowledge the staff of Springer for their timely and accurate assistance and for their kindness.

Finally, we would like to express our sincere thanks to all the people who helped us, lightening some of our daily duties or simply putting up with us patiently, during the writing of this book.

Contents

Part I

Part I

Chapter 1
From the Nuclear Fusion to the Radiated Energy on the Earth

1.1 Inside the Universe

The Sun is the central body of the solar system. It is informally designed as a *yellow dwarf star*. It belongs to spectral type *G2V* where *G2* is referred to its temperature surface (approximately 5778 K) and *V* means that the Sun is a main sequence star and generates its energy by fusion of Hydrogen into Helium. Other *GV* stars are Alpha Centauri A, Tau Ceti, and 51 Pegasi. It should be noted that the Sun is white, however, it appears as yellow from the Earth for Rayleigh scattering effect through the atmosphere.

The Sun shares the galaxy's rotating motion, moreover, it shows a further rotating motion around its axis that is inclined of 7° 15′ on the ecliptic plane. Since the Sun has to be considered as a gas sphere rather than a solid body, the angular speed cannot be constant, it is lower on the poles and greater on the equator, the rotation period is about 25 days.

The Sun is a typical star whose properties have been studied more deeply compared to other stars for the exceptional closeness to the Earth. Fortunately, solar physics coincides with the physics of the great part of our galaxy stars and, with good chance, with that of the great part of the other galaxies.

Due to the turn of the Earth (that is our point of observation) on its axis, to the revolution round the sun and finally to the distance from the Earth to the Sun and from the Sun to the other stars, we experience an apparent motion of the Sun and of the stars. This can be explained by considering that even if stars are always in the same direction from us all the day and all the year (this is the reason for which they are referred as fixed stars), the Sun slightly changes its direction from us all day, from its rising to its setting during the year as a consequence of the Earth revolution around it.

Figure 1.1 shows (not in scale) the fixed stars, the Sun and the Earth with its motion around the Sun in two position named A and B. From the point of observation A, the Sun appears between the Aries and Pisces constellation; a month later with a rotation of $360°/12 = 30°$, the Earth will be situated in the point B and the

M. C. Di Piazza and G. Vitale, *Photovoltaic Sources*, Green Energy and Technology, DOI: 10.1007/978-1-4471-4378-9_1, © Springer-Verlag London 2013

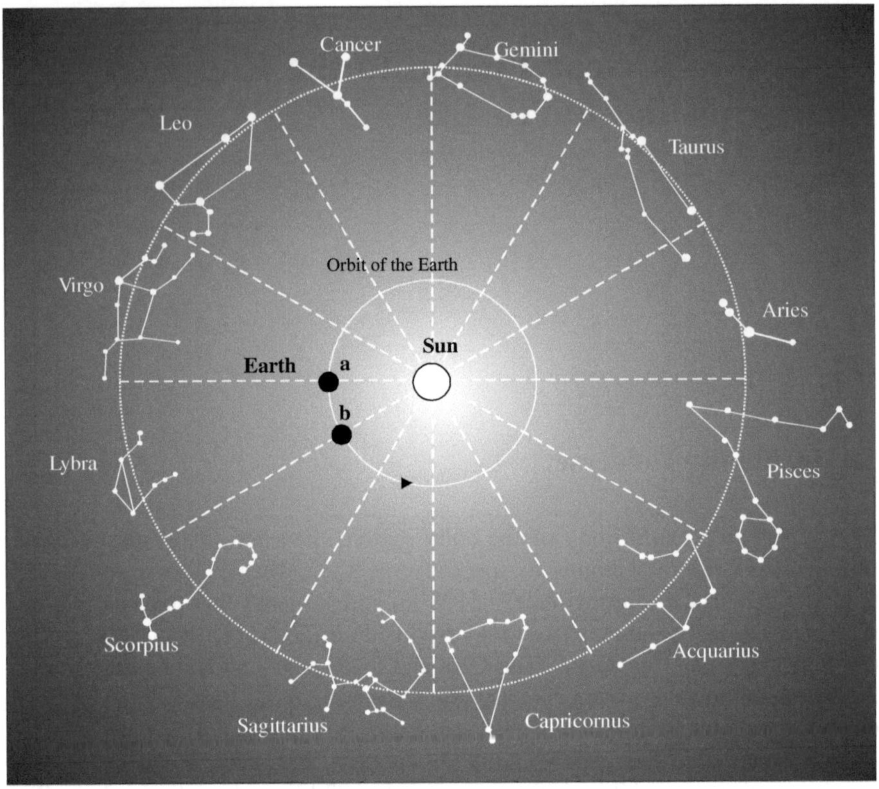

Fig. 1.1 Motion of the Earth around the Sun with constellations on the background

Sun will appear between Aries and Taurus; and so on. As soon as the Earth's rotation will be completed, the Sun will have apparently covered a complete circle around the celestial sphere. The circle thus described by the Sun on the celestial sphere is known as the *ecliptic*.

The ecliptic plane and the plane of the Earth's equator are inclined each other of an angle of 23½°. This inclination is crucial for Earth climate behavior, it defines seasons, it implies that the Sun is higher in the sky (and gives higher sunlight hours) during the summer than during winter.

As it is shown in Fig. 1.2, the ecliptic and the equatorial plane have two points in common that define the Vernal and Autumnal equinoxes.

When the Sun is at the Vernal Equinox, it rises exactly east, and sets exactly west. The days are as long as the nights. In the next 6 months, when the Sun passes from Aries to Virgo it rises north of east and sets north of west; in the northern hemisphere, the days are longer than the nights.

When the Sun passes from Gemini into Cancer, it is on the Summer Solstice, it has reached its greatest north declination and appears at 23.45° N latitude of the equator. Then, it moves toward Libra and there it crosses the equator on Autumnal

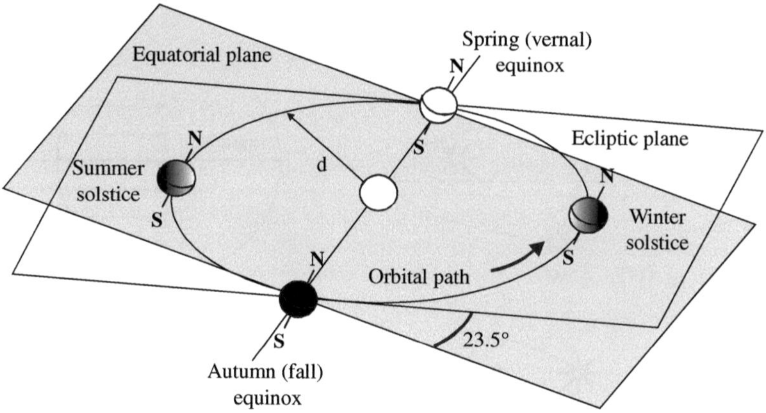

Fig. 1.2 Motion of the Earth around the Sun

Equinox. During the next 6 months the Sun goes from Libra to Pisces, it rises south of east and sets south of west. In the northern hemisphere the nights are then longer than the days. When the Sun passes from Sagittarius into Capricornus, it reaches the Winter Solstice in which it has the greatest south declination, appearing at 23.45° S latitude of the equator and begins to return toward the equator.

The different tracks of the Sun during the year are sketched in Fig. 1.3. It should be noted that in any case the Sun culminates always at south and covers an angular displacement of 15° per hour.

1.2 The Sun

1.2.1 Inside the Sun

The Sun has not exhibited variation in the last billion of years (3×10^{16} s); it means the presence of a force that is opposite to its force of gravity. As a matter of fact the collapse time can be evaluated as $^1/_{\sqrt{G\bar{\rho}}}$ where G is the Newton's gravitational constant)[1] and $\bar{\rho}$ is the mean density. For $G \approx 6{,}7 \times 10^{-11}$ N (m/kg)^2 and $\bar{\rho} = 1{,}4 \times 10^3$ kg/m^3 the collapse time is lower than 1 h. The antagonist force is due to thermal pressure gradient between its center and the surface. This thermal gradient is sustained by nuclear reaction that converts hydrogen to helium.

[1] The symbol G has been used in Sect. 1.1 to identify the Sun spectral class. It is commonly used for Newton's gravitational constant as well, however, it should not confuse the reader because the context in which the two symbols are used is different. In the next chapters, G will denote the solar irradiance and neither the Sun spectral class nor Newton's gravitational constant will be utilized any more.

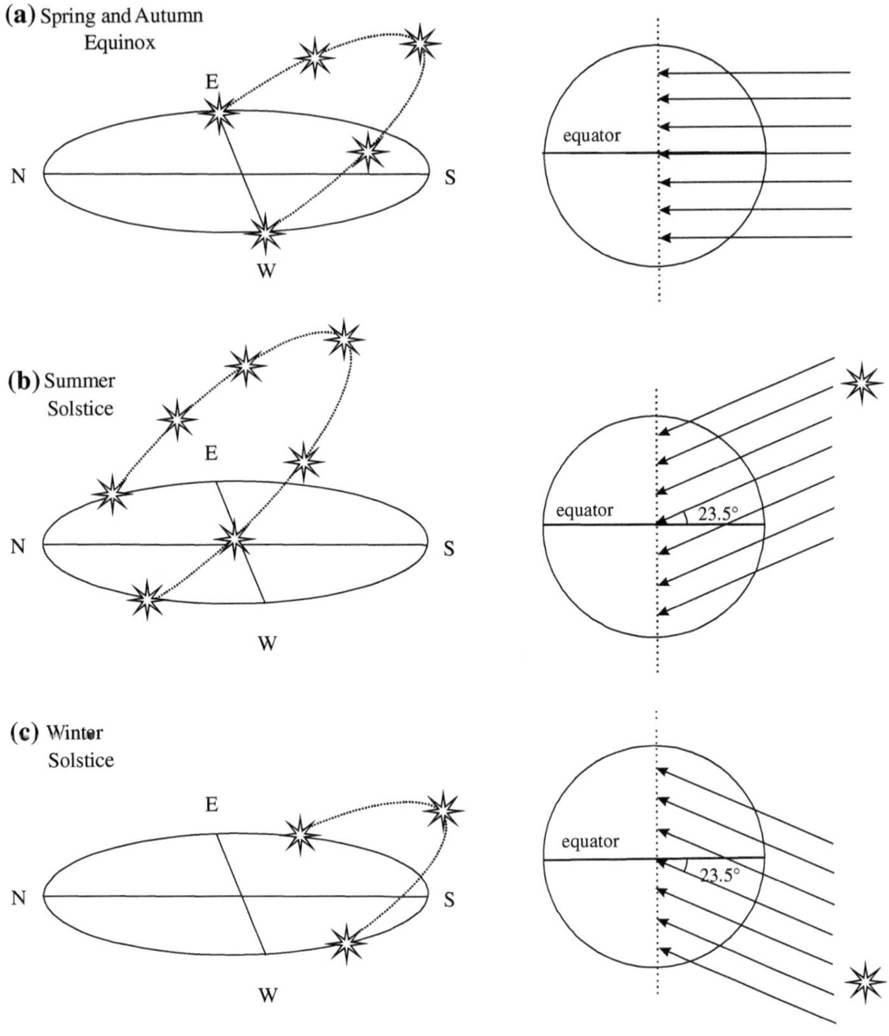

Fig. 1.3 Different positions of the Sun during the year. **a** spring and autumn equinox: the Sun rises at east and goes down at west, day and night have the same length. **b** summer solstice: the Sun rises at north-east and goes down at north-west, days are longer than the nights. **c** winter solstice: the Sun rises at south-east and goes down at south-west, nights are longer than the days. In any case, the Sun culminates at south

As a matter of fact, the Sun is formed by hydrogen (75 %), helium (23 %), and some heavy elements (2 %). Inside the Sun, the nuclear fusion of four hydrogen nuclei (protons) into a helium nucleus occurs.

In particular, first, a couple of protons are transformed into a deuterium nucleus and a positive electron. Then, the deuterium nucleus is aggregated with a proton resulting in a helium nucleus. This last nucleus is formed by two protons and one

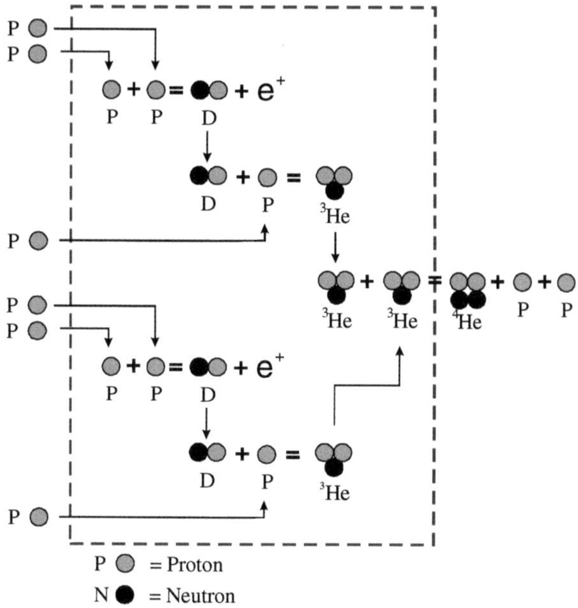

Fig. 1.4 Nuclear fusion inside the Sun

neutron and it is indicated as ^{3}He. As soon as this transformation occurs twice, the two ^{3}He nuclei form a helium nucleus with two neutrons or ^{4}He and a couple of protons are obtained.

The whole reaction has six protons as inputs and gives a ^{4}He nucleus and two protons as output. By comparing the mass of the four protons (4×1.00731 amu) where 1 amu $= 1.66 \times 10^{-27}$ kg, and the mass of ^{4}He nucleus, equal to 4.002063 amu, it can be noted that the resulting mass is reduced by a quantity named the mass defect of the Helium nucleus. The lacking mass has been transformed into energy according to the Einstein's law $E = mc^2$.

Each second, 6×10^{11} kg of hydrogen are transformed into Helium with mass loss of 4×10^3 kg that gives 4×10^{20} J. This energy raises the internal temperature of the Sun up to about 5×10^6 K.

The nuclear fusion steps are drawn in Fig. 1.4.

The temperature decreases toward the surface with a gradient of about 10^{-4} K/cm.

The thermal radiation inside the Sun is not visible, during the way to external surface it is scattered and absorbed. Considering the scattering of free electrons inside the Sun (Thompson effect), the free length path is $\lambda \cong 10^{24}/n$ cm, where n is the density of free electrons equal to 10^{24}. As a consequence ($\lambda = 1$ cm) the Sun appears very opaque. In spite of its low value, the temperature gradient supplies the energy flux against the Sun opacity and maintains the external surface at a temperature of about 5800 K. Now, energy is radiated uniformly in all directions following the Plank's blackbody radiation formula:

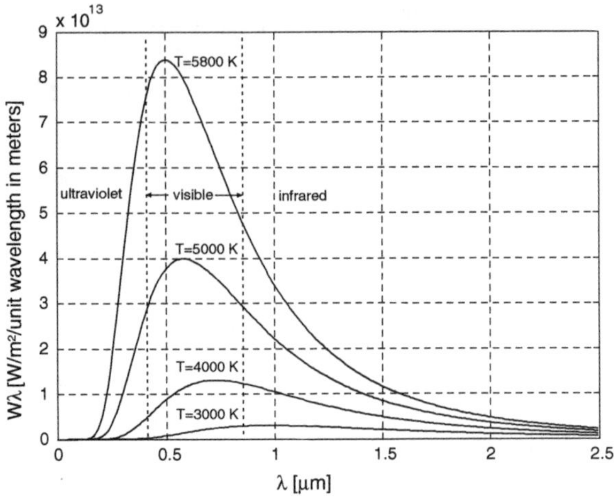

Fig. 1.5 Plank's blackbody radiation versus wavelength for different temperatures

$$W_\lambda = \frac{2\pi hc^2}{\lambda^5\left(e^{\frac{hc}{\lambda kT}} - 1\right)} \tag{1.1}$$

where h is the Plank's constant equal to 6.63×10^{-34} J.s and k is the Boltzmann's constant equal to 1.38×10^{-23} J/mol·K. It should be noted that Eq. 1.1 is expressed in [W/m^2/unit wavelength in meters], its shape depends on temperature and wavelength. Figure 1.5 shows some different curves representing the Plank's blackbody radiation formula obtained for different values of temperature, T. For T = 5800 K, a great amount of energy is inside the visible spectrum, this gives the characteristic white color of the Sun. Higher temperature implies a blue shift, on the contrary lower temperature gives a red shift. For T = 3000 K, nearly all energy is in the infrared spectrum. This last aspect will be taken again into consideration in Sect. 1.6. Finally, integrating from $\lambda = 0$ to infinity (Eq. 1.1) for T = 5800 K the power density coming from the Sun is obtained; this value, calculated on the outside atmosphere surface, is equal to 1376 W/m^2.

1.2.1.1 How Long Does it Occur?

The value of 1367 W/m^2 before entering into the atmosphere, considering an Earth–Sun distance of 1.5×10^{11} m, corresponds to the power supplied by the Sun equal to 3.86×10^{26} J/s.

On the other hand, the energy corresponding to the transformation of 1/10 of the Sun's mass, calculated by the Einstein law E = mc^2, is equal to 9.2×10^{43} J. Hence, the Sun will consume 1/10th of its hydrogen mass within 7,5 billion of years. This last value is greater that the expected life of the Earth. The solar energy, therefore, can be successfully exploited without to be afraid of depletion.

1.2.1.2 Clean Energy Issues

Photovoltaic energy is, without any doubt, a form of clean energy. However, the reader should be borne in mind that it is based on another form of clean energy available by the Sun. As a matter of fact, the nuclear fusion that occurs in the Sun requires a couple of hydrogen nuclei and produces one deuterium nucleus, one electron and releases about 1.4 MeV. This value corresponds to about 2×10^6 times the energy given by combustion of the same mass of coal.

Unfortunately, in normal conditions, two hydrogen nuclei cannot be in touch for the electric repulsion. On the contrary, inside the Sun, the extremely high pressure and temperature conditions make possible this nuclear reaction.

An alternative form of clean energy is given by annihilation in which a couple of particle–antiparticle becomes energy transforming their mass.

Considering, as example, a couple proton–antiproton, annihilation gives an energy equal to $E = mc^2 = 2 \times 1.67 \times 10^{-27}(3 \times 10^8)^2 = 3\ 10^{-10}J = 1872$ MeV. This value is greater compared to that obtained by nuclear fusion of about 10^3.

Both nuclear fusion and annihilation are common in the Universe but improbable on the Earth and it is not foreseeable their exploitation in the next future for energy generation.

1.3 From the Sun to the Earth

Energy coming from the Sun travels for 150×10^6 km before encountering the Earth's atmosphere. Before this it is practically unchanged, but from this boundary several phenomena occur.

Inside the atmosphere, a part of sunlight is absorbed from molecules that raise their own energy. It means that some wavelengths are lessened and the curve of Fig. 1.5 exhibits some hole known as absorption lines. For example, the presence of ozone produces absorption in the ultraviolet region while water vapor and carbon dioxide absorb mainly in the visible and infrared parts of the spectrum.

Scattered sunlight makes blue the sky color and gives us sunlight from north when the Sun is in the south. Without sunlight scattering, the sky would appear black as during the night with the moon. Even with a light shadowing, sunlight is scattered and these rays cannot be focused as the direct rays.

Finally, at ground level, sunlight can be either absorbed or reflected.

Nomenclature for some of the above described situations has been introduced in the literature. The *direct* or *beam* radiation is the sunlight that reaches Earth without scattering while scattered sunlight is defined as *diffuse* radiation. Sunlight reflected from ground is called *albedo* radiation and the sum of direct, diffuse, and albedo radiation is the *global* radiation.

It is clear that the path length through atmosphere plays an important role in the spectrum of sunlight at Earth level. Conventionally, the vertical path length directed to the sea level is defined as *air mass* = 1 (AM1). If Sun rays are inclined,

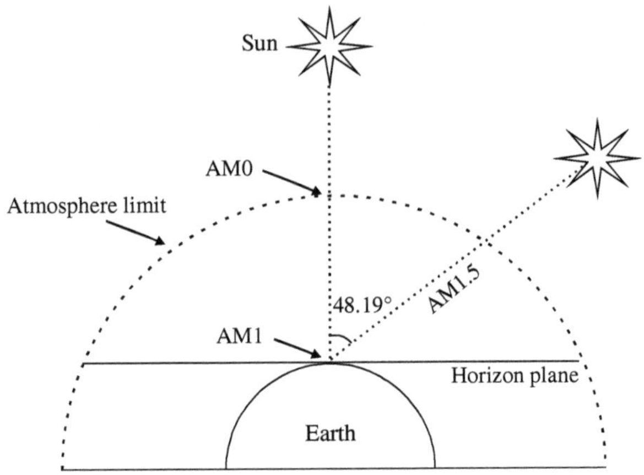

Fig. 1.6 Different values of AM vs. solar position

the AM value is greater than one. Considering the absorption at AM1, the power density coming from the Sun is diminished of about 70 %; it is slightly greater than 1000 W/m^2.

The ratio of monthly average daily total horizontal insulation on the Earth's surface $\bar{H}$ to the monthly average daily total horizontal insulation outside the atmosphere $\bar{H}_{ext}$ is defined as *clearness index* K_T. It is a value that varies with the place and during the year. It includes the effects of the weather, diffuse, reflected direct radiation as well.

$$K_T = \frac{\bar{H}}{\bar{H}_{ext}} \tag{1.2}$$

For AM different from unity, power density I [W/m^2] can be deduced by:

$$I = 1367(0.7)^{AM^{0.678}} \tag{1.3}$$

The value of AM1.5 is used as a standard calibration spectrum for photovoltaic cells. It corresponds to the spectrum obtained when the Sun forms an angle of 48.19° with the line perpendicular to the earth in the considered point. Figure 1.6 shows different values of AM.

If photovoltaic performance is evaluated under conditions that differ from AM1.5, a calibration factor must be adopted. In general, this correction factor gives the short-circuit current taking into account environmental conditions as AM and water vapor and turbidity. It should be borne in mind that the calibration factor value is influenced by the semiconductor used to set up the photovoltaic generator, for example amorphous silicon has a much higher sensitivity to water and turbidity than crystalline Si, CdTe, or $CuInSe_2$. In any case, the calibration factor is more accurate for conditions near to AM1.5.

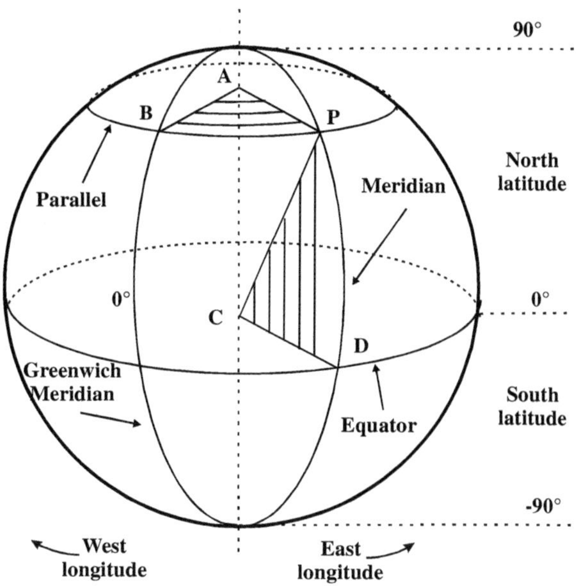

Fig. 1.7 Representation of latitude and longitude

1.3.1 Finding One's Bearings on the Earth

The Earth seen by the Sun appears always as a dot. However, the effects of the Sun are very different depending on the point of observation on the Earth surface where a dedicate coordinate system is fundamental to define the position and the related effects tied to solar radiation.

Two parameters have to be introduced: latitude and longitude.

Latitude (abbreviation: Lat., φ, or phi) is the angle between the equatorial plane and a line that is normal to the reference ellipsoid, which approximates the shape of the Earth. The loci of the points with the same latitude are called *parallels*. Parallels trace circles on the surface of the Earth, parallel to the equator. Conventionally, the north pole is 90° N; the south pole is 90° S. The equator defines 0° parallel and the fundamental plane of all geographic coordinate systems. The equator divides the globe into Northern and Southern Hemispheres.

The half circles joining the two poles that cut perpendicularly the equator are defined as *meridians*. A line passing to the rear of the Royal Observatory, Greenwich (near London in the UK) has been chosen as the international zero-longitude reference line, the Prime Meridian.

Longitude (abbreviation: Long., λ, or lambda)[2] is the angle east or west of a reference meridian between the two geographic poles to another meridian that passes through an arbitrary point. All meridians are halves of great circles, they converge at the north and south poles. Places to the east are in the eastern

[2] The Longitude symbol should not be confused with the wavelength.

hemisphere, and places to the west are in the western hemisphere. The antipodal meridian of Greenwich is both 180°W and 180°E.

The latitude/longitude "webbing" is known as the conjugate graticule. Figure 1.7 shows a representation of latitude and longitude on the Earth.

1.3.2 The Greenhouse Effect

Energy received by Earth during the day is re-radiated during the night into space. For maintaining its constant temperature, a thermal equilibrium is necessary. However, the Earth has a mean temperature of 300 K, hence it can reradiate at long wavelength compared to the Sun. The atmosphere has a crucial role in this equilibrium. It should be borne in mind that, without atmosphere (like on the moon), the Earth's temperature would fall about 255 K, corresponding to −18 °C. With the atmosphere, a natural background of 270 ppm of CO_2 maintain the mean temperature of 300 K.

The balance can be compromised by "greenhouses gases" as carbon dioxide, methane, and chlorofluorocarbons (CFCs) that are mostly transparent to Sun's wavelength but, absorbing the short wavelength radiation, prevent Earth to reradiate during the night. As a consequence, Earth's temperature rises; this phenomenon is known as global warming or greenhouse effect.

1.4 Tracking the Sun

As soon as solar energy has reached the Earth, the target is to use the greatest possible part of it to produce energy under other form as electric energy.

As it is shown in the above sections, the Sun position and height are continuously varying during the day and during the year and depend on the position of the observer.

To this aim a mathematical description of the motion of the Sun has to be set up.

The power density of sunlight is defined as *irradiance.* The irradiance received by the Earth from the Sun at the top of the atmosphere (AM0) is defined as *solar constant* and it is equal to 1367 W/m^2.

The energy density of sunlight is defined as *irradiation* and it is expressed in [kWh/m^2]. The irradiation is obtained by integrating the irradiance over time, usually the time interval of integration is referred to one day and it is performed on daylight hours. Another way to express irradiation consists on *peak sun hours* (psh). It is the time interval, in hours, that is necessary to an irradiance of 1 kW/m^2 to produce the daily irradiation obtained by integrating the real irradiance over the daylight hours of the same day. The daily irradiation is numerically equal to the daily psh.

Figure 1.8 shows the irradiance plot measured in a real situation (photovoltaic plant located at 38°06′15.17″ N, 13°20′49.24′ E.) and corresponding irradiation. It

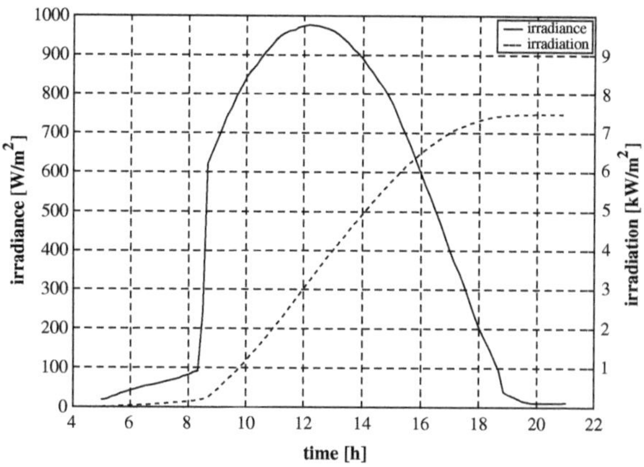

Fig. 1.8 Irradiance and irradiation plot

should be noted that the typical "bell shaped" curve is deformed due to the presence of a wall.

With reference to Fig. 1.2, the distance from the Earth to the Sun d [m] can be described by Eq. 1.4 that represents an elliptic orbit with the Sun placed in one of the foci.

$$d = 1{,}5 \times 10^{11}\left\{1 + 0.017\sin\left[\frac{360(n-93)}{365}\right]\right\} \tag{1.4}$$

The parameter n represents the number of the day, where the first of January is set for $n = 1$. It should be noted that the distance varies slightly, as a consequence its mean value of 1.5×10^{11} m is assumed as the distance.

The angle formed by a line from the center of the Earth to the Sun and the equatorial plane is defined as *declination* δ. The declination, whose value is null on Spring and Autumn equinox, is equal to 23.45° on Summer solstice (assuming as positive angles in north direction respect to the equator) and it is equal to −23.45° on Winter solstice.

During the year the declination varies as:

$$\delta = 23.45^\circ \sin\left[\frac{360(n-80)}{365}\right] \tag{1.5}$$

Even if Eq. (1.4) is an approximation (the year is assumed 365 days long and $n = 80$ is the first day of Spring), it can be used successfully.

A line perpendicular to the Earth is defined as *zenith*. The zenith angle θ_z is the angle between a straight line passing from the center of the Earth to the Sun and the zenith.

The *solar noon* is defined as the time in which the Sun is at its highest point in the sky. For given time zone, solar noon occurs at 12 only for one point at a given latitude; at longitude east (west) of this point solar noon occurs before (after) 12. It can be noted from Fig. 1.3 that at solar noon a shadow points directly to north or south, depending on latitude.

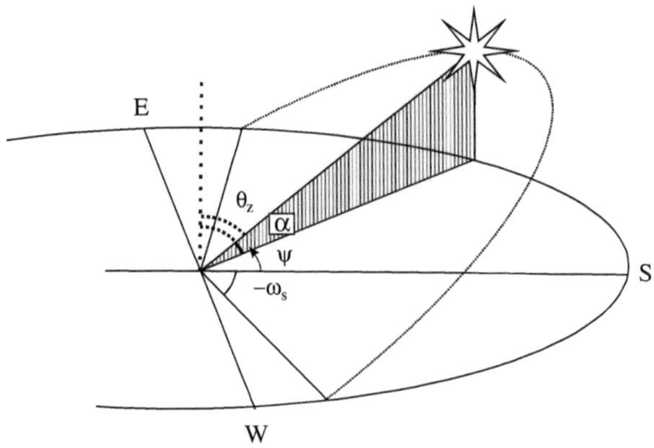

Fig. 1.9 Representation of solar angle and azimuth

The latitude and declination are related to the zenith angle at solar noon by the following relationship:

$$\theta_z|_{\text{solar noon}} = \phi - \delta \tag{1.6}$$

where ϕ is the latitude and δ is the declination.

Several information can be deduced by Eq. 1.6. It depends on the place by means of latitude and on the day of the year by the declination. $\theta_z|_{\text{solar noon}} = 0$ for $\phi = \delta$, but being $-23.45° \leq \delta \leq 23.45°$ only places with $-23.45° \leq \phi \leq 23.45°$ will be able to see the Sun on a straight line perpendicular to the Earth in a day of the year. Moreover, if $\theta_z|_{\text{solar noon}} \leq 90°$ the highest point falls below the horizon and there will be no sunset. It can occur if $\varphi \leq -66.55°$. In particular in the south pole where $\phi \cong -90°$ for $0 \leq \delta \leq 23.45°$(from Spring to Autumn equinox) six months of darkness are expected.

The *solar altitude* α is defined as the angle formed by the horizon and a straight line connecting the Earth and the Sun in a plane perpendicular to horizon. The solar altitude is complementary to the zenith angle.

The *azimuth angle* ψ is the Sun angular position east or west to south. Conventionally, azimuth is null at solar noon and it increases toward the east. However, some papers refer the azimuth to north, in such case at solar noon $\psi = 180°$. Figure 1.9 gives a representation of solar angles and azimuth.

The same information given by azimuth can be expressed by *hour angle* ω as the difference between noon and the time. Considering a rotation of 360° in 24 h

$$\omega = \frac{12 - \text{time}|_{24\,\text{h}}}{24} 360 = 15(12 - \text{time}|_{24\,\text{h}}) \tag{1.7}$$

where $\text{time}|_{24h}$ is the time referred to solar midnight on a 24-h clock.

The sunrise angle can be deduced by knowing the latitude and declination by:

$$\omega_s = ar\cos(-\tan\phi\tan\delta) \tag{1.8}$$

and the sunset angle is given by $-\omega_s$

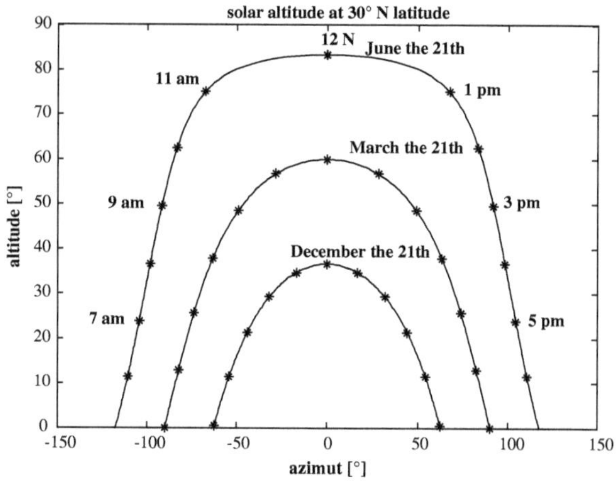

Fig. 1.10 Solar altitude versus azimuth angle during the year at a latitude of 30° N

By observing that in a day from sunrise to sunset the displacement angle is equal to $2\omega_s$ and that a complete rotation requires 24 h, it is possible to calculate the daylight hours *DH*

$$DH = \omega_s \frac{48}{360} = \frac{ar\cos(-\tan\phi\tan\delta)}{7.5} \tag{1.9}$$

Finally, the position of the Sun can be obtained using the solar altitude and the azimuth angle versus latitude, declination, and hour angle.

$$\sin\alpha = \sin\delta\sin\phi + \cos\delta\cos\phi\cos\omega \tag{1.10}$$

$$\cos\psi = \frac{\sin\alpha\ \sin\phi - \sin\delta}{\cos\alpha\ \cos\phi} \tag{1.11}$$

It should be remembered that all angles are expressed in degrees.

Solar altitude versus azimuth angle during the year at a latitude of 30° N and 65° N are respectively sketched in Figs. 1.10 and 1.11, by comparing these figures, the different Sun altitude can be noted.

1.5 Measuring Sunlight

In Sect. 1.4, it has been explained how irradiance can be predicted versus day of the year and latitude. Irradiance can be also measured with dedicated systems. The advantages consist, for example, on taking into account the real irradiance value in that moment to maximize the collected energy.

Precise sensors are: the pyranometer and the pyrheliometer.

The pyranometer measures global radiation. It can be mounted horizontally or on the collector surface. In this last case, it measures global irradiance perpendicular to

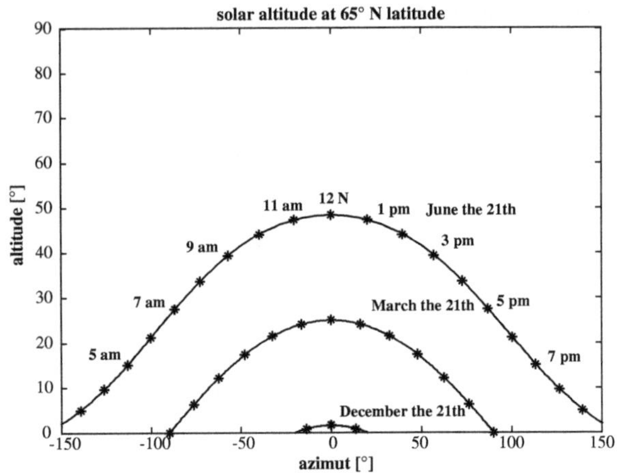

Fig. 1.11 Solar altitude versus azimuth angle during the year at a latitude of 65° N

the collector. On the outside, there is a lens that is transparent for wavelength from 0.285 to 2.8 μm. The sensor is a circular multijunction, wire–wound thermopile, it has as accuracy of 1 % for altitudes higher than 20° and its output voltage is 9 μV/W/m^2.

The black and white pyranometer is less accurate and operates on the basis of different heating of black and white wedges. Each wedge has a thermocouple that measures temperature and a voltage dependent on temperature difference is given as output.

The normal incidence pyrheliometer operates on the basis of the collected beam irradiance into a small solid angle (about 5.5°). It is obtained by means of a tube blackened inside. The tube is sealed with dry air to eliminate absorption of incident beam by vapor. The sensor is a wire–wound thermopile and its sensitivity is about 8 μV/W/m^2. To perform continuing measurements a tracker is needed.

In order to measure diffuse component of irradiance, a pyranometer can be used with a shadow band stand.

Finally, some less precise and cheaper measurement systems exist as cadmium sulfide photocells and silicon photodiodes. These devices are not sensitive to the whole solar spectrum and cannot be calibrated to measure total energy.

1.6 Sunlight Emulation

In Fig. 1.5, it has been highlighted that a blackbody radiation corresponding to 3000 K falls mainly in the infrared spectrum. This is the typical case of incandescent lamp filaments. It has been demonstrated that these kind of lamps have a poor efficiency as lighting systems, moreover they are unable to emulate sunlight due to their different spectrum compared to the Sun.

Other lighting systems as gas discharge lamps differ from blackbody radiation due to presence in their spectrum of characteristic lines of the adopted gas, as in the case of fluorescent tubes, high pressure sodium, and high intensity discharge lamps.

As it will be described in the following chapter, a photovoltaic device can generate electricity depending on the energy received under form of radiated waves. As a consequence, standardized spectral test condition has to be defined in order to evaluate the goodness of photovoltaic generator.

1.7 Collecting Sunlight

Once arrived on the Earth, it is necessary to optimally collect sunlight before transforming it into electric energy. The best situation is to have the collector orthogonal to beam, unfortunately, as it has been described before, the Sun's position varies continuously during the day and day of the year.

Under the hypothesis of a fixed collector, a good choice can be done considering that at solar noon the Sun will be perpendicular to a collector which forms an angle of θ_z with the horizon plane (tilt angle). In this situation the Sun is highest in the sky and lowest AM is obtained. Considering that the Sun covers 15° per hour, the optimal position will be maintained for about 2 h. A satisfactorily efficiency, however, is assured between 10 a.m. and 2 p.m. as shown in Fig. 1.9 and 1.10. It should be noted that the use of Eq. 1.6 implies the choice of a declination value. A mean value referred to the season can be adopted. If, for example, best performance is expected during the summer on northern hemisphere, using Eq. 1.5 a value of $-14.93°$ is obtained which implies that the collector has to be mounted at about $\varphi - 15°$

Some installations could require different orientation. For example, in a grid connected plant, collected energy has to be maximum when the consumer request is higher and it occurs after solar noon. In this case, hence, a west facing orientation shifts the peak of collected energy toward afternoon.

Of course, the collector can have one o two degree of freedom to track Sun beam. In general, efficiency gain is greater in summer rather than in winter. Moreover, increased cost and energy required to tracking system have to be taken into account by designer.

In presence of diffused sunlight, the beam irradiance is a small fraction of the global irradiance.

1.8 Conclusions

Energy is created by nuclear fusion process into the Sun: four hydrogen nuclei are transformed into a helium nucleus, the whole mass is decreased and this lack becomes energy radiated into the universe. This energy reaches the Earth, it is partially scattered by the atmosphere, and it can be utilized to be converted into electric energy by photovoltaic devices.

Through this chapter, the energy generation process, how it can be tracked, collected, and measured to be transformed and utilized as electric energy are explained.

Bibliography

Emery K, Del Cueto J, Zaaiman W (2002) Spectral corrections based on optical air mass. In: Proceedings. 29th IEEE PV Specialists conference, pp 1725–1728

Gueymard CA (2004) The sun's total and spectral irradiance for solar energy applications and solar radiation models. Sol Energy 76:423–453

Messenger RA, Ventre J (2003) Photovoltaic systems engineering, 2nd edn. CRC Press, Boca Raton

Sciama DW (1971) Modern cosmology. Cambridge University Press, Cambridge

Wentham S, Green MA, Watt ME, Corkish R (2007) Applied photovoltaics, Earthscan

Chapter 2
From Radiated Energy to Electrical Energy: Physics of Photovoltaic Cells

2.1 Prologue: The Photoelectric Effect

The transformation of the radiated energy coming from the Sun into electrical energy implies the study of the interaction of electromagnetic waves with matter.

This mechanism can be understood starting from the photoelectric effect in which electrons are emitted from a material when it is exposed to electromagnetic radiation.

In particular, it was observed that (using visible light for alkali metals, near-ultraviolet for other metals, and extreme ultraviolet for non-metals) the energy of emitted electrons increased with the frequency and did not depend on the intensity of the radiation.

This effect was first observed by Heinrich Hertz in 1887 and for several years it was apparently in contrast with James Clerk Maxwell's wave theory of light; according to this theory, the electron energy would be proportional to the intensity of the radiation.

The following main experimental results, for given material, were observed:

1. the rate at which photoelectrons are ejected is directly proportional to the intensity of the incident light;
2. a threshold frequency, below which no photoelectrons are emitted, exists;
3. above the threshold frequency, if the intensity of light is increased, the number of emitted electrons is increased as well but their maximum energy does not vary; moreover very low intensity of incident light, with frequency greater that the threshold, is able to extract electrons;
4. above the threshold frequency, if the frequency of incident light is increased, the maximum energy of photoelectrons is also increased;

Albert Einstein theorized, in 1905, that light is composed of discrete quanta, now called photons, and that the energy of a quantum of light is given by the product of the frequency of the corresponding wave multiplied by a constant, later called Planck's constant.

M. C. Di Piazza and G. Vitale, *Photovoltaic Sources*, Green Energy and Technology,
DOI: 10.1007/978-1-4471-4378-9_2, © Springer-Verlag London 2013

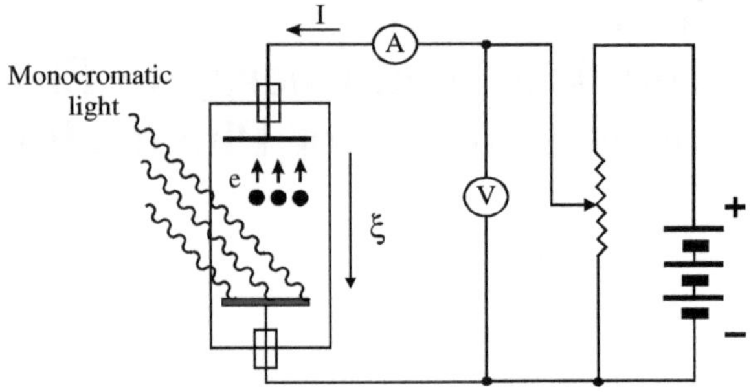

Fig. 2.1 Experimental setup for evaluation of the photoelectric effect

$$E = h\nu \tag{2.1}$$

where E is the energy of a single quantum of light, $h = 6.62 \times 10^{-34}$ J·s is the Planck's constant and ν is the frequency of the electromagnetic wave (incident photon). For this discovery, Einstein earned the Nobel Prize in Physics in 1921 and the quantum revolution in physics began.

If a photon has enough energy, it can give its energy to an electron and it is ejected.

The maximum kinetic energy $K_{\max}$ of an ejected electron is given by:

$$K_{\max} = h\nu - \varphi \tag{2.2}$$

where $\varphi = hf_0$ is the work function (sometimes denoted W), which is the minimum energy required to remove a delocalized electron from the surface of any given metal. It follows that the frequency ν of the incident photon must be greater than f_0 to extract electrons.

The experimental setup to verify the photoelectric effect is sketched in Fig. 2.1. The light beam hits the material that is situated inside a bulb. Two plates, subjected to a variable potential, generate an electric field (ξ) that is able to break the photoelectrons. The galvanometer measures the current generated by photoelectrons.

When the frequency of the incident beam is greater that the threshold frequency, by applying a potential such that $K_{\max} = eV$, the measured current is null.

It should be noted that it is convenient to measure energy in electronvolt [eV] instead of Joule (1 eV = $1.6 \cdot 10^{-19}$ J).

It is easy to calculate that the energy associated with a photon emitted by Sun having $\lambda = 0.5$ μm (see Fig. 1.5) is equal to:

$$E\big|_{\lambda=0.5\,\mu\text{m}} = h\nu = h\frac{c}{\lambda} = 6.62 \cdot 10^{-34}\frac{3 \cdot 10^{8}}{0.5 \cdot 10^{-6}} = 3.972 \cdot 10^{-19} J = 2.48\,\text{eV}$$

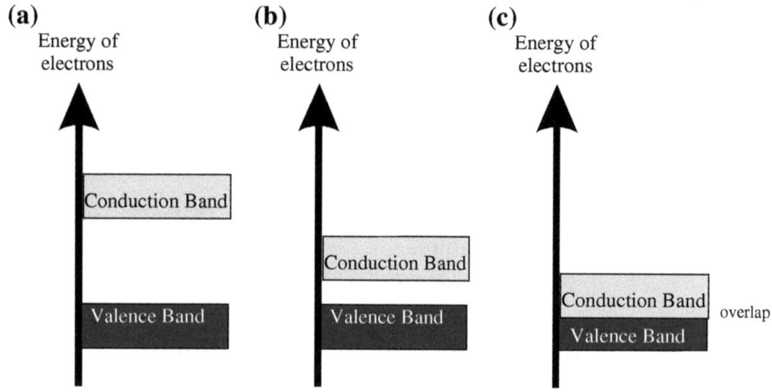

Fig. 2.2 Schematic representation of the band structure for: **a** insulator, **b** semiconductor, **c** conductor

2.2 Metals, Semiconductors, Insulators

In a single isolated atom, energy levels of electrons are discrete. For hydrogen atom, the Bohr's model gives:

$$E_H = \frac{-m_o q^4}{8\,\varepsilon_o^2 h^2}\,\frac{1}{n^2} \tag{2.3}$$

where m_o is the free electron mass and q its charge, ε_o^2 is the free space permittivity and n is a positive integer known as principal quantum number. The fundamental level corresponds to $n = 1$ and the related energy is $E_{H(n=1)} = -13.6\,\text{eV}$.[1]

If N atoms interact (for example in a crystal), N outer levels have energy only slightly different and thermal energy allows electrons to pass from one level to another (the energy corresponding to $T = 300$ K is $kT \approx 0.026$ eV). Resulting energy levels are grouped in bands. Two main bands are recognizable: conduction band and valence band. These two bands are separated by a forbidden region that is characterized by an energy value E_g. This value makes the difference among insulators, conductors, and semiconductors.

In an insulator, the forbidden band has a wide energy (for example $E_g = 9$ eV for SiO_2) neither thermal energy nor an electric field is able to raise the energy of an electron to send it into the conduction band. Due to the absence of free electrons for conduction, the material behaves as an insulator.

On the contrary, in a conductor the conduction band is partially superimposed to the valence band. As a consequence, there are many electrons available for conduction and an electric field can give them sufficient energy to perform conduction.

In a semiconductor, the two bands are separated but the energy of the forbidden band is low ($E_g = 1.12$ eV for Si at T = 300 K) and it is easy to give energy to an

[1] The negative value of this energy corresponds to a tied electron

electron to go into the conduction band. In this case, the hole in the valence band contributes to the conduction as well as the electron in the conduction band (Fig. 2.2).

The forbidden band amplitude varies with temperature, for Si the amplitude is:

$$E_g(T) = 1.17 - \frac{(4.73 \cdot 10^{-4})T^2}{T + 636} \tag{2.4}$$

The temperature coefficient is negative, it means that the forbidden band amplitude decreases with temperature.

2.3 Inside the Band Structure of a Semiconductor

The motion of electrons inside a crystal is influenced by the periodical crystalline structure of the semiconductor that determines a periodic potential field due to the atom's nucleus and the tightly bound core electrons.

The dynamic behavior is obtained by solving the Schrödinger equation.

$$\Delta^2\psi + \frac{2m_0}{(h/2\pi)^2}[E - U(\vec{r})]\,\psi = 0 \tag{2.5}$$

where m_0 is the free electron mass, E is the energy of the electron, and $U(\vec{r})$ is the periodic potential inside the semiconductor. The solution of Schrödinger equation defines the band structure, the allowed energy, and the crystal momentum $p = \hbar k$ where k is the wave vector and $\hbar = h/2\pi$.

The relationship between energy and momentum near the minimum of the conduction band and near the maximum of the valence band is nearly parabolic; as a consequence, the effective mass is defined as the inverse of the second derivative of the energy respect to the momentum[2]:

$$m* \equiv \left[\frac{d^2E}{dp^2}\right]^{-1} \tag{2.6}$$

This definition allows a particle to be considered as a classical particle with an effective mass $m*$. The effective mass $m*$ depends on the band.

Near the top of the valence band this value is negative. These places are usually empty due to electrons thermally excited (as it will be explained in the next section) that are in the conduction band.

It is easier to consider these states as occupied by another kind of particles said *holes* that behaves as classical free particles with a positive effective mass m_p^*, while electrons are considered as classical free particles with effective mass m_n^*.

Figure 2.3 describes a simplified energy band diagram for a semiconductor at $T > 0$ in energy versus momentum diagram.

[2] It should be borne in mind that $E = p^2/2m$

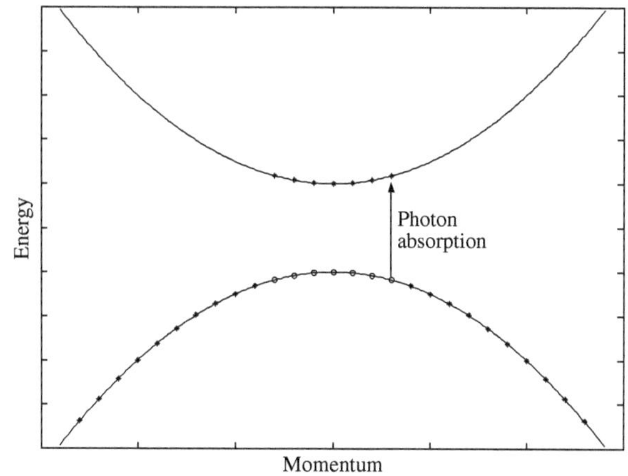

Fig. 2.3 Energy versus momentum representation of the energy band structure for a direct band-gap semiconductor

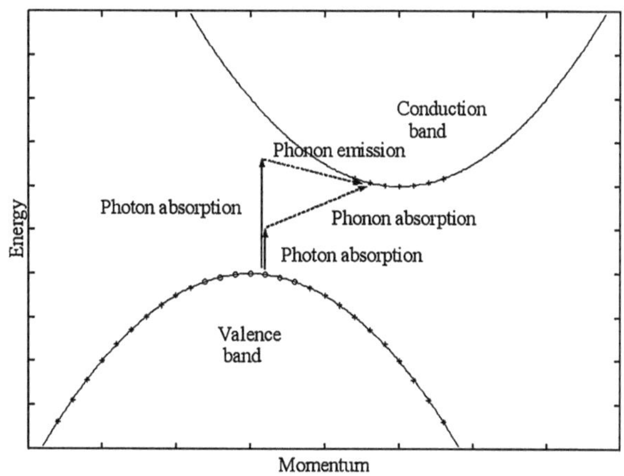

Fig. 2.4 Energy versus momentum representation of the energy band structure for indirect band-gap semiconductor

It should be noted that the minimum of the conduction band occurs for the same momentum values of the maximum of the valence band. As a consequence, an electron can do a transition maintaining constant the momentum. This kind of semiconductor is a *direct band-gap* semiconductor.

The behavior of an *indirect band-gap* semiconductor and the energy band diagram is illustrated in Fig. 2.4.

The difference between these two kinds of semiconductors is important for the study of absorption and emission of light in a semiconductor.

2.4 Absorption of Light

As explained in Sect. 2.1, radiated energy interacts with the matter, including semiconductors, as photons, whose energy is $E = h\nu$, and momentum $p_\lambda = h/\lambda$.

The excitation of an electron from the valence band to the conduction band is called *fundamental absorption* and, as a consequence, a hole appears in the valence band.

Both the total energy and the momentum must be conserved; in particular, for direct band-gap semiconductors (GaAs, GaInP, CdTe, and $CU(InGa)Se_2$) a transition can occur remaining constant the momentum of the photon as shown in Fig. 2.3. The crystal momentum is equal to $p_\lambda = h/l$ where l is the lattice constant and it is bigger than the photon momentum. Being the wavelength of sunlight of the order of 10^{-4} cm and the lattice constant of 10^{-8} cm, it can be assumed that the conservation law can be applied only to the photon momentum.

The probability of an induced transition from a level E_1 into the valence band to a level E_2 into the conduction band for a photon with energy $E_2 - E_1 = h\nu$ is given by a coefficient α that depends on the difference between the photon energy and the forbidden band gap.

$$\alpha(h\nu) \approx \left(h\nu - E_g\right)^{1/2} \tag{2.7}$$

Some semiconductors allow only transitions with $p \neq 0$, in such cases:

$$\alpha(h\nu) \approx \frac{1}{h\nu}\left(h\nu - E_g\right)^{3/2} \tag{2.8}$$

In indirect band-gap semiconductor, like Si and Ge, the maximum of the valence band and the minimum of conduction band occur for different values of the momentum as sketched in Fig. 2.4. Conservation of the momentum implies in this case the emission or the absorption of a phonon.[3] In particular, if the photon energy is greater than the difference between the starting electron energy level in the valence band and the final level in conduction band, a phonon is emitted. On the contrary, if the photon energy is lower than the difference between the starting electron energy level in the valence band and the final level in conduction band, a phonon is absorbed.

The absorption coefficient is different depending on absorption ($\alpha_a(h\nu)$) or emission ($\alpha_e(h\nu)$) phenomenon.

$$\alpha_a(h\nu) \propto \frac{\left(h\nu - E_g + E_{ph}\right)^2}{e^{E_{ph}/kT} - 1} \tag{2.9}$$

$$\alpha_e(h\nu) \propto \frac{\left(h\nu - E_g - E_{ph}\right)^2}{1 - e^{E_{ph}/kT}} \tag{2.10}$$

where E_{ph} is the phonon energy. It should be noted that for indirect band-gap semiconductor, the absorption of a photon depends on the availability of energy states, and on the absorbed/emitted phonons as well. This makes the absorption

---

[3] The phonon is considered as a particle representation of a lattice vibration in the semiconductor, it is a low energy particle with a relatively high momentum

coefficient for indirect transition smaller than the corresponding one for direct transition. As a result, light is able to penetrate more inside an indirect band-gap semiconductor.

2.5 Allowable States for Holes and Electrons

By the Schrödinger equation, it is possible to obtain the density of states both for electrons and holes. The influence of periodic potential is incorporated into the effective mass. The density of available states for unit volume and energy gives the energy levels allowable that could be occupied if the corresponding particle has sufficient energy.

In the conduction band ($E > E_C$), the density of states is given by:

$$g_C(E) = \frac{m_n^* \sqrt{2m_n^*(E - E_C)}}{\pi^2 \hbar^3} \quad \left[\mathrm{cm^{-3}\,eV^{-1}}\right] \tag{2.11}$$

And in the valence band ($E < E_V$) the density of states is given by:

$$g_V(E) = \frac{m_p^* \sqrt{2m_p^*(E_V - E)}}{\pi^2 \hbar^3} \quad \left[\mathrm{cm^{-3}\,eV^{-1}}\right] \tag{2.12}$$

where E_C and E_V are the energy of the conduction and valence band, respectively.

2.6 Energy Distribution for Holes and Electrons

Considering a semiconductor in thermal equilibrium, with no generation of carrier and insulated from the external, the probability that a state is occupied by an electron with energy E is given by the Fermi function.

$$f(E) = \frac{1}{1 + e^{(E - E_F)/kT}} \tag{2.13}$$

where E_F is the Fermi energy. This value depends on the material.

At $T = 0$, the Fermi function is a step function. All electrons have energy below or equal to E_F. When temperature rises there is a non-null probability that electrons could have energy above E_F. In Fig. 2.5 Fermi function is plotted for $T = 0$ K, $T = 300$ K, and $T = 3000$ K; Fermi level is set to 3 eV and the band gap is 1.1 eV as in the Silicon. It should be noted that at environmental temperature, the number of electrons with energy greater than the lower limit of conduction band is negligible.

In the conduction band, a state is occupied if there is an available state and if an electron has sufficient energy. As a consequence, to obtain the equilibrium electron

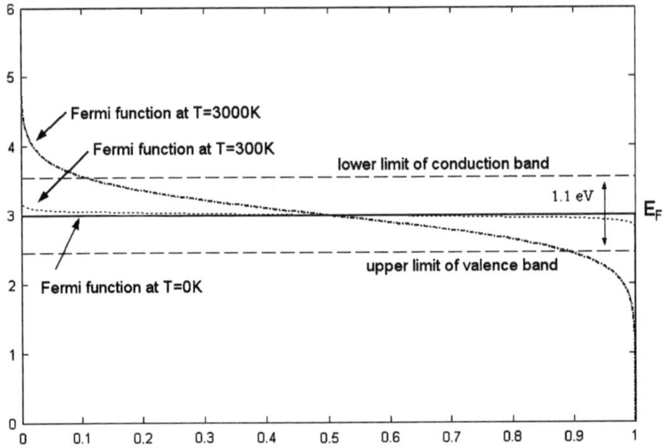

Fig. 2.5 Fermi level and state density

density, it is necessary to multiply the corresponding density of state (Eq. 2.11) for the probability that the electron has the corresponding energy (Eq. 2.13) and to integrate it from the minimum level to be occupied (E_C) to infinity.

$$n_0 = \int_{E_C}^{\infty} g_C(E)f(E)dE = 2\left(\frac{2\pi m_n^* kT}{h^2}\right)^{3/2} e^{-(E_C - E_F)/kT} \tag{2.14}$$

where the following position can be done:

$$N_C = 2\left(\frac{2\pi m_n^* kT}{h^2}\right)^{3/2} \tag{2.15}$$

N_C is the effective density of state in the conduction band. In the same way, the holes concentration at the equilibrium is given by:

$$p_0 = \int_{-\infty}^{E_V} g_V(E)[1 - F(E)]dE = 2\left(\frac{2\pi m_p^* kT}{h^2}\right)^{3/2} e^{(E_V - E_F)/kT} \tag{2.16}$$

and the following position can be done:

$$N_V = 2\left(\frac{2\pi m_p^* kT}{h^2}\right)^{3/2} \tag{2.17}$$

Equations (2.14) and (2.16) are obtained under the hypothesis that E_F is sufficiently far from E_C and E_V ($E_C - E_F > 3kT$ and $E_F - E_V > 3kT$) in such a case, the semiconductor is said *nondegenerate* and the product of electrons and hole concentration does not depend on Fermi energy. In an undoped (intrinsic)

Fig. 2.6 Periodic table

Period ↓ / Group →	13	14	15	16
2	5 **B** $2s^22p^1$	6 **C** $2s^22p^2$	7 **N** $2s^22p^3$	8 **O** $2s^22p^4$
3	13 **Al** $3s^23p^1$	14 **Si** $3s^23p^2$	15 **P** $3s^23p^3$	16 **S** $3s^23p^4$
4	31 **Ga** $3d^{10}4s^23p^1$	32 **Ge** $3d^{10}4s^23p^2$	33 **As** $3d^{10}4s^23p^3$	34 **Se** $3d^{10}4s^23p^4$
5	49 **In** $4d^{10}5s^25p^1$	50 **Sn** $4d^{10}5s^25p^2$	51 **Sb** $4d^{10}5s^25p^3$	52 **Te** $4d^{10}5s^25p^4$

semiconductor, the electron and hole concentration is equal and this value is the intrinsic carrier concentration $\underline{n_i}$.

$$n_i = \sqrt{n_0 p_0} = \sqrt{N_C N_V} e^{(E_V - E_C)/2kT} = \sqrt{N_C N_V} e^{-E_g/2kT} \tag{2.18}$$

This last equation represents the *law of mass action:* regardless of doping, the product of hole and electron densities is a material property that does not depend on temperature. It should be noted that the wider is the forbidden band the lower is the intrinsic carrier concentration.

Finally, the Fermi level for an intrinsic semiconductor can be evaluated by equating the electrons and hole concentration given by Eqs. (2.14) and (2.16), respectively.

$$E_{F_i} = \frac{E_C + E_V}{2} + \frac{kT}{2} \ln\left(\frac{N_V}{N_C}\right) \tag{2.19}$$

The Fermi level is very close to the middle of band gap.

The intrinsic carrier concentration for Silicon ($n_i \approx 10^{10}$ cm^{-3}) makes it an insulator, to exploit it as a semiconductor the introduction of specific impurities or dopants is necessary.

2.7 Doping

The conductivity of a semiconductor can be varied by introducing specific dopants. With reference to the part of periodic table sketched in Fig. 2.6, it can be noted that phosphorous has five valence electrons ($3s^23p^3$) whereas boron has three valence electrons ($3s^23p^1$).

If phosphorous atoms are introduced in a silicon crystal, one of its five valence electrons becomes available for conduction, the remaining four electrons are tied with covalence bonds of silicon lattice (see Fig. 2.7a). This kind of dopant is said *donor.* In the same way by introducing boron, its three valence electrons are tied

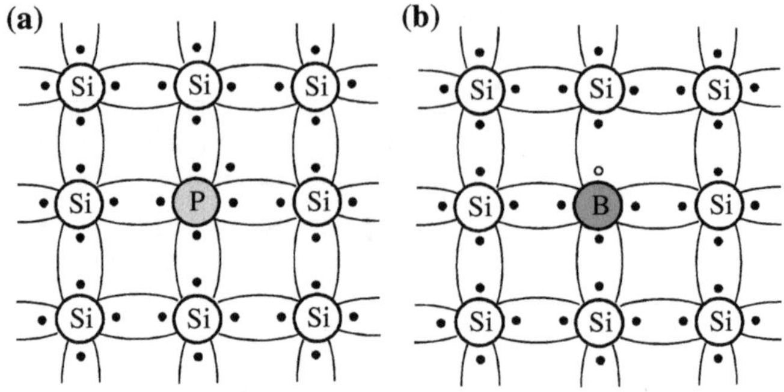

Fig. 2.7 **a** *n-type* doping with Phosphorous As, **b** *p-type* doping with Boron

with covalence bonds of silicon lattice and a hole remains (see Fig. 2.7b). This kind of dopant is said *acceptor.*

From the point-of-view of energy levels, the presence of donor introduces additional energy levels near the conduction band (within few *kT*), hence thermal energy can allow the added electron to move to the conduction band. In the same way, the presence of an acceptor introduces additional energy levels near the valence band.

In case of donor introduction, electrons are the primary source of conduction and the semiconductor is said *n-type*, on the contrary, if an acceptor is introduced, conduction is due to hole, and the semiconductor is said *p-type*.

The atoms of donors (N_D) or acceptors (N_A) are usually completely ionized, as a consequence for *n-type* semiconductor $n_0 \approx N_D$ and for *p-type* semiconductor $p_0 \approx N_A$. This hypothesis will be maintained in the following, throughout the chapter.

The presence of dopant changes the Fermi level compared to an intrinsic semiconductor, this value can be recalculated by using Eq. (2.19); for an *n-type* semiconductor:

$$E_{Fn} = E_{F_i} + kT \ln\left(\frac{N_D}{n_i}\right) \tag{2.20}$$

Compared to an intrinsic semiconductor the Fermi level is increased.

For a *p-type* semiconductor:

$$E_{Fp} = E_{F_i} - kT \ln\left(\frac{N_A}{n_i}\right) \tag{2.21}$$

and the Fermi level is lower compared to the intrinsic semiconductor.

The Eq. (2.18) holds even for doped semiconductors, for an *n-type*:

$$n_i^2 = n_0 \cdot p_0 = p_0 \cdot N_D \tag{2.22}$$

and the donors' concentration can be expressed versus the Fermi level for an intrinsic semiconductor:

$$n_0 = N_C e^{-(E_C - E_{Fn})/kT} = N_C e^{-(E_C - E_{Fi})/kT} e^{(E_{Fn} - E_{Fi})/kT} = n_i e^{(E_{Fn} - E_{Fi})/kT} \quad (2.23)$$

where the n_i is obtained by using Eqs.(2.18) and (2.19) with the position $N_C \approx N_{V.}$

For a *p-type*:

$$n_i^2 = n_0 \cdot p_0 = n_0 \cdot N_A \quad (2.24)$$

$$p_0 = n_i e^{-\left(E_{Fp} - E_{Fi}\right)/kT} \quad (2.25)$$

In an *n-type* semiconductor, electrons represent majority carriers and holes minority carriers. Usually, if necessary, their concentration symbol includes a pedex to indicate the semiconductor type. Hence, in an *n-type* semiconductor there are n_n majority carriers and p_n minority carriers. In a *p-type* semiconductor, there are p_p majority carriers and n_p minority carriers. If necessary, to specify the equilibrium conditions a further pedex "*o*" can be added.

When double doping with both donors and acceptors is performed, the type of the semiconductor is determined by the greatest impurity concentration.

Supposing the total ionization of both donors and acceptors, if $N_D > N_A$ an *n-type* results with $n_n \approx N_D - N_A$ electrons and p_n holes concentration, on the contrary if $N_A > N_D$ a *p-type* results with $p_p \approx N_A - N_D$ holes and n_p electrons concentration. As a matter of fact, considering, for example, the case of $N_D > N_A$, being the intrinsic semiconductor and both the dopants neutral after doping, the final positive charge (due to holes and ionized donors) must be equal to the negative charge (due to electrons and ionized donors):

$$n_n + N_A = p_n + N_D \quad (2.26)$$

using the action mass law written in the form $n_i^2 = n_n p_n$ it is possible to obtain:

$$\begin{cases} n_n = \frac{1}{2}\left[N_D - N_A + \sqrt{(N_D - N_A)^2 + 4n_i^2}\right] \\ p_n = \dfrac{n_i^2}{n_n} \end{cases} \quad (2.27)$$

usually the difference between donors and acceptors is greater than the intrinsic concentration, it means $|N_D - N_A| > n_i^2$.

2.8 Carrier Transport

Until now, the semiconductor has been considered in equilibrium status with n_o electron and p_0 holes concentration. In this condition, electrons move continuously inside the lattice in all directions but, if observed for sufficiently long time, its displacement is null. The mean distance between two collisions is known as the

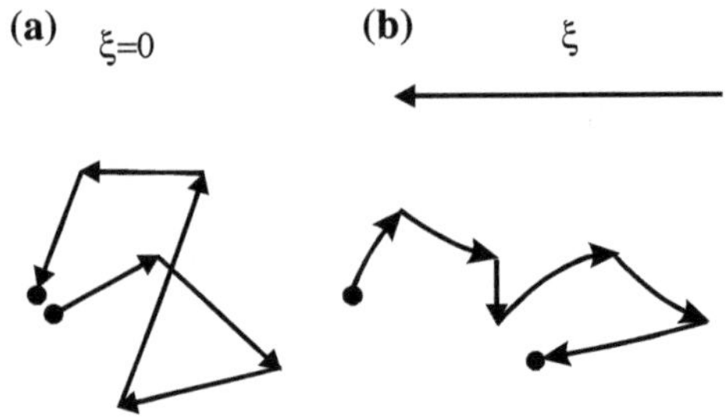

Fig. 2.8 **a** Random thermal motion, **b** thermal motion under electric field

free mean path (L_c) and the corresponding time interval between two collisions as the *mean free time* (τ_c). This situation is schematically sketched in Fig. 2.8a.

By using the effective mass m_n^*, the electrons' kinetic energy can be expressed by assigning *kT/2* for each degree of freedom.

$$\frac{1}{2} m_n^* v_{th}^2 = \frac{3}{2} kT \tag{2.28}$$

where v_{th} is the mean free speed, at environment temperature ($T = 300$ K) it is equal to 10^7 cm/s both for Si and GaAs. Considering a typical value of 10^{-5} cm for the free mean path it follows $\tau_c = L_c/v_{th} = 10^{-12}$ s = 1 ps.

If an electric field is applied to the semiconductor, a carriers' displacement occurs. The motion is schematically sketched in Fig. 2.8b: the electron (hole) is continuously scattered but there is a non-null mean displacement.

The motion of electrons and holes can be changed by applying an electric field or by varying locally their concentration. In the former case the process is called *drift*, in the latter *diffusion*.

2.8.1 Drift Current

With regard to drift current, if an electric field is applied to a uniformly doped semiconductor, the energy of the bands does not remain constant but it is raised in the direction of the field. Electrons are moved to the opposite direction of the field toward lower energy of conduction band and holes toward higher energy of the valence band. The motion is schematically sketched in Fig. 2.9: the electron (hole) is continuously scattered but there is a non-null mean displacement.

Electrons have a drift velocity in the opposite direction of the electric field.

$$\vec{v}_{\text{drift\_}n} = -\mu_n \vec{\xi} = \mu_n \nabla \phi \tag{2.29}$$

where ϕ is the electrostatic potential and μ_n is the electrons mobility coefficient. As a consequence a drift current occurs; its density is equal to the product of drift velocity for the density of electrons (or holes) and for the unit charge.

$$\vec{J}_{\text{drift\_}n} = -qn\vec{v}_{\text{drift\_}n} = qn\mu_n \vec{\xi} = -qn\mu_n \nabla \phi \tag{2.30}$$

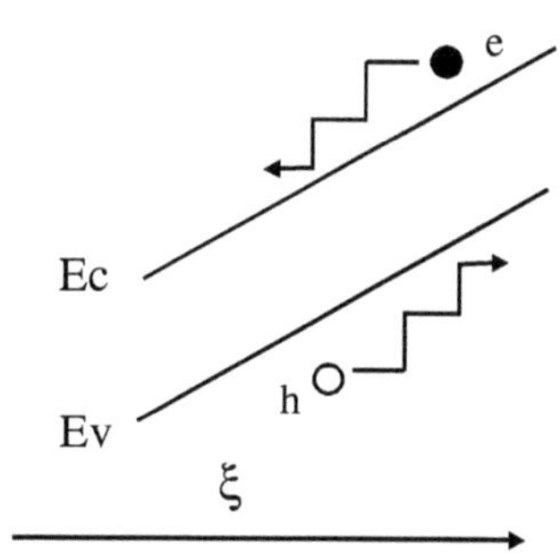

Fig. 2.9 Energy levels and electrons (holes) motion in presence of an electric field

For a semiconductor whose length (corresponding to the x direction) is bigger than the other two dimensions and the electric field is in the x direction. A one-dimensional (1D) representation can be used, as depicted in Fig. 2.10; Eq. (2.30) becomes:

$$J_{drift\_n} = -qnv_{drift\_n} = -qn\mu_n\xi = -qn\mu_n\frac{\partial\phi}{\partial x} \tag{2.31}$$

where $\nabla\phi$ reduces to the potential derivative along x.

The drift velocity is proportional to the electric field ξ or to the derivative along x direction of the electrostatic potential φ.

$$v_{drift\_n} = -\mu_n\xi = \frac{\partial\phi}{\partial x} \tag{2.32}$$

In the same way for holes:

$$\vec{J}_{drift\_p} = qp\vec{v}_{drift\_p} = qp\mu_p\vec{\xi} = qp\mu_p\nabla\varphi \tag{2.33}$$

$$\vec{v}_{drift\_p} = \mu_p\vec{\xi} = \mu_p\nabla\phi \tag{2.34}$$

The coefficients that tie velocity to electric field can be evaluated by equating the impulse of the force ($-q\xi$) multiplied for time interval to the quantity of motion achieved in the same time. For electrons:

$$-q\xi\tau_n = m_n^* v_{drift\_n} \tag{2.35}$$

$$\mu_n = \frac{q\tau_n}{m_n^*} \tag{2.36}$$

And for holes:

$$q\xi\tau_p = m_p^* v_{drift\_p} \tag{2.37}$$

$$\mu_p = \frac{q\tau_p}{m_p^*} \tag{2.38}$$

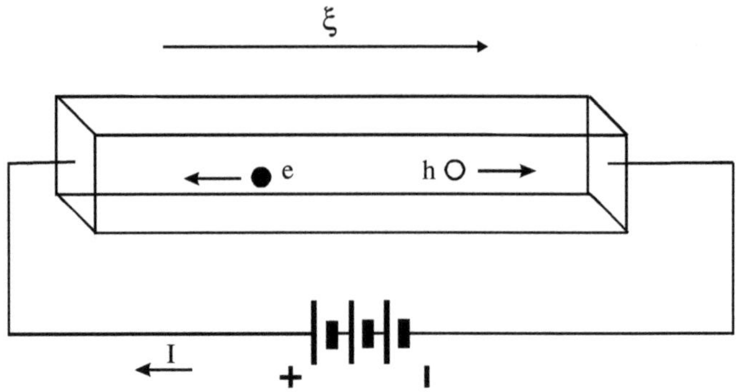

Fig. 2.10 Semiconductor with electric field applied: a drift current formed by holes and electron occurs

Mobility coefficient takes into account scattering mechanism for lattice interaction and ionized impurity

2.8.2 *Diffusion Current*

When concentration is not uniform, carriers tend to move from regions with high concentration to regions with low concentration. A diffusion current occurs and its density is given respectively for electrons and holes by:

$$\vec{J}_{\text{diff\_}n} = qD_n \nabla n \tag{2.39}$$

$$\vec{J}_{\text{diff\_}p} = -qD_p \nabla p \tag{2.40}$$

where D_n and D_p are the diffusivity coefficients for electrons and holes respectively, and q is the electron charge.

Considering a long semiconductor where the injection is performed on a lateral surface, the current occurs in x direction:

$$J_{\text{diff\_}n} = qD_n \frac{\partial n}{\partial x} \tag{2.41}$$

$$J_{\text{diff\_}p} = -qD_p \frac{\partial p}{\partial x} \tag{2.42}$$

The total current induced in a semiconductor is the sum of diffusion and drift currents for electrons and holes:

$$\vec{J}_n = \vec{J}_{\text{drift\_}n} + \vec{J}_{\text{diff\_}n} \tag{2.43}$$

$$\vec{J}_p = \vec{J}_{\text{drift\_}p} + \vec{J}_{\text{diff\_}p} \tag{2.44}$$

In thermal equilibrium, diffusion current and drift current must balance. Comparing these two terms, the *Einstein relationship* is obtained.

$$D_n = \left(\frac{kT}{q}\right)\mu_n \tag{2.45a}$$

$$D_p = \left(\frac{kT}{q}\right)\mu_p \tag{2.45b}$$

By using Eqs. (2.28), (2.36), and (2.38), the mean free path is obtained versus diffusivity and mean free time:

$$L_n = \sqrt{D_n \tau_n} \tag{2.46a}$$

$$L_p = \sqrt{D_p \tau_p} \tag{2.46b}$$

Finally, a further term due to dielectric displacement field should be added.

$$\vec{J}_{\text{disp}} = \frac{\partial \vec{D}}{\partial t} = \varepsilon \frac{\partial \vec{\xi}}{\partial t} \tag{2.47}$$

where ε is the electric permittivity of the semiconductor. However, it should be considered that in case of photovoltaic cell that is operated in DC, this term can be neglected.

For Silicon at $T = 300$ K and impurity concentration of 10^{16} cm^{-3}, electron mobility is about 1400 cm^2/Vs and consequently diffusivity is about 35 cm^2/s. Lower values are exhibited by holes for which mobility is about 490 cm^2/Vs and consequently diffusivity is about 10 cm^2/s. Another semiconductor as GaAs has values greater of about five times. Mobility and diffusivity decrease with impurity concentration.

2.8.3 Semiconductor Resistivity

Considering Eqs. (2.30) and (2.33), the total current density due to an electric field is given by:

$$J = J_n + J_p = \left(qn\mu_n + qp\mu_p\right)\xi \tag{2.48}$$

This equation defines the resistivity of the semiconductor as:

$$\rho = \frac{1}{\sigma} = \frac{1}{\left(qn\mu_n + qp\mu_p\right)} \tag{2.49}$$

2.9 Semiconductor Fundamental Equations

The semiconductor obeys to two fundamental laws: the Poisson's equation and the continuity equation.

The Poisson's equation comes from Gauss' law for electricity, it says that in a control volume the divergence of electric displacement field $\vec{D}$ is equal to the free charge density ρ_s. For a linear, isotropic, and homogeneous medium $\vec{D} = \varepsilon\vec{E}$.

The continuity equation is a local conservation law. It affirms that a quantity (no matter what: particles, charges or people) in a given volume can change in the time only for a flux variation of the quantity through the surface that contains the volume.

2.9.1 *The Poisson's Equation*

The general form for a semiconductor was proposed by Van Roosbroeck in 1950:

$$\nabla \cdot \varepsilon\vec{\xi} = \rho_s \tag{2.50}$$

where ρ_s is the spatial charge density. For a semiconductor, ρ_s can be expressed as the contribution of positive charge (holes density plus the ionized donors) and the negative charge (electron density and ionized acceptors).

$$\nabla \cdot \varepsilon\vec{\xi} = q(p + N_D - n - N_A) \tag{2.51}$$

This last equation can be particularized along x direction:

$$\varepsilon\frac{d\xi}{dx} = q(p + N_D - n - N_A) \tag{2.52}$$

2.9.2 *Continuity Equation*

The continuity equation can be written both for electrons and holes. The general form for electrons is:

$$\nabla \cdot \vec{J}_n = q\left(R_n - G_n + \frac{\partial n}{\partial t}\right) \tag{2.53}$$

The first member is the divergence of electron current density. It represents the flux of electrons through an arbitrary volume.

The second member is the variation of the electrons' number during the time. It is composed of a first term that gives the velocity of the recombination R_n. The

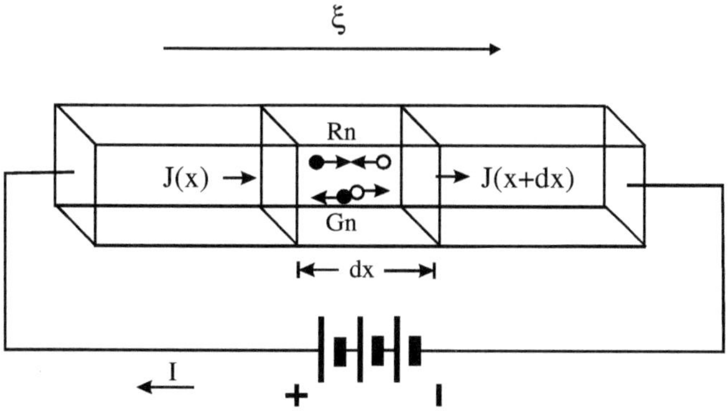

Fig. 2.11 Semiconductor with electric field applied: application of the continuity equation to a volume with length *dx*

recombination process occurs when an electron falls from conduction band to valence band and a couple electron–hole disappears.

Several recombination mechanisms are possible: by defects in the forbidden band, radiative recombination, and a further term called Auger recombination.

The first type of recombination occurs when an electron or a hole makes a transition to a parasitic energy level inside the forbidden band and phonons are emitted. In radiative recombination, a transition from conduction band to valence band origins a photon emission.

In Auger recombination, an electron and a hole recombine in a band-to-band transition, but in this case the resulting energy is given off to another electron or hole, in this last case three particles are involved.

The second term of Eq. (2.53), G_n, gives the velocity of generation, this occurs for light absorption or carrier injection.

The third term is the time variation of electron density.

In 1D representation, as depicted in Fig. 2.11, the continuity equation for electrons is:

$$\frac{\partial J_n}{\partial x} = q\left(R_n - G_n + \frac{\partial n}{\partial t}\right) \tag{2.54}$$

For holes:

$$\nabla \cdot \vec{J}_p = q\left(G_p - R_p - \frac{\partial p}{\partial t}\right) \tag{2.55}$$

$$\frac{\partial J_p}{\partial x} = q\left(G_p - R_p + \frac{\partial p}{\partial t}\right) \tag{2.56}$$

2.10 Minority Carrier Diffusion Equations

A 1D representation of the semiconductor is considered in the following.

For steady state conditions, Eq. (2.54) can be used neglecting the time derivative. Moreover, current density derivative considering Eqs. (2.30) and (2.39) is given by:

$$\frac{d}{dx}J_n = q(R_n - G_n) = \frac{d}{dx}\left(-qn\mu_n\xi + qD_n\frac{\partial n}{\partial x}\right) \tag{2.57}$$

and for holes

$$\frac{d}{dx}J_p = q(R_p - G_p) = \frac{d}{dx}\left(-qn\mu_p\xi + qD_p\frac{\partial p}{\partial x}\right) \tag{2.58}$$

For small electric field, the contribution of electric field space derivative can be neglected. It corresponds to neglect the drift current compared to diffusion current.

Under low level injection, the recombination rate for electrons can be written as the difference between the minority carriers in that position and their equilibrium value divided by the minority carrier lifetime.

$$R_n = \frac{p_n - p_{n0}}{\tau_p} = \frac{\Delta p_n}{\tau_p} \tag{2.59}$$

It should be noted that electron recombination depends on the corresponding minority carriers. As a matter of fact, the recombination can occur when a minority carrier is available. For an *n-type* material, from Eq. (2.58) the corresponding minority carrier diffusion equation is obtained:

$$D_p\frac{d^2}{dx^2}\Delta p_n - \frac{\Delta p_n}{\tau_p} = -G(x) \tag{2.60}$$

For a *p-type* material the corresponding minority carrier diffusion equation is:

$$D_n\frac{d^2}{dx^2}\Delta n_p - \frac{\Delta n_p}{\tau_n} = -G(x) \tag{2.61}$$

When an n-doped semiconductor is lighted on a lateral surface, a carrier injection is performed, the minority carrier concentration is given by Eq. (2.57) imposing $G(x) = 0$ with boundary conditions:

$$\begin{cases} p_n(x=0) = p_n(0) \\ p_n(x \Rightarrow \infty) = p_{n0} \end{cases} \tag{2.62}$$

At the surface, the injection causes a constant value of minority carriers' concentration $p_n(0)$, on the contrary, for x sufficiently far from the injection surface, the concentration returns to the equilibrium value p_{n0}.

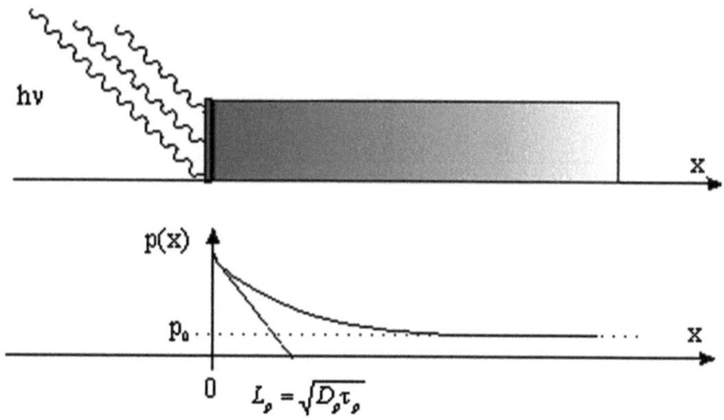

Fig. 2.12 Representation of a semiconductor bar with an incident light on the lateral surface that causes charge injection: the charge concentration relaxes with spatial constant L_p, for x sufficiently far from the injection zone the equilibrium concentration is restored

The solution is:

$$p_n(x) = p_{n0} + [p_n(0) - p_{n0}]e^{-x/L_p} \tag{2.63}$$

The minority carriers' concentration relaxes with exponential law. The relaxation space constant is given by Eqs. (2.46a, b).

The resultant diffusion current can be obtained by Eq. (2.41).

It should be noted that, if the semiconductor length is about L_p, a different boundary condition must be imposed on the corresponding surface where the minority concentration carriers are different from equilibrium value.

Figure 2.12 shows a typical situation in which the incident light on the lateral surface causes charge injection, their concentration relaxes with spatial constant L_p, if the distance is sufficiently far from the injection zone, the equilibrium concentration is restored and the charge concentration equals the equilibrium value p_0.

2.11 P–N Junction

A p–n junction can be conceptually conceived as a two doped semiconductor of *n-type* and *p-type* that have a surface in common. When both semiconductors are separated, they are electrically neutral. As soon as they get in touch, majority carriers of *n-type* semiconductors (the electrons) begin to diffuse into the *p-type* semiconductor and vice versa. As a result, near the surface of separation between the two semiconductors, in *n-type* semiconductor, holes coming from *p-type* semiconductor tend to combine with electrons and the positive charge of the corresponding ionized donors is not more compensated by majority carriers. Inside the *n-type* region, near the junction, where there are no more majority

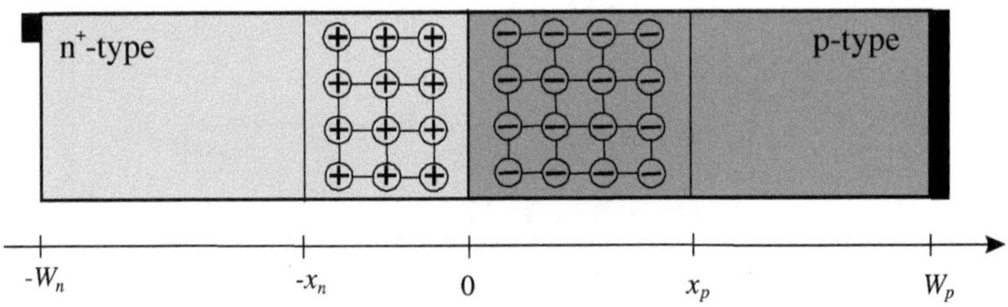

Fig. 2.13 Schematic representation of a p–n junction

charges, a depletion is observed and the corresponding zone remains with fixed positive charges.

In the same way, in *p-type* side, electrons coming from *n-type* semiconductor tend to combine with holes and the negative charge of the ionized acceptors is not more compensated by majority carriers. Inside the *p-type* region, near the junction, where there are no more majority charges, a depletion is observed and the corresponding zone remains with fixed negative charges.

As the fixed charges are uncovered, an electric field is produced and the diffusion process is slowed down. A p–n junction is drawn in Fig. 2.13 in 1D representation; the origin ($x = 0$) is the junction surface, x_p and W_p are the depletion boundary at the end of *p-type* region, while $-x_n$ and $-W_n$ are the depletion boundary at the end of *n-type* region. It should be noted that, if a semiconductor is more doped than the other (usually indicated with apex +), the greater quantity of free carrier diffused in the other semiconductor cause a more extended depletion.

It is assumed a uniformed and nondegenerated doping and that dopants are fully ionized.

The whole zone in which there are fixed uncompensated charge is called *depletion region* or *space charge region*. The remaining zones can be considered as neutral (often called *quasineutral*). The electric field due to the fixed charges origins an electrostatic potential difference called *built-in voltage*.

The Poisson's Eq. (2.50) can be rewritten as:

$$\frac{d\xi}{dx} = -\nabla^2\phi = \frac{q}{\varepsilon}(p_O + N_D - n_O - N_A) \tag{2.64}$$

where ϕ is the electrostatic potential, p_O and n_o are the hole and electron equilibrium concentration, N_D is the concentration of ionized donors (positive fixed charges), and N_A is the concentration of ionized acceptors (negative fixed charges).

Equation (2.64) can be particularized for each zone. In quasineutral zone, the net charge is null and it follows $\nabla^2\phi = 0$, inside the positive charged depletion region $\nabla^2\phi = -qN_D/\varepsilon$ and inside the negative charged depletion region $\nabla^2\phi = qN_A/\varepsilon$.

The built-in voltage is the potential difference between points $x = -x_n$ and $x = x_p$.

$$V_{bi} = \phi(-x_n) - \phi(x_p) \tag{2.65}$$

Assuming that $\phi(x_p) = 0$ The process of the potential along x is:

$$\phi = \begin{cases} V_{bi} & for \quad x \leq -x_n \\ V_{bi} - \frac{qN_D}{2\varepsilon}(x + x_n)^2 & for \quad -x_n < x < 0 \\ \frac{qN_A}{2\varepsilon}(x - x_p)^2 & for \quad 0 \leq x < x_p \\ 0 & for \quad x \geq x_p \end{cases} \tag{2.66}$$

Imposing that the second and the third relationship in Eq. (2.66) give the same values of ϕ for $x = 0$ and the continuity of the electric field in the same point, that gives:

$$x_n N_D = x_p N_A \tag{2.67}$$

the length of depletion zone is obtained versus built-in voltage.

$$W_D = x_n + x_p = \sqrt{\frac{2\varepsilon}{q}\left(\frac{N_A + N_D}{N_A N_D}\right) V_{bi}} \tag{2.68}$$

Equation (2.68) says that the length of depletion zone depends on the built-in voltage. For diode operation, this value can be lessened by applying a positive voltage on p-side (forward bias) or widened by a positive voltage on n-side (reverse bias).

In particular, in case of forward bias, the built-in voltage V_{bi} is lessened by a quantity equal to the applied voltage and the corresponding electric field changes the equilibrium conditions allowing holes belonging to p zone to pass through the junction and reaching the n zone and electrons belonging to n zone to pass through the junction and reaching the p zone, as well.

The depletion zone length can be calculated by Eq. (2.68) using the new value of $(V_{bi} - V)$ where V is the applied voltage.

For low level charge injection, a diffusion current due to minority charges as described by Eq. (2.63) for holes in n zone occurs. The corresponding current is calculated by (2.42)

$$J_{diff\_p} = qD_p \frac{\partial p}{\partial x} = q\frac{D_p}{L_p}[p_n(0) - p_{n0}]e^{-x/L_p} \tag{2.69}$$

Now, taking into account that the total current remains the same along x because inside the quasineutral region there is no appreciable generation–recombination rate, it is easy to evaluate Eq. (2.69) for $x = 0$.

$$J_{diff\_p}(0) = q\frac{D_p}{L_p}[p_n(0) - p_{n0}] = q\frac{D_p}{L_p}\Delta p_n \tag{2.70}$$

It should be noted that, because current is constant along x direction, decreasing the minoritary carriers current, a complementary current due to majority carriers occurs along x. For $x \gg L_p$ the current is due only to majority carriers.

Next step consists on evaluating the term inside square brackets. Considering the Eq. (2.58), its second member contains two terms with a high value. As a matter of fact, there is a big value of the electric field and a great variation of the concentration carriers. Being the first member small, as it represents a low level injection the following approximated relationship can be used:

$$qp\mu_p\vec{\xi} = qD_p\frac{dp}{dx} \tag{2.71}$$

By using Einstein relationship (2.45a) and assuming $V_T = kT/q$, Eq. (2.71) becomes:

$$\xi = \frac{V_T}{p}\frac{dp}{dx} = -\frac{dV}{dx} \tag{2.72}$$

As for the holes concentration, it should be observed that, inside the p zone it assumes the value corresponding to the thermal equilibrium p_{p0}, at the edge of the n zone, near the depletion zone it is equal to $p_n(0)$. Rewriting Eq. (2.72) and integrating over the depletion zone:

$$\int_{p_{p0}}^{p_n(0)} \frac{dp}{p} = -\frac{1}{V_T}\int_0^{V_{bi}-V} dV \tag{2.73}$$

it follows that:

$$p_n(0) - p_{p0}e^{-(V_{bi}-V)/V_T} \tag{2.74}$$

Finally, the relationship between p_{p0} and p_{n0} can be found by integrating Eq. (2.72) under no bias condition (it means $V = 0$) obtaining:

$$p_{n0} = p_{p0}e^{-(V_{bi})/V_T} \tag{2.75}$$

from which

$$p_n(0) - p_{n0} = p_{n0}\left[e^{V/V_t} - 1\right] \tag{2.76}$$

and the Eq. (2.70) becomes:

$$J_{diff\_p}(0) = q\frac{D_p p_{n0}}{L_p}\left[e^{V/V_t} - 1\right] \tag{2.77}$$

Finally, the junction current is determined by adding the contribution of electrons and multiplying for the section area A.

$$I = I_s(e^{V/V_t} - 1) \tag{2.78}$$

where

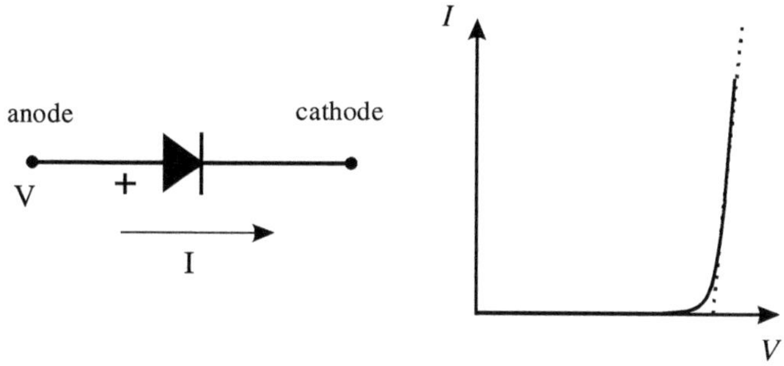

Fig. 2.14 Electric symbol and voltage versus current diode characteristic

$$I_s = Aq\left(\frac{D_p p_{n0}}{L_p} + \frac{D_n n_{p0}}{L_n}\right) \tag{2.79}$$

Equation (2.78) is known as the Shockley diode equation.

When the forward bias voltage approaches V_{bi} the depletion zone tends to vanish and the current is limited by the semiconductor and ohmic contact, as well. In this case, the voltage versus current characteristic is approximated by straight line.

When a reverse bias is applied, it means that a positive voltage is applied to the n zone contact, and Eq. (2.78) can be still utilized. As a matter of fact, the exponential term is negligible and a reverse saturation current given by Eq. (2.79) is obtained. In this case, the obtained small current is given only by carriers generated inside the junction and it does not depend on the applied reverse bias.

Figure 2.14 shows the voltage versus current characteristic of a diode.

From what explained above, it is clear that a diode allows the current to pass from p zone to n zone when it is forward biased. The ohmic contact belonging to the p zone is called *anode* while the ohmic contact belonging to the n zone is called *cathode*.

2.12 P–N Junction Capacitance

In Sect. 2.11, it has been explained that current near the junction is due to minority carrier injection. When the bias voltage changes, the corresponding charge concentration varies as well, but similarly to a capacitor, this concentration requires a finite time interval to reach a new equilibrium.

The equivalent capacitance can be calculated by considering the minority carrier diffusion in Eq. (2.63) in which injection is due to junction direct voltage polarization for which electrons are injected in p region. If the p region is more doped than the n region, it can be assumed that current is given only by holes injected into the n region.

The whole charge variation compared to equilibrium condition can be calculated as:

$$Q = A\int_0^{\infty} q[p_n(x) - p_{n0}]dx = A\int_0^{\infty} q[p_n(0) - p_{n0}]e^{-x/L_p}dx = \\ = Aq[p_n(0) - p_{n0}]L_p = AqL_p\Delta p_n \tag{2.80}$$

where A is the junction cross-sectional area. By comparing Eqs. (2.80) with (2.70) and using Eq. (2.46b), it follows that:

$$J_{\text{diff\_}p} = \frac{Q}{\tau_p} \tag{2.81}$$

It means that the current is proportional to the excess of minority charges.

For direct polarization the capacitance of the junction is called *diffusion capacitance*, for inverse junction polarization the capacitance is called *transition capacitance* and it has lower value than the diffusion capacitance; moreover under PV cells operation the inverse polarization of the junction does not occur, on the contrary, diffusion capacitance evaluation gives information about dynamic behavior.

By using Eq. (2.81), under static condition:

$$C_D = \frac{dQ}{dV} = \tau_p\frac{dI}{dV} = \tau_p g \tag{2.82}$$

where g is the differential conductance of the junction; it can be obtained on the basis of Eq. (2.78) neglecting the terms corresponding to the inverse saturation current.

$$g = \frac{dI}{dV} = \frac{d}{dV}\left[I_s\left(e^{V/V_T} - 1\right)\right] = \frac{I_s}{V_T}e^{V/V_t} = \frac{I}{V_T} \tag{2.83}$$

then:

$$C_D = \tau_p\frac{I}{V_T} \tag{2.84}$$

The diffusion capacity is proportional to the current. If both holes and electrons contribute to the current, the diffusion capacity is the sum of two terms due to electrons and holes respectively. Finally, comparing Eqs. (2.82) and (2.84), it can be seen that the time constant of the *RC* equivalent circuit is equal to the minority charges lifetime:

$$\frac{C_D}{g} = \tau_p \tag{2.85}$$

If applied voltage varies with time, the dynamic diffusion capacity is defined on the basis of charge variation in time interval dt as:

$$i = \frac{dQ'}{dt} = \frac{dQ'}{dv} \cdot \frac{dv}{dt} = C'_D \frac{dv}{dt} \tag{2.86}$$

During the time interval dt, only the charge near the junction varies, as a matter of fact, carrier diffusion requires more time to reach a new equilibrium condition. For this reason, the charge variation is lower compared to static condition and a lower value of diffusion capacity is expected. In any case, the dynamic value of diffusion capacitance depends on the applied voltage waveform; an expression of this voltage can be obtained for a sinusoidal case with amplitude V_m and pulsation ω superimposed to a fixed bias V_I.

$$v(t) = V_1 + V_m e^{j\omega t} \tag{2.87}$$

if V_m is a small signal, nonlinear effect of I–V curve can be neglected and the resulting current has the form

$$i(t) = I_1 + gV_m e^{j\omega t} + j\omega C'_D V_m e^{j\omega t} \tag{2.88}$$

The excess minority carrier concentration contains, in addition to the term due to static polarization, a further time dependent term:

$$\Delta p_n(x,t) = \Delta p_n(x) e^{j\omega t} \tag{2.89}$$

By utilizing Eqs. (2.57) and (2.58) where there is no charge generation and neglecting the electric field derivative, the differential equation for the excess minority charges is obtained:

$$D_p \frac{d^2}{dx^2} \Delta p_n - \frac{\Delta p_n}{\tau_p} - \frac{d(\Delta p_n)}{dt} = 0 \tag{2.90}$$

By substituting Eq. (2.89) in Eq. (2.90) and taking into account that from Eq. (2.46a) $L_p = \sqrt{D_p \tau_p}$:

$$\frac{d^2}{dx^2} \Delta p_n = \frac{1 + j\omega\tau_p}{L_p^2} \Delta p_n \tag{2.91}$$

that gives:

$$\Delta p_n = ke^{-\frac{(1 + j\omega\tau_p)^{\frac{1}{2}} x}{L_p}} e^{j\omega t} \tag{2.92}$$

the resulting excess minority carriers' concentration is given by the time constant term defined by Eq. (2.63) plus the time varying term defined by Eq. (2.92).

$$\Delta p_n = p_{no}\left(e^{\frac{V_1}{V_T}} - 1\right)e^{-x/L_p} + ke^{-\dfrac{(1+j\omega\tau_p)^{\frac{1}{2}}x}{L_p}}e^{j\omega t} \tag{2.93}$$

where k is a constant that can be found by equating the junction law Eq. (2.76) considering that the junction voltage is defined by Eq. (2.87)

$$\Delta p_n = p_{no}\left(e^{\dfrac{V_1 + V_m e^{j\omega t}}{V_T}} - 1\right)e^{-x/L_p} \tag{2.94}$$

with the resulting excess minority carriers' concentration expression Eq. (2.93) both evaluated for $x = 0$.

Since, for $y \ll 1$, $e^y \approx 1 + y$ assuming $(V_m/V_T)e^{j\omega t} = y$, the following expression of k is obtained:

$$k = p_{n0}\frac{V_m}{V_T}e^{\frac{V_1}{V_T}} \tag{2.95}$$

Finally, the junction current is obtained by using Eq. (2.42) for $x = 0$.

$$J_{diff\_p} = -qD_p\left.\frac{\partial p}{\partial x}\right|_{x=0}$$

$$I_p(0) = \frac{AqD_pp_{n0}}{L_p}\left(e^{\frac{V_1}{V_T}} - 1\right) + \frac{AqD_pp_{n0}V_m e^{V_1/V_T}}{V_T}\cdot\frac{\sqrt{(1+j\omega\tau_p)}}{L_p}e^{j\omega t} \tag{2.96}$$

It should be noted that the first term of the second member of Eq. (2.96) corresponds to the static current of the p–n junction, I_1.

For low frequencies, corresponding to $\omega\tau_p \ll 1$, the approximation $\sqrt{1 + j\omega\tau_p} \approx 1 + j\omega\tau_p/2$ is valid, then Eq. (2.96) gives:

$$I_p(0) = I_1 + \frac{AqD_pp_{n0}V_m e^{V_1/V_T}}{V_TL_p}\left(1 + \frac{j\omega\tau_p}{2}\right)e^{j\omega t} \tag{2.97}$$

$$\begin{cases} g = g_0 = \frac{AqD_pp_{n0}e^{V_1/V_T}}{V_TL_p} \\ C_D = \frac{g\tau_p}{2} \end{cases} \tag{2.98}$$

It should be noted that the conductance is the same one of static value g_0 but capacitance is divided by two. They both depend on the static polarization voltage by exponential term.

For high frequencies corresponding to $\omega\tau_p \gg 1$

$$\sqrt{(1 + j\omega\tau_p)} = \sqrt{(j\omega\tau_p)} = \sqrt{(\omega\tau_p)e^{j\frac{\pi}{4}}} = \sqrt{\left(\frac{\omega\tau_p}{2}\right)}(1 + j)$$

and Eq. (2.96) gives:

$$\begin{cases} g = g_0\sqrt{\left(\frac{\omega\tau_p}{2}\right)} \\ C_D = g_0\sqrt{\left(\frac{\tau_p}{2\omega}\right)} \end{cases} \tag{2.99}$$

In this last case, both diffusion capacitance and conductance depend on frequency.

2.13 The PV Cell

The photovoltaic (PV) cell is basically a p–n junction with a central depletion region and two lateral quasineutral zone. At the end of each zone an electrical contact is placed. The more heavily doped zone is called the *emitter* zone and the other is the *base* zone. This last region is also called the *absorber region* because the great part of incident light is absorbed here. Differently from a diode, the PV cell is designed so to allow holes–electrons couples to be generated inside the junction due to incident light.

The aim of this section is to define the law that ties voltage and current of a PV cell including the dependence on incident light.

In a word, the total current due to both electrons and holes has to be found.

$$I = A[J_p(x) + J_n(x)] \tag{2.100}$$

where A is the semiconductor's surface. To this aim, the electrons continuity equation (2.54) in steady state condition can be integrated over the depletion region obtaining:

$$\int_{-x_N}^{x_P} \frac{\partial J_n}{\partial x} dx = J_n(x_P) - J_n(-x_N) = q \int_{-x_N}^{x_P} [R_n(x) - G_n(x)]\, dx \tag{2.101}$$

Now, evaluating Eq. (2.100) for $x = -x_N$ and using Eq. (2.101) to deduce $J_n(-x_N)$, the following equation is obtained:

$$I = A\left[J_p(-x_N) + J_n(x_P) + q\int_{-x_N}^{x_P} G_n(x)dx - q\int_{-x_N}^{x_P} R_n(x)dx\right] \tag{2.102}$$

This last equation contains four terms. The first one is the hole density current evaluated at the end of depletion zone in *n*-type region, the second is the electron density current evaluated at the end of depletion zone in *p*-type region. These two terms are referred to minority carriers. The third term is the optical generation rate and the last term represents the recombination rate in the space charge region.

2.13.1 Minority Carriers Current Density

The minority carrier diffusion equations in the quasineutral n-type and p-type regions have to be solved with suitable boundary conditions. In particular, for n-type semiconductor Eq. (2.60) with boundary condition at $x = -W_N$ and $x = -x_{N;}$ and for *p-type* semiconductor Eq. (2.61) with boundary condition at $x = W_P$ and $x = x_P$.

It has been explained in Sect. 2.8 that doping changes Fermi energy level. Thus, the law of mass action can be written as:

$$n_i^2 = pne^{-(E_{Fn}-E_{Fp})/kT} = pne^{-(\Delta E_{Fnp})/kT} \tag{2.103}$$

where ΔE_{Fnp} is the difference between Fermi level due to n doping and Fermi level due to p doping.

Assuming that the majority carriers maintain a constant concentration in their respective quasineutral zones, for n-type it means that Fermi energy level is constant for $-W_n \leq x \leq -x_n$ and for p-type that Fermi energy level is constant for $x_p \leq x \leq W_p$, therefore it can be assumed everywhere inside the semiconductor:

$$\Delta E_{Fnp} = E_{Fn}(-W_n) - E_{Fp}(-W_p) = E_{Fn}(x) - E_{Fp}(x) = qV \tag{2.104}$$

Now, being in n-type semiconductor quasineutral region $n = N_D$, the first boundary condition is obtained:

$$p_n(-x_n) = \frac{n_i^2}{N_D} e^{qV/kT} \tag{2.105}$$

And, being in p-type semiconductor quasineutral region $p = N_A$, it follows that:

$$n_p(x_p) = \frac{n_i^2}{N_A} e^{qV/kT} \tag{2.106}$$

It should be noted the presence of the factor $e^{qV/kT}$ in Eqs. (2.105) and (2.106), that will be present in the final expression of the current generated by the PV cell.

The last two boundary conditions are obtained considering that the front contact at the n-type zone is built by a metallic grid with an effective front surface $S_{F,eff}$, and the effective recombination is given by:

$$\frac{d}{dx}\Delta p = \frac{S_{F,eff}}{D_p} \Delta p(-W_n) \tag{2.107}$$

This situation is different from an ideal ohmic contact in which $\Delta p = 0$.

For the *p-type* semiconductor the back contact is quite an ohmic contact, however, if a thin more heavily doped region near this contact is introduced to increase the number of collected minority carriers, a condition similar to Eq. (2.107) can be imposed introducing the effective back surface S_{BSF}:

$$\frac{d}{dx}\Delta n = \frac{S_{BSF}}{D_n}\Delta p(W_p) \tag{2.108}$$

2.13.2 Optical Generation Rate

As explained before, only photons with wavelength $\lambda \leq hc/E_g$ can contribute to generate holes–electrons couples. The generation rate depends on a grid shadowing factor s, on the reflectance $r(\lambda)$, on the absorption coefficient $\alpha(\lambda)$ and on incident photon flux $f(\lambda)$ according to the Eq. (2.109).

$$\int_{-x_N}^{x_P} G_n(x)dx = (1-s)\int_{\lambda} [1 - r(\lambda)]f(\lambda)(e^{-\alpha(W_N - x_N)} - e^{-\alpha(W_N - x_P)})d\lambda \tag{2.109}$$

2.13.3 Recombination Rate

This term depends on the length of the depletion zone and on intrinsic carrier concentration, it is inversely proportional to the effective lifetime in the depletion region τ_D. Moreover, a further term is present in which the exponential contains the voltage multiplied for $(q/2kT)$ it is equal to about 19.23 J/C for $T = 300$ K.

$$\int_{-x_N}^{x_P} R_n(x)dx = \frac{W_D n_i}{\tau_D}(e^{(qV/2kT)} - 1) \tag{2.110}$$

2.13.4 Current Versus Voltage Law of Photovoltaic Cell

By solving minority carrier diffusion equation with boundary conditions and considering the generation and recombination, the current versus voltage law of photovoltaic cell is obtained after non-trivial calculations, not included here.

$$I = I_{ph} - I_{s1}(e^{qV/kT} - 1) - I_{s2}(e^{qV/2kT} - 1) \tag{2.111}$$

The first term is the short circuit current and collects the contributes of the two quasineutral zones and of the depletion zone.

The second term contains the dark saturation current due to the recombination in the quasineutral region I_{s1}. It should be observed that in Eq. (2.111) this

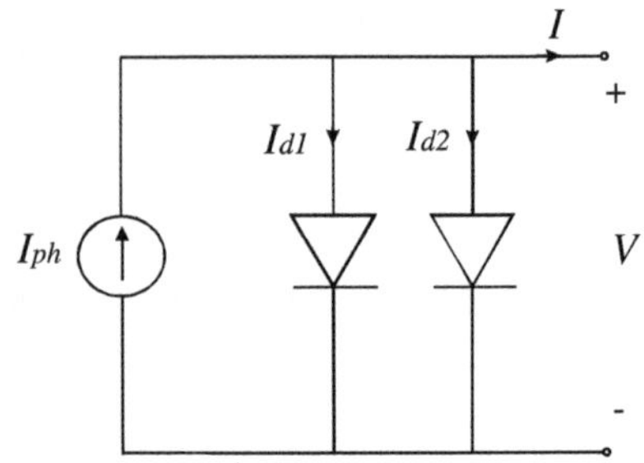

Fig. 2.15 Physical circuit model of a PV cell

phenomenon is considered ideal. When the non-ideality of the diffusion is taken into account, an ideality factor should be introduced in the exponential, as will be discussed in Chap. 3.

Finally, the last term contains the dark saturation current due to the recombination in the space charge region I_{s2}. This last term is bias dependent because it is proportional to the length of the depletion zone.

2.14 Physical Model of a PV Cell

Equation (2.111) reproduces the physical phenomena of a PV cell. It can be regarded as a Kirchhoff's current law (KCL) written for a node in an electrical circuit for which the sum of currents flowing into that node is equal to the sum of currents flowing out of that node. On the basis of this law, an equivalent circuit can be deduced. It represents a physical circuit model of a PV cell. This circuit is drawn in Fig. 2.15.

It should be noted that the output current is the sum of a current given by a generator that depends on solar irradiance minus the current that flows through the two diodes. The first current corresponds to I_{ph} in Eq. (2.111), the second current corresponds to $I_{s1}(e^{qV/kT}-1)=I_{d1}$ and the third current corresponds to$I_{s2}(e^{qV/2kT}-1)=I_{d2}$.

As a matter of fact, the second and the third term of Eq. (2.111) can be considered as Shockley diode equations. Finally, the output voltage is obtained by the diodes direct bias due to the current generator.

During operating conditions, when solar radiation occurs, the generator current flows through the diodes and a voltage appears at the terminals. If no load is applied this voltage is an "open circuit" voltage, i.e., the voltage of a directly polarized p–n junction and it is the maximum value achievable by a PV cell. If a load is connected, a part of the current of the generator flows into the load, voltage decreases and electric power is supplied to the load. The conversion process is completed.

Starting from solar radiation, electric energy has been obtained.

It should be noted that if the load is raised (it corresponds to a lower resistance) current rises too and voltage decreases; the supplied power reaches a maximum and then decreases until the short circuit condition.

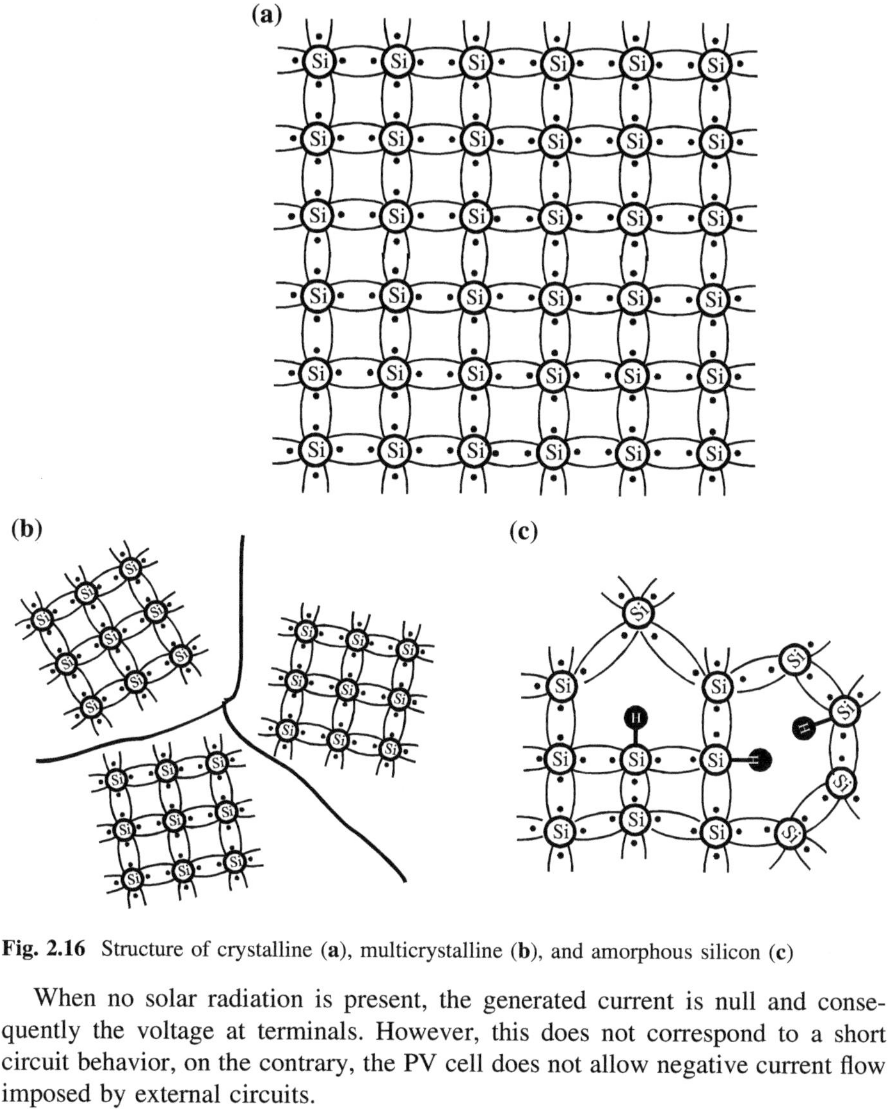

Fig. 2.16 Structure of crystalline (**a**), multicrystalline (**b**), and amorphous silicon (**c**)

When no solar radiation is present, the generated current is null and consequently the voltage at terminals. However, this does not correspond to a short circuit behavior, on the contrary, the PV cell does not allow negative current flow imposed by external circuits.

2.15 Semiconductor Types

Silicon can be arranged in different forms to obtain a photovoltaic cell depending on crystal structure. Following the classification on the basis of planar grain size, microcrystalline, polycrystalline, and multicrystalline can be distinguished. In particular, microcrystalline material exhibits a grain smaller than 1 μm, polycrystalline smaller than 1 mm, and multicrystalline smaller than 10 cm.

A schematic representation of structure of crystalline, multicrystalline, and amorphous silicon is drawn in Fig. 2.16.

2.15.1 Crystalline Silicon

Crystalline silicon is considered as an ideal structure where the pattern is regular throughout the whole surface. All theory explained above is developed with reference to this structure. The main advantage consists of highest ratio solar irradiance—produced electric power. With Monocrystalline silicon, power conversion efficiency ranging from 20 to 24 % is expected, with GaAs, power conversion efficiency ranging from 20 to 29 % is expected.

Crystalline silicon, on the other hand, is expensive owing to manufacturing process. For this reason, several alternative cheaper silicon structures have been developed.

2.15.2 Multicrystalline

Multicrystalline and polycrystalline silicon can be produced by a less sophisticated technique compared with crystalline. However, in this case, the presence of grain boundaries must be taken into account. In particular, cell performance is reduced because at the boundaries the carriers flow is blocked, the level structure is altered, and the current that would flow across p–n junction is shunted away.

Some remedies have been devised as, for example, the use of grains of few millimeters to cover the entire distance from the back to the front of the cell with minimum number of grains. With Polycrystalline silicon, a power conversion efficiency ranging from 13 to 18 % is expected.

2.15.3 Amorphous

Amorphous silicon presents a less regular structure with unsatisfied bonds. These "dangling" bonds are passivated by hydrogen by allowing doping (otherwise impossible) and raising the band gap form 1.1 eV of crystalline silicon to 1.7 eV; in this way, photons of higher energy can be absorbed and the required thickness of the material is lower. As a consequence, amorphous silicon can be used as a "thin film" form deposited on glass or other substrates for low cost applications.

The band structure of amorphous materials is similar to the crystalline material over short distance and a mobility gap, in which conduction occur, can be defined. However, there are a great number of localized energy states within mobility gap, corresponding to band tails and gangling bonds that make different the amorphous behavior compared to crystalline silicon.

2.15.4 Thin Film

The so called thin-film technologies, where thin-film cells are deposited on foreign substrate, are indicated for a transition from other technologies for module manufacturing.

It is possible, by increasing the ratio to hydrogen silane in the gas from which amorphous silicon is deposited, to obtain a microcrystalline material in which columns of crystallites are separated by amorphous region.

Actually, established thin-film technologies use amorphous silicon and nano-crystalline Si film (a-Si), cadmium telluride (CdTe), and copper indium/gallium diselenide (CIGs), they have efficiency of 8–13 % for a-Si, of 10–17 % for CdTe, and for 10–19 % for CIGs. Moreover, a reduced cost is expected. As a matter of fact, in a crystalline Si module about 40 % of cost is due to wafer or solar cell ribbon manufacturing.

2.15.5 Polymer Solar Cell

In addition to the previously cited inorganic material as Silicon (Si), Gallium Arsenide (GaAs), Cadmium Telluride (CdTe), and Cadmium-Indium-Selenide (CIS), endeavors to reduce costs due to solar grade silicon and fabrication processes have led to the exploitation of organic thin-film materials.

In particular, conjugated polymers and molecules are easy to be produced, they have mechanical flexibility as plastics, and can exhibit electronic properties similar to conductors and semiconductors. Moreover, recent progresses in soft lithography techniques as inkjet printing and microcontact printing have increased the interest toward the fabrication of integrated devices over both rigid and flexible substrates having a large area.

However, properties of organic and inorganic semiconductors are significantly different. In an organic semiconductor, as explained above, the absorption of a photon generates a hole–electron couple and they are transported for their high mobility and internal electric field.

On the contrary, in organic semiconductor dissociation into free charges carriers does not occur at room temperature. As a consequence, two different materials are necessary and charges are created as an effect of the transfer of photoinduced electrons from one to the other material.

In particular, after the absorption of solar irradiance, for which a matching between solar spectrum and absorption spectrum is desirable, a donor material (D) is required to give the excited electron to the acceptor material (A). As a result, a radical cation of the donor (D^+) and a radical anion of the acceptor (A^-) occur.

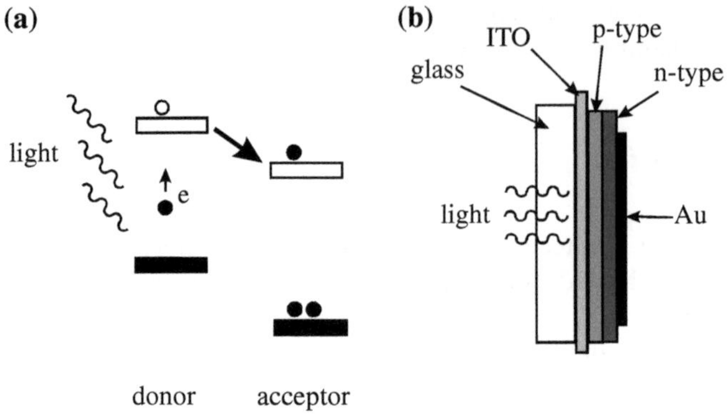

Fig. 2.17 Electron transfer from donor to acceptor material (**a**), section of an organic cell (**b**)

In addition, competitive processes as fluorescence and non-radiative decay must be inhibited and the transfer of photogenerated charges must be aided. For these reasons, the transferred electron should be slowed down as much as possible.

These last considerations are schematized in Fig. 2.17 (a). Here, an electron is excited by light absorption to the highest occupied molecular orbital (HOMO) in the donor material and then it is transferred to the lowest unoccupied molecular orbital (LUMO) in the acceptor material. Now, an extra electron is present in acceptor material and a hole appears in the donor material. Figure 2.17 (b) sketches the photovoltaic cell. Light passes through a glass and a transparent electrode (ITO); the electrodes are dissimilar but both are metallic to collect charges and to transfer them to external circuits.

Among organic cells the most successful appear to be: Dye-sensitized cells, Double layer cells, and Bulk heterojunction cells.

Dye-sensitized cells use an organic dye adsorbed at the surface of an inorganic wide band-gap semiconductor for absorption of light and injection of photoelectrons into the conduction band of the semiconductor.

Double layer cells consist on a sandwich of a single layer of organic dye between two dissimilar electrodes.

Finally, in bulk heterojunction cells, *p*- and *n*-type materials are mixed together, but, because of the intrinsic tendency of polymer materials to remain separate at nanometer dimension, junctions throughout the material are created.

2.16 Conclusions

The energy under the form of an electromagnetic wave is transformed in the electrons motion inside a semiconductor. Different semiconductor types allows a p–n junction to be obtained and used as a photovoltaic cell from which electric power, in terms of generated voltage and current at its terminal, is obtained.

Analyzing the electrons behavior inside a semiconductor, the fundamental equations of the PV cell and a physical circuit model, using a current generator and two diodes, is obtained.

Bibliography

Ashcroft NW, Mermin ND (1976) Solid state physics. Saunders College Publishing, USA

Basore PA (1994) Defining terms for cristalline silicon solar cells. Prog Photovolt: Res Appl 2:177–179

Grätzel M (2003) Photochemistry reviews. J Photochem Photobiol C 4:145–153

Kittel C (2004) Introduction to solid state physics. Wiley, London

Spol'skii EV (1969) Atomic physics. lliffe Books, London

Sze SM (2001) Semiconductor devices, physics and technology. Wiley, London

Von Roeden B (2006) Thin-film PV module a review, solar energy, focus July/August

Wöhrle D, Meissner D (1991) Organic solar cells. Adv Mater 3:129–138

Millman J (1988) Microelectronics, 2nd edn. McGraw Hill Higher Education, London

Jassen R Introduction to polymer solar cells, http://user.chem.tue.nl/janssen/SolarCells/Polymer%20solar%20cells.pdf

Chapter 3
Photovoltaic Source Models

3.1 Introduction

Nowadays, the more and more growing interest in applications of photovoltaic (PV) generation, with all the related problems of optimal exploitation, environmental impact, and grid stability, has determined a speedup of the research in this field. In particular, several issues are currently under consideration such as the prediction of PV energy production, the optimal choice and design of the power converters interfacing the PV generator to utility or load, and the study of all the problems related to the power electronic control. In order to suitably face these problems, an accurate modeling of the PV source is necessary. As a matter of fact, such modeling allows the electrical characteristics of a PV source to be defined, for varying load and environmental conditions, so to obtain an evaluation of the PV source behavior in any operating condition.

The mathematical model of a PV source is, therefore, useful when dealing with dynamic analysis of power converters, assessing the most suitable maximum power point tracking (MPPT) algorithms and, above all, when developing simulations tools for PV systems, either at software or hardware level. Some terms used, henceforth, need an explanation. A PV source is in general a device converting the light of sun into electricity. The elementary PV source is the PV *cell*. A device formed of series/parallel connected PV cells is called PV *module*. Generally, the PV *module* is the basic element of larger PV systems. A PV *field* may be either a module or a set of series/parallel connected modules.

Summarizing the contents of this chapter, the fundamentals of PV sources modeling are given, starting from the static double diode model of a PV cell up to the simplified single diode representation of a PV field, including parasitic and nonideality effects, obtained under uniform illumination. This last model will be used in the following for further investigations on the PV system emulation.

The dynamic model of a PV source is then described. Furthermore, some solutions to extend the modeling to the case of nonuniform solar irradiance are outlined too.

M. C. Di Piazza and G. Vitale, *Photovoltaic Sources*, Green Energy and Technology,
DOI: 10.1007/978-1-4471-4378-9_3, © Springer-Verlag London 2013

3.2 Static Model

3.2.1 Circuit Model of a PV Cell

A mathematical expression of the current/voltage (*I–V*) terminal characteristics for PV cells, coming from the theory of semiconductors, has been described in Chap. 2, Sect. 2.13.4.

The basic equation that analytically describes the *I–V* characteristic of the ideal PV cell is derived by the solutions of the minority carrier diffusion equations. In particular, these solutions can be used to evaluate the minority carrier current densities equations, as shown in Chap. 2. Introducing these last equations in the expression of the PV cell total current, the following relationship is obtained:

$$I = I_{ph} - I_{s1}(e^{qV/kT} - 1) - I_{s2}(e^{qV/2kT} - 1) \tag{3.1}$$

where:

- I is the PV cell terminal current
- V is the PV cell terminal voltage
- k is the Boltzman constant $= 1.38 \times 10^{-23}$ J/K
- q is the electronic charge $= 1.6 \times 10^{-19}$ C
- T is the PV cell temperature
- I_{s1} is the dark saturation current due to recombination in the quasi-neutral region (diffusion)
- I_{s2} is the dark saturation current due to recombination in the space charge region
- I_{ph} is the photo-generated current, linearly depending on solar irradiance

The photo-generated current and the dark saturation currents are given by rather complicated expressions depending on the cell structure, material properties, and operating conditions, as described in Chap. 2.

An in-depth understanding of the PV cell operation requires a detailed study of these terms. On the other hand, the basic form of Eq. (3.1) gives important information about the cell's behavior.

From a circuit point of view, Eq. (3.1) can be considered as a Kirchhoff current law; it shows that a PV cell can be modeled as an ideal current source in parallel with two diodes, reflecting the physical behavior of the *p-n* junction.

The first diode has an unitary ideality factor, whereas an ideality factor equal to two is assigned to the second diode as exhibited by denominators of the exponentials in Eq. (3.1).

Figure 3.1 shows the circuit model of a PV cell where the whole current flowing through the diodes is:

$$I_d = I_{d1} + I_{d2} = I_{s1}(e^{qV/kT} - 1) + I_{s2}(e^{qV/2kT} - 1) \tag{3.2}$$

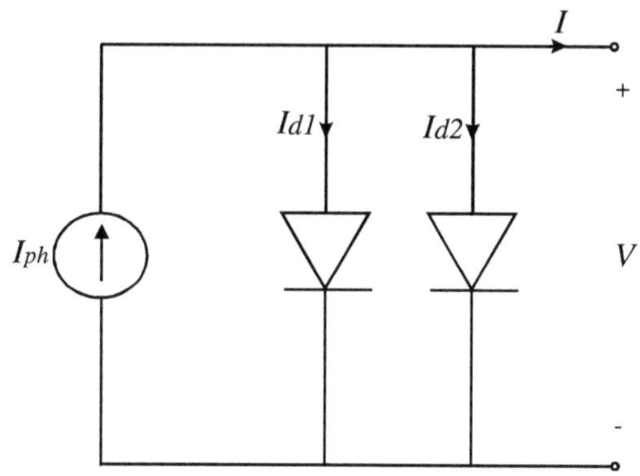

Fig. 3.1 PV cell circuit model

The PV cell circuit model is based on the assumption of linearity, i.e., the assumption that the current flowing through the PV cell is the superposition of two currents, one due to the junction bias, and the other due to illumination.

From this assumption, it is possible to obtain the PV cell I–V characteristic according to the composition schematized in Fig. 3.2. Considering that the current generator and the diode are parallel connected, the I–V curve is obtained by algebraically summing, point-by-point, the devices current having the same voltage.

From the I–V curve shape it is possible to assess that a PV cell, especially if a practical device is considered, can be regarded as an hybrid current–voltage source depending on the operating point. In particular, two regions can be distinguished on the I–V curve, namely a nearly constant current region and a nearly constant voltage region. These regions will be referred, hereinafter, as constant current region and constant voltage region, for convenience.

Once the electrical characteristic of a PV cell is derived, some remarkable points associated to the PV cell behavior can be defined. They are: the *short-circuit current point*, the *open circuit voltage point*, and the *maximum power point*. As well as the above cited points, an important parameter relevant to the PV cell efficiency to be defined is the *fill factor*.

3.2.1.1 Short-Circuit Current

If the PV source is connected to a load with low impedance, nearly all the photo-generated current flows through the load, the diodes are not biased and the voltage at the PV source terminals is near to zero. In this case, the terminal current equals the photo-generated current that is well approximated by the short-circuit current I_{sc}, i.e., the terminal current for $V = 0$. I_{sc} is, therefore, the ordinate of the point of the PV cell characteristic intersecting the I-axis.

The short-circuit current in a PV cell is proportional to the solar irradiance, for a wide range of cell design and irradiance values, provided that the temperature of

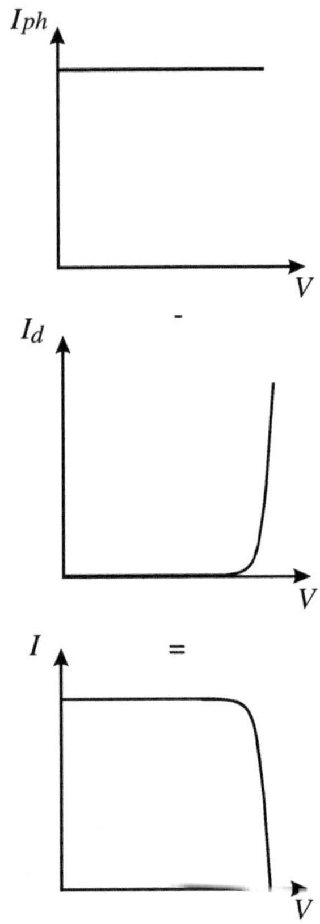

Fig. 3.2 I–V characteristic of the PV cell, obtained by composition of photo generated and diode currents

the cell is the same and that the PV cell does not receive high irradiance values, as in the case of concentrating PV systems.

In a practical PV device, datasheets only provide the nominal short-circuit current, which is the maximum current available at the terminals of the device for a given solar irradiance.

Even if in many cases the assumption of the independence of the short-circuit current on temperature is acceptable, according to literature it can be removed introducing a temperature coefficient for the short-circuit current. For a silicon PV cell, a typical value of 6.4×10^{-6} A/cm^2 °C is frequently used.

3.2.1.2 Open Circuit Voltage

When the PV source is connected to a load with high impedance, nearly all the photo-generated current flows through the diodes, the diodes are directly biased and the PV cell terminal current drops rapidly up to the open circuit condition $I = 0$.

When the open circuit condition is reached, all the photo-generated current is flowing through diodes, thus, considering that $I_{ph} \approx I_{sc}$, the open circuit voltage V_{oc}, can be expressed as:

$$V_{oc} = \frac{2}{3}\frac{kT}{q}\ln\left(\frac{I_{sc} + I_{d1} + I_{d2}}{I_{d1}I_{d2}}\right) \approx \frac{2}{3}\frac{kT}{q}\ln\left(\frac{I_{sc}}{I_{d1}I_{d2}}\right) \tag{3.3}$$

since $I_{sc} \gg I_{d1} + I_{d2}$.

Equation (3.3) demonstrates that the value of the open circuit voltage depends on the logarithm of $I_{sc}/I_{d1}I_{d2}$. Therefore, provided that the cell temperature is constant, the open circuit voltage values scale logarithmically with the short-circuit current which, in turn shows a linear dependence on solar irradiance. This means that the open circuit voltage has a logarithmic dependence on solar irradiance. As a result, it is possible to assess that solar irradiance has a much larger effect of on short-circuit current than on open circuit voltage.

On the other hand, a temperature effect on the open circuit voltage is noticeable. In particular, for typical silicon PV cells in a wide range of operating temperatures, a linear dependence of V_{oc} on T is observed with a temperature coefficient of about −2.3 mV/ °C.

3.2.1.3 Maximum Power Point

The power generated by a PV cell is expressed by the product of the terminal voltage and current. This output power is zero both at the short circuit and open voltage points since the output voltage and the output current are zero, respectively, in these two cases. Between these two extreme points, the power generated by the PV cell is positive.

A remarkable point of the PV cell electrical characteristic is the point where the produced power is at a maximum. This point is referred as maximum power point (MPP) and the corresponding voltage and current are indicated as V_{MP} and I_{MP}, as a rule.

Figure 3.3 shows the PV cell I–V characteristic with the short circuit, open voltage and maximum power points highlighted. As depicted in Fig. 3.3, the MPP defines the rectangle whose area, $P_{MP} = V_{MP}I_{MP}$, is the largest obtainable for any I–V couple on the cell characteristic. The coordinates of the V_{MP} are found by solving the following equation:

$$\left.\frac{dP}{dV}\right|_{V=V_{MP}} = \left.\frac{d(IV)}{dV}\right|_{V=V_{MP}} = \left.\left(I + V\frac{dI}{dV}\right)\right|_{V=V_{MP}} = 0 \tag{3.4}$$

then I_{MP} is determined by evaluating Eq. (3.1) at $V = V_{MP}$.

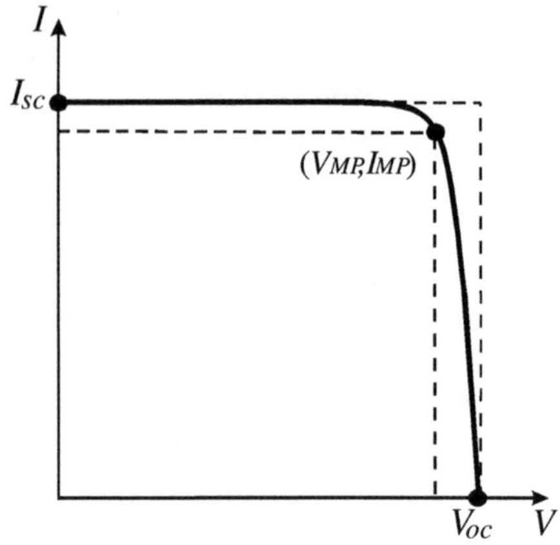

Fig. 3.3 I–V characteristic of the PV cell with remarkable points highlighted

3.2.1.4 Fill Factor

The fill factor (FF) of a PV cell is an important figure of merit whose meaning is relevant to the cell efficiency. It is defined as the ratio between the maximum power $P_{\mathrm{MP}} = I_{\mathrm{MP}}V_{\mathrm{MP}}$, and the product $I_{\mathrm{sc}}V_{\mathrm{oc}}$, as expressed in Eq. (3.5).

$$FF = \frac{P_{\mathrm{MP}}}{I_{\mathrm{sc}}V_{\mathrm{oc}}} = \frac{I_{\mathrm{MP}}V_{\mathrm{MP}}}{I_{\mathrm{sc}}V_{\mathrm{oc}}} \tag{3.5}$$

The FF is a measure of the squareness, in a word of the ideality, of the PV cell I–V characteristic; in particular, it represents the ratio between the areas of the rectangles highlighted in Fig. 3.3.

The literature gives an approximation of the FF by the following empirical expression:

$$FF = \frac{V_{\mathrm{oc}} - \frac{kT}{q}\ln(qV_{\mathrm{oc}}/kT + 0.72)}{V_{\mathrm{oc}} + kT/q} \tag{3.6}$$

The relationship between the FF and the PV cell efficiency η is:

$$\eta = \frac{P_{\mathrm{MP}}}{P_i} = FF\frac{I_{\mathrm{sc}}V_{\mathrm{oc}}}{P_i} \tag{3.7}$$

where P_i is the power incident on the PV cell, whose value is tied to the characteristic of the light spectrum hitting the PV cell. Therefore, the closer the FF is to one the higher will be the cell efficiency. In the ideal case FF = 1.

3.2.2 *Diffusion Diode Non-Ideality*

In the mathematical model of the PV cell expressed by Eq. (3.1), a unitary ideality factor of the diffusion diode has been assigned.

In a practical device, a *I–V* characteristic with ideality factor equal to 1 is unlikely. Therefore, an additional parameter A_0 is usually introduced in the PV cell *I–V* equation to take into account the nonidealities in the junction behavior. With this assumption, Eq. (3.1) can be reformulated as follows:

$$I = I_{\text{ph}} - I_{s1}\left(e^{qV/A_0kT} - 1\right) - I_{s2}\left(e^{qV/2kT} - 1\right) \tag{3.8}$$

3.2.3 *Parasitic Resistance Effects*

In the operation of a practical PV cell several phenomena, affecting the cell response, occur. Most of these effects have not been modeled until now, since the solar cell has been described as a nearly ideal device.

It should be observed that, for an accurate representation of the PV cell electrical behavior, the cited effects should be taken into account.

First of all the so-called parasitic resistance effects have to be introduced in the model. In particular, two resistive terms have to be accounted for: the series resistance and the shunt resistance.

3.2.3.1 Series Resistance

The series resistance R_s in a PV cell is a lumped element used to model the series power losses due to the current circulation through different parts of the device. It can be placed between the cathode of the diodes and the corresponding output terminal.

R_s results from the sum of several structural resistances. It mainly depends on the contact of the metal base with the *p*-doped semiconductor layer, the resistance of the base and emitter layers, the contact resistance of the *n*-doped layer with the top metal grid, the resistance of the grid, and other contact resistances.

The effect of the series resistance on the PV cell I–V curve is represented in Fig. 3.4. A quantitative evaluation of this effect is given in Sect. 5.4.2.

It should be observed that R_s has a stronger influence on the part of the electrical characteristic at the right of the MPP, i.e., the constant voltage region. Moreover, it has no effect on the open circuit voltage, whereas it lessens the short-circuit current.

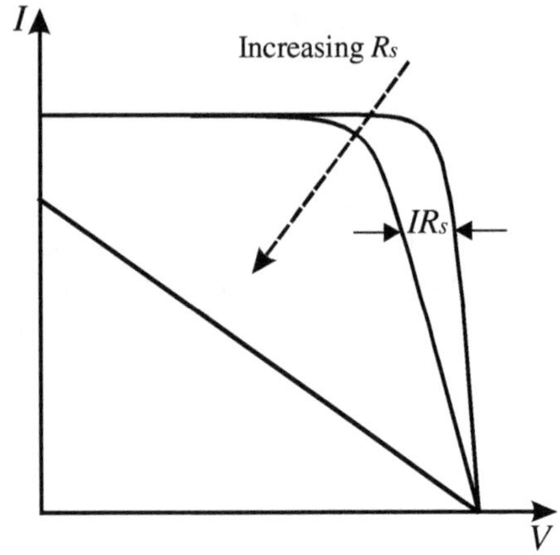

Fig. 3.4 Effect of the series resistance on the I–V characteristic

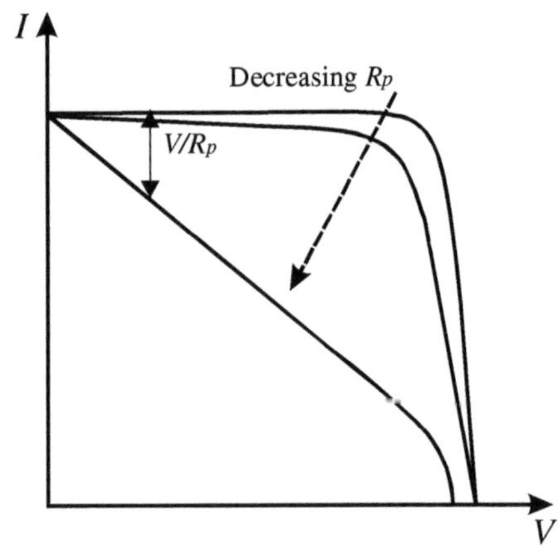

Fig. 3.5 Effect of the shunt resistance on the I–V characteristic

3.2.3.2 Shunt Resistance

The shunt resistance R_p in a PV cell is a lumped element used to model the effects of leakage currents of the *p-n* junction. It can be placed in parallel with the diodes.

Several shunt resistive losses are found in a practical PV device, for example local short circuits in the emitter layer or perimeter shunts along the semiconductor borders. Therefore, R_p mainly depends on the fabrication process of the PV cell.

The effect of the shunt resistance on the PV cell *I–V* curve is represented in Fig. 3.5. A quantitative evaluation of this effect is given in Sect. 5.4.2.

The shunt resistance has a stronger influence on the part of the electrical characteristic at the left of the MPP, i.e., the constant current region. Furthermore, it has no effect on the short-circuit current, whereas it affects the open circuit voltage.

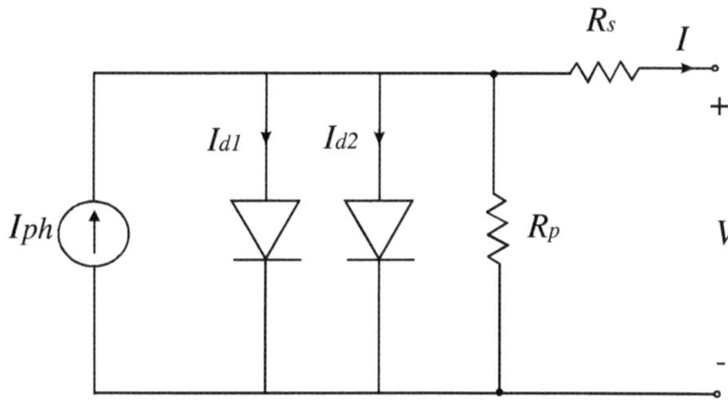

Fig. 3.6 Circuit model of a practical PV cell including series and shunt parasitic resistances

3.2.4 Generalized Double Diode Model

The circuit model referred to a practical PV cell, including the parasitic resistances, is shown in Fig. 3.6.

The corresponding equation, taking into account the nonideality of the diffusion diode as well, is:

$$I = I_{\mathrm{ph}} - I_{s1}\left(e^{q(V+IR_s)/A_0kT} - 1\right) - I_{s2}\left(e^{q(V+IR_s)/2kT} - 1\right) - \frac{(V + IR_s)}{R_p} \quad (3.9)$$

Equation (3.9) represents the generalized double diode mathematical model of a real PV cell.

The double diode or double exponential model of a PV cell is generally accepted as reflecting the electrical behavior of real PV cells, especially those constructed from polycrystalline silicon. This representation of solar cell behavior, incorporating the space charge recombination effects by a separate current component with its own exponential voltage dependence, is considered by the literature as the most accurate, especially when dealing with low illuminations. This consideration is important when modeling is oriented to the optimization of solar cells manufacturing processes.

On the other hand, when the study is focused on a PV modules or fields, the implications of low irradiance are less important since the PV electric power generation occurs at high irradiance values.

In the following subsection the simplified single diode representation, which is the most commonly used when the PV modeling is aimed to power generation issues, is given.

3.2.5 Simplified Single Diode Model

The single diode model is a simplified representation of the PV cell *I–V* characteristic. It is derived from the Shockley diode equation including a diode quality

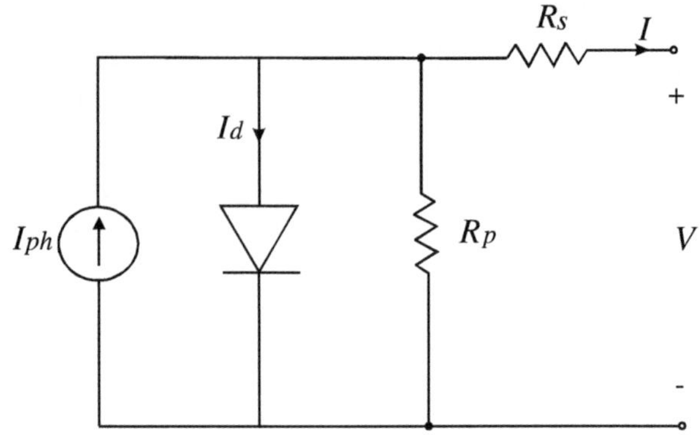

Fig. 3.7 Single diode circuit model of a practical PV cell

factor A_q to approximate the effect of both diffusion and recombination in the space charge layer.

3.2.5.1 The Shockley Diode Equation

The Shockley diode equation relates the current of a *p-n* junction diode, I_D to the diode voltage V_D, according to the so-called *I–V* characteristic:

$$I_D = I_s(e^{V_D/A_q V_T} - 1) \tag{3.10}$$

where I_s is the saturation current or scale current of the diode, $V_T = kT/q$ is the thermal voltage, and A_q is the diode ideality factor.

Equation (3.10) can be regarded as equivalent to Eq. (2.78) but, in this case, the dependence on the fabrication process and on the semiconductor material is considered by introducing the ideality factor of the diode.

A_q varies from 1 to 2; in particular, $A_q \approx 1$ for diodes dominated by the recombination in quasi-neutral region and $A_q \approx 2$ for diodes dominated by the recombination in the depletion region.

3.2.5.2 PV Cell Single Diode Equivalent Circuit

On the basis of the Shockley diode representation, the *I–V* equation of the PV cell becomes:

$$I = I_{\text{ph}} - I_s(e^{q(V+IR_s)/A_q kT} - 1) - \frac{(V + IR_s)}{R_p} \tag{3.11}$$

The corresponding equivalent circuit is that sketched in Fig. 3.7.

The obtained model is quite general, it can be used for different solar cells where the parameters represent physical phenomena of the source. For example, in case of organic cells (see Sect. 2.15.5), the current source I_{ph} corresponds to the

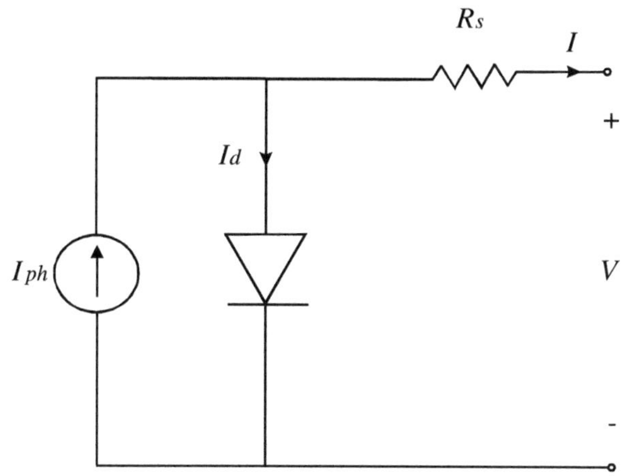

Fig. 3.8 Simplified single diode circuit model of a practical PV cell

number of free electrons/holes pairs after generation and before any recombination process, the shunt resistor R_p is due to the recombination at the interface between the donor and the acceptor material, and may include other recombination effects far away from the dissociation site, the series resistor R_s considers conductivity of each carrier in the transport medium and it is affected by space charges and traps.

If the shunt resistor R_p can be neglected, the last term in Eq. (3.11) is null and the so called four parameter model, given by Eq. (3.12), is obtained.

$$I = I_{\mathrm{ph}} - I_s\left(e^{q(V+IR_s)/A_q kT} - 1\right) \tag{3.12}$$

On the basis of such a simplification, a final simplified single diode equivalent circuit is derived. This equivalent circuit is shown in Fig. 3.8; it will be considered as the reference model both for the simulated analysis of the PV system and for the set up of the PV source emulator henceforth.

As previously explained, the presence of the shunt resistor in the circuit model is mainly due to the leakage current of the *p-n* junction depending on the fabrication process of the PV cell. Considering PV modules or fields as a part of an electric generation system, the shunt resistance has a stronger influence in an uncommon region of operation, i.e., the constant current region and for low solar irradiance values. Therefore, the simplification introduced by neglecting R_p does not affect significantly the validity of the model and its use for the scope of emulating the PV source, since the power generation occurs at high irradiance values and in the neighborhood of the MPP. This can be deduced by the example shown in Fig. 5.18.

It is a common practice to neglect the term '−1' in Eq. (3.12) because, in silicon devices, the dark saturation current is much smaller than the exponential term. With this further assumption, the final PV cell model formulation is:

$$I = I_0 - e^{[(V+IR_s)K_1 + K_2]} \tag{3.13}$$

with the positions:

$$\begin{aligned} I_{\mathrm{ph}} &= I_0 \\ \frac{q}{A_q kT} &= K_1 \\ I_s &= e^{K_2} \end{aligned} \tag{3.14}$$

Using Eqs. (3.13) and (3.14), the PV cell electrical behavior can be studied once the four model parameters K_1, K_2, R_s, and I_0 are known.

3.2.5.3 PV Voltage versus Current Model Explicit Formulation

The PV model inversion, i.e., the model representation as $V = f(I)$, is particularly useful for the source emulation purpose. As a matter of fact, it is convenient to obtain the voltage on the basis of the actual current, as it will be explained in detail in the chapters dedicated to emulation.

When the single diode model neglecting R_p is considered, the analytical solution of the equation $V = f(I)$ is straightforward. In particular, this solution is obtained by the inversion of Eq. (3.12) in which the dark saturation current is neglected respect to the exponential term or by the inversion of Eq. (3.13).

The following equations are obtained, respectively:

$$V - \frac{A_q kT}{q} \ln\left(\frac{I_{\mathrm{ph}} - I}{I_s}\right) - IR_s \tag{3.15}$$

$$V = \frac{\ln(I_0 - I) - K_2}{K_1} - IR_s \tag{3.15a}$$

It should be observed that the logarithmic term is null for $I = I_{\mathrm{ph}} - I_s$ and it becomes negative for $I_{\mathrm{ph}} - I_s < I < I_{\mathrm{ph}}$. Since I_s is much smaller than the other currents, it is necessary to put particular care to acquire experimental values of current near to the short-circuit condition because the presence of a superimposed noise could result in a negative voltage value or a calculation error evaluating the logarithm.

When the considered PV model is that described in Eq. (3.11), the inversion is more complex and can be managed using the Lambert W function.

The Lambert W function is defined to be the multivalued inverse of the function $f(w) = we^w$.

For our purpose, the attention can be focused on the real-valued Lambert W$(f(w))$, then the relation is defined only for $f(w) \geq -1/e$. If the equation is written under the form $Y = Xe^X$, the X value is obtained as $X =$ Lambert W(Y).

In Fig. 3.9 the real branch of the Lambert W function for $x > -1/e$ is plotted, according to the following Matlab® commands:

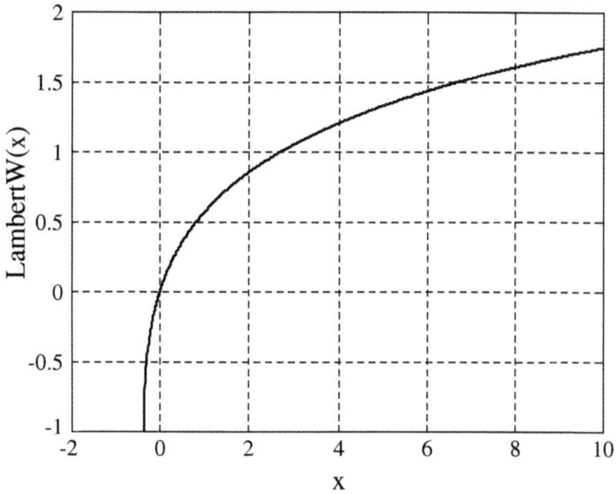

Fig. 3.9 Real branch of the Lambert W function

≫*x* = (−*1/2.7183:0.01:10*);
≫*L* = *lambertw*(*x*);
≫*plot*(*x*, *L*);

On this basis, the following explicit expressions of $V = f(I)$ are obtained:

$$V = -IR_s + (A_q V_T)\ln\left(\frac{A_q V_T}{I_s R_p}\text{Lambert }W\left\{\frac{I_s R_p}{A_q V_T}\exp\left[\frac{R_p}{A_q V_T}(I_{\text{ph}} - I + I_s)\right]\right\}\right) \tag{3.16}$$

$$\begin{aligned} V &= I(R_p + R_s) - I_{ph}R_p + I_s R_p \\ &\quad - (A_q V_T)\text{Lambert }W\left\{\frac{I_s R_p}{A_q V_T}\exp\left[\frac{R_p}{A_q V_T}(I_{\text{ph}} - I + I_s)\right]\right\} \end{aligned} \tag{3.17}$$

The demonstration of (3.16) and (3.17) are given in the next section "Calculus of $I = f(V)$ by Lambert W Function".

3.2.5.4 Calculus of $V = f(I)$ by Lambert W Function

Equation (3.11) can be rewritten as:

$$I = I_{\text{ph}} - I_s\left(e^{\frac{V+IR_s}{A_q V_T}} - 1\right) - \frac{(V + IR_s)}{R_p} \tag{3.18}$$

where $V_T = \frac{kT}{q}$

$$I_{\text{ph}} - I + I_s = I_s e^{\frac{V+IR_s}{A_q V_T}} + \frac{(V + IR_s)}{R_p} \tag{3.19}$$

multiplying both member for $\frac{R_p}{A_q V_T}$

$$\frac{R_p}{A_q V_T}\left(I_{ph} - I + I_s\right) = \frac{R_p I_s}{A_q V_T} e^{\frac{V+IR_s}{A_q V_T}} + \frac{(V + IR_s)}{A_q V_T} \tag{3.20}$$

taking the exponential of both members

$$\exp\left[\frac{R_p}{A_q V_T}\left(I_{\text{ph}} - I + I_s\right)\right] = \exp\left[\frac{R_p I_s}{A_q V_T} e^{\frac{V+IR_s}{A_q V_T}}\right] \cdot \exp\left[\frac{(V + IR_s)}{A_q V_T}\right] \tag{3.21}$$

multiplying both member for $\frac{I_s R_p}{A_q V_T}$

$$\frac{I_s R_p}{A_q V_T} \exp\left[\frac{R_p}{A_q V_T}\left(I_{\text{ph}} - I + I_s\right)\right] = \frac{I_s R_p}{A_q V_T} e^{\frac{V+IR_s}{A_q V_T}} \cdot \exp\left[\frac{R_p I_s}{A_q V_T} e^{\frac{V+IR_s}{A_q V_T}}\right] \tag{3.22}$$

It should be noted that the second member of Eq. (3.22) is in the form we^w where:

$$w = \frac{I_s R_p}{A_q V_T} e^{\frac{V+IR_s}{A_q V_T}} \tag{3.22a}$$

Finally, the Lambert W function can be applied to both members

$$\text{Lambert } W\left\{\frac{I_s R_p}{A_q V_T} \exp\left[\frac{R_p}{A_q V_T}\left(I_{\text{ph}} - I + I_s\right)\right]\right\} = \frac{I_s R_p}{A_q V_T} e^{\frac{V+IR_s}{A_q V_T}} \tag{3.23}$$

and the $V = f(I)$ formulation is obtained as:

$$V = -IR_s + \left(A_q V_T\right) \ln\left(\frac{A_q V_T}{I_s R_p} \text{Lambert } W\left\{\frac{I_s R_p}{A_q V_T} \exp\left[\frac{R_p}{A_q V_T}\left(I_{\text{ph}} - I + I_s\right)\right]\right\}\right) \tag{3.24}$$

or by using Eq. (3.18) as:

$$I_s R_p e^{\frac{V+IR_s}{A_q V_T}} = -I\left(R_p + R_s\right) + I_{\text{ph}} R_p - I_s R_p - V \tag{3.25}$$

$$\begin{aligned} &- I\left(R_p + R_s\right) + I_{ph} R_p - I_s R_p - V = \\ &\left(A_q V_T\right) \text{Lambert } W\left\{\frac{I_s R_p}{A_q V_T} \exp\left[\frac{R_p}{A_q V_T}\left(I_{\text{ph}} - I + I_s\right)\right]\right\} \end{aligned} \tag{3.26}$$

$$\begin{aligned} V = {} & I\left(R_p + R_s\right) - I_{ph} R_p + I_s R_p \\ & - \left(A_q V_T\right) \text{Lambert } W\left\{\frac{I_s R_p}{A_q V_T} \exp\left[\frac{R_p}{A_q V_T}\left(I_{\text{ph}} - I + I_s\right)\right]\right\} \end{aligned} \tag{3.27}$$

3.2.5.5 PV Current versus Voltage Model Explicit Formulation

The Lambert W function can be used to express I versus V as an explicit function, as well. In this case, the following expression is obtained:

$$I = \frac{A_q V_T}{R_s} \text{Lambert } W\left[\frac{I_s R_p R_s}{A_q V_T (R_s + R_p)} \exp\left(\frac{R_p (I_{ph} R_s + I_s R_s + V)}{A_q V_T (R_s + R_p)}\right)\right] + \frac{V - (I_{\text{ph}} - I_s) R_p}{R_s + R_p} \tag{3.28}$$

The demonstration of Eq. (3.28) is given in the next section “Calculus of $I = f(V)$ by Lambert W Function”.

3.2.5.6 Calculus of $I = f(V)$ by Lambert W Function

From (3.16)

$$IR_p = I_{\text{ph}} R_p - I_s R_p e^{\frac{V+IR_s}{A_q V_T}} + I_s R_p - V - IR_s \tag{3.29}$$

$$I = \frac{I_{\text{ph}} R_p - I_s R_p e^{\frac{V+IR_s}{A_q V_T}} + I_s R_p - V}{R_p + R_s} \tag{3.30}$$

By multiplying both members for R_s

$$IR_s = \frac{R_s}{R_p + R_s}\left(I_{\text{ph}} R_p - I_s R_p e^{\frac{V+IR_s}{A_q V_T}} + I_s R_p - V\right) \tag{3.31}$$

Then, by adding V to both members,

$$V + IR_s = \frac{R_s}{R_p + R_s}\left(I_{\text{ph}} R_p - I_s R_p e^{\frac{V+IR_s}{A_q V_T}} + I_s R_p - V \frac{R_p}{R_s}\right) \tag{3.32}$$

Now, dividing for $A_q V_T$,

$$\frac{V + IR_s}{A_q V_T} = \frac{R_s}{A_q V_T (R_p + R_s)}\left(I_{\text{ph}} R_p - I_s R_p e^{\frac{V+IR_s}{A_q V_T}} + I_s R_p - V \frac{R_p}{R_s}\right) \tag{3.33}$$

$$\frac{V + IR_s}{A_q V_T} + \frac{I_s R_p R_s}{A_q V_T (R_p + R_s)} e^{\frac{V+IR_s}{A_q V_T}} = \frac{R_s\left(I_{\text{ph}} R_p + I_s R_p - V \frac{R_p}{R_s}\right)}{A_q V_T (R_p + R_s)} \tag{3.34}$$

Taking the exponential of both members

$$e^{\frac{V+IR_s}{A_qV_T}} \cdot \exp\left[\frac{I_sR_pR_s}{A_qV_T(R_p+R_s)} e^{\frac{V+IR_s}{A_qV_T}}\right] = \exp\left[\frac{R_s\left(I_{\mathrm{ph}}R_p + I_sR_p - V\frac{R_p}{R_s}\right)}{A_qV_T(R_p+R_s)}\right] \tag{3.35}$$

By multiplying both members for $\frac{I_sR_pR_s}{A_qV_T(R_p+R_s)}$

$$\frac{I_sR_pR_s}{A_qV_T(R_p+R_s)} e^{\frac{V+IR_s}{A_qV_T}} \cdot \exp\left[\frac{I_sR_pR_s}{A_qV_T(R_p+R_s)} e^{\frac{V+IR_s}{A_qV_T}}\right] = \frac{I_sR_pR_s}{A_qV_T(R_p+R_s)} \exp\left[\frac{R_s\left(I_{\mathrm{ph}}R_p + I_sR_p - V\frac{R_p}{R_s}\right)}{A_qV_T(R_p+R_s)}\right] \tag{3.36}$$

now, the first member is in the form we^w, then taking the Lambert W of both members:

$$\frac{I_sR_pR_s}{A_qV_T(R_p+R_s)} e^{\frac{V+IR_s}{A_qV_T}} = \text{Lambert } W\left\{\frac{I_sR_pR_s}{A_qV_T(R_p+R_s)} \exp\left[\frac{R_s\left(I_{\mathrm{ph}}R_p + I_sR_p - V\frac{R_p}{R_s}\right)}{A_qV_T(R_p+R_s)}\right]\right\} \tag{3.37}$$

$$I_sR_pe^{\frac{V+IR_s}{A_qV_T}} = \frac{A_qV_T(R_p+R_s)}{R_s} \text{Lambert } W\left\{\frac{I_sR_pR_s}{A_qV_T(R_p+R_s)} \exp\left[\frac{R_s\left(I_{\mathrm{ph}}R_p + I_sR_p - V\frac{R_p}{R_s}\right)}{A_qV_T(R_p+R_s)}\right]\right\} \tag{3.38}$$

The first member can be written by using Eq. (3.16) as:

$$I_sR_pe^{\frac{V+IR_s}{A_qV_T}} = -I(R_p+R_s) + I_{\mathrm{ph}}R_p - I_sR_p - V \tag{3.39}$$

$$-I(R_p+R_s) + I_{\mathrm{ph}}R_p - I_sR_p - V = \frac{A_qV_T(R_p+R_s)}{R_s} \text{Lambert } W\left\{\frac{I_sR_pR_s}{A_qV_T(R_p+R_s)} \exp\left[\frac{R_s\left(I_{\mathrm{ph}}R_p + I_sR_p - V\frac{R_p}{R_s}\right)}{A_qV_T(R_p+R_s)}\right]\right\} \tag{3.40}$$

$$I(R_p + R_s) = -\frac{A_q V_T (R_p + R_s)}{R_s} \text{Lambert } W \left\{ \frac{I_s R_p R_s}{A_q V_T (R_p + R_s)} \exp\left[\frac{R_s \left(I_{\text{ph}} R_p + I_s R_p - V \frac{R_p}{R_s} \right)}{A_q V_T (R_p + R_s)} \right] \right\} + I_{\text{ph}} R_p - I_s R_p - V \tag{3.41}$$

$$I = -\frac{A_q V_T}{R_s} \text{Lambert } W \left\{ \frac{I_s R_p R_s}{A_q V_T (R_p + R_s)} \exp\left[\frac{R_s \left(I_{\text{ph}} R_p + I_s R_p - V \frac{R_p}{R_s} \right)}{A_q V_T (R_p + R_s)} \right] \right\} + \frac{I_{\text{ph}} R_p - I_s R_p - V}{(R_p + R_s)} \tag{3.42}$$

3.3 Cell-Module/Field

The models described until now have been developed for a single PV cell; however, they can be extended to the case of series/parallel connection of cells inside a PV module or field.

Starting from the double diode model, expressed by Eq. (3.9), an analogous model and equivalent circuit of a PV module/field is obtained provided that a suitable scaling of the parameters is performed. In particular, considering a module/field formed of N_P cells in parallel and N_S cells in series, if solar irradiance and temperature are assumed to be uniform, the following relationships are valid:

$$\begin{cases} I_{\text{ph,tot}} = N_P I_{\text{ph}} \\ I_{\text{tot}} = N_P I \\ V_{\text{tot}} = N_S V \end{cases} \tag{3.43}$$

where the subscript '*tot*' stands for a quantity referred to the whole module or field.

The resulting analytical model is:

$$I_{tot} = I_{\text{ph,tot}} - N_P \left[I_{s1} \left(\frac{q}{e^{A_0 kT}} \left(\frac{V_{\text{tot}}}{N_S} + \frac{I_{\text{tot}} R_s}{N_P} \right) - 1 \right) + I_{s2} \left(e^{\frac{q}{2kT} \left(\frac{V_{\text{tot}}}{N_S} + \frac{I_{\text{tot}} R_s}{N_P} \right)} - 1 \right) - \frac{\frac{V_{\text{tot}}}{N_S} + \frac{I_{\text{tot}} R_s}{N_P}}{R_p} \right] \tag{3.44}$$

On the other hand, starting from the single diode model expressed by Eq. (3.12), considering the simplification introduced in Eq. (3.13) and with the following positions:

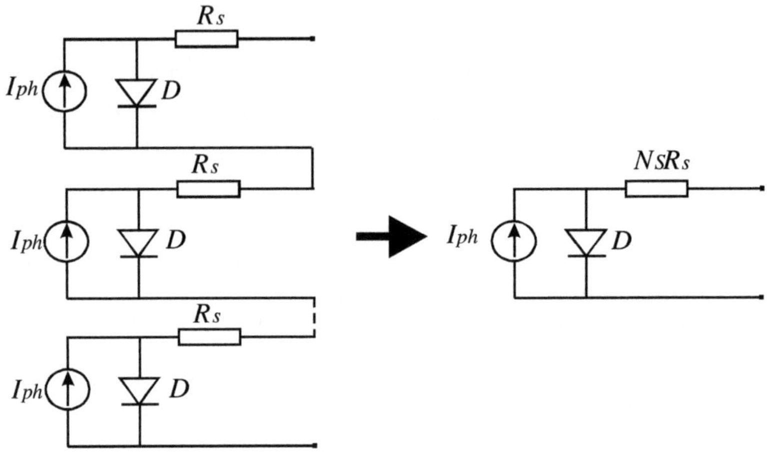

Fig. 3.10 Equivalent circuit model of N_S series connected cells

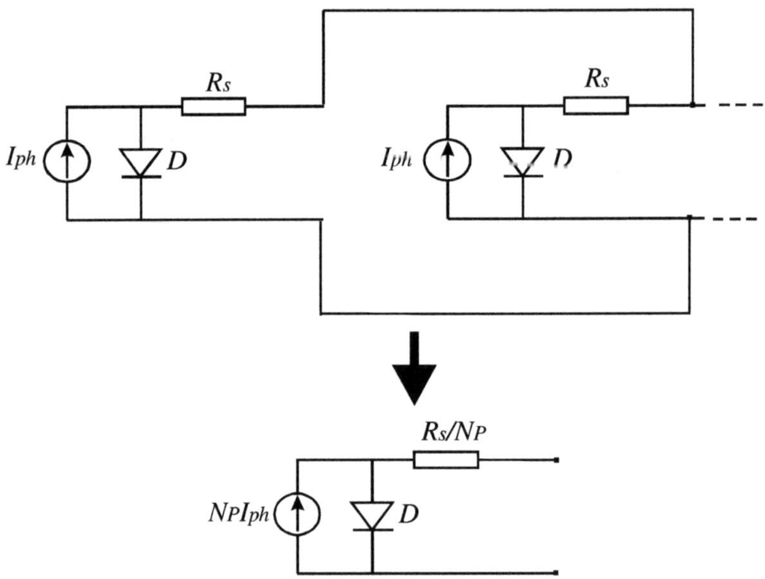

Fig. 3.11 Equivalent circuit model of N_P parallel connected cells

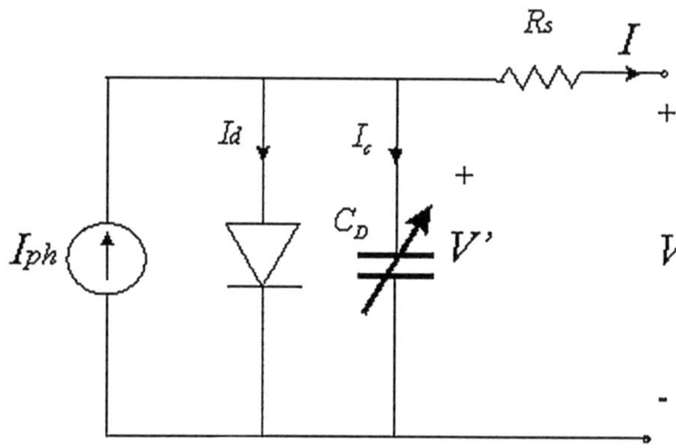

Fig. 3.12 Dynamic circuit model containing nonlinear diffusion capacitance

$$\begin{aligned} I_{\text{ph,tot}} &= I_{0,\text{tot}} \\ \frac{q}{A_q kT} &= K_1 \\ N_P I_s &= e^{K_2} \end{aligned} \tag{3.45}$$

the final expression of the module/field model is:

$$I_{\text{tot}} = I_{0,\text{tot}} - e^{\left[\left(\frac{V_{\text{tot}}}{N_S} + \frac{I_{\text{tot}} R_s}{N_P}\right) K_1 + K_2\right]} \tag{3.46}$$

The equivalent circuit model of the PV module/field can be deduced in this case according to the schemes represented in Figs. 3.10 and 3.11.

3.4 Dynamic Model

The considered models are suited for slow variations of solar irradiance and load. Several situations, however, need a precise knowledge of solar cell behavior during transients. In these cases, a dynamical model is necessary.

3.4.1 Parallel Capacitance

In Sect. 2.12, it has been explained that the *p-n* junction includes a capacitive effect due to the effect of excess minority carriers stored in the quasi-neutral region. This effects can be modeled by a voltage depending capacitance that, for low frequency variations, can be expressed as:

$$C_D = \tau_p \frac{I}{V_T} = \tau_p \frac{I_d}{V_T} e^{b\frac{V'}{V_T}} = C_0 e^{b\frac{V'}{V_T}} \tag{3.47}$$

where C_O is the so-called base capacitance and b is a fitting parameter.

To consider this dynamic effect in the PV cell model, the variable capacitance, defined by Eq. (3.47), must be included in the circuit model. As a consequence, during transients, a discrepancy between current and voltage obtained by static and dynamic model is expected. This difference depends on the capacitive current I_c that changes the effect of photo-generated current.

The dynamic circuit model containing the nonlinear diffusion capacitance C_D is shown in Fig. 3.12.

For the circuit scheme in Fig. 3.12, the following current balance is valid:

$$I = I_{\text{ph}} - I_d - I_C \tag{3.48}$$

where

$$I_C = \frac{\mathrm{d}}{\mathrm{d}t}(C_D V') = C_D \frac{\mathrm{d}V'}{\mathrm{d}t} + V' \frac{\mathrm{d}C_D}{\mathrm{d}t} \tag{3.49}$$

using (3.47)

$$\frac{\mathrm{d}C_D}{\mathrm{d}t} = C_0 \frac{b}{V_T} e^{b\frac{V'}{V_T}} \tag{3.50}$$

Therefore,

$$I_C = C_0 e^{b\frac{V'}{V_T}} \left(b\frac{V'}{V_T} + 1 \right) \frac{\mathrm{d}V'}{\mathrm{d}t} \tag{3.51}$$

Finally, taking into account that $V' = V - IR_S$, the expression of the capacitive current versus electrical parameters measured at the PV cell terminals can be achieved.

$$I_C = C_0 e^{b\frac{(V-IR_S)}{V_T}} \left(b\frac{V - IR_S}{V_T} + 1 \right) \frac{\mathrm{d}(V - IR_S)}{\mathrm{d}t} \tag{3.52}$$

if the variation of the external voltage V is negligible, the following equation is obtained:

$$I_C = -R_S C_0 e^{b\frac{(V-IR_S)}{V_T}} \left(b\frac{V - IR_S}{V_T} + 1 \right) \frac{\mathrm{d}I}{\mathrm{d}t} \tag{3.53}$$

By Eq. (3.53) it is possible to notice that the nonlinear junction capacitance influences the current balance in Eq. (3.48) with a term proportional to the series resistance and the base capacitance that is relevant only when a fast load current variation occurs. Finally, being this capacitance parallel connected to the diode and the current generator, when N_S cells are series connected, the capacitance equivalent value is divided by N_S.

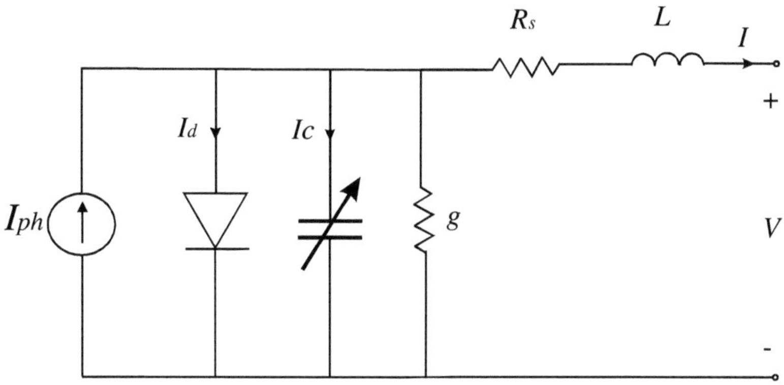

Fig. 3.13 Dynamic circuit model containing nonlinear diffusion capacitance and parasitic inductance

3.4.2 Series Inductance

For sudden photovoltaic current or load variation at first instants, time domain waveforms of external voltage and current may differ from those obtained by the model described in Sect. 3.4.1. This is due to both the facts that for fast variations a more complicated capacitance model should be used, as explained in 2.12, for high frequencies and to the presence of a low parasitic inductance of electrical wires connecting the cell to the electric terminals.

The dynamic circuit model containing nonlinear diffusion capacitance and parasitic inductance is shown in Fig. 3.13. When N_S cells are series connected, the parasitic equivalent inductance is multiplied by N_S.

3.5 Modeling PV Fields under Nonuniform Illuminating Conditions

As previously stated, the PV modules are composed of matrices of solar cells interconnected in series and parallel and PV fields are given by a set of series/parallel connected modules.

The overall electrical characteristic of a PV field is obtained by the composition of the electrical characteristics of its modules.

Figures 3.14 and 3.15 schematize, for the case of two identical modules under the same solar irradiance, the composition process used to obtain the *I*–*V* characteristic of the PV field, from the knowledge of the *I*–*V* characteristics of the modules.

The overall series connection *I*–*V* curve is obtained by summing the voltage point-by-point for each value of the common current. As a result, the open circuit voltage is twice the open circuit voltage of the single module.

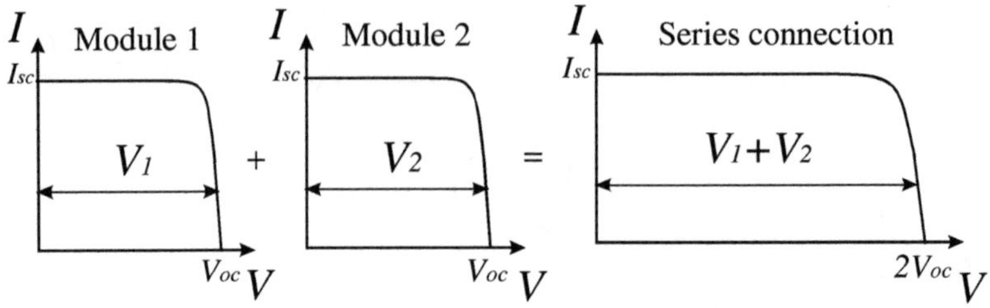

Fig. 3.14 Composition of *I–V* curves for series connected modules under the same solar irradiance in a PV field

Fig. 3.15 Composition of *I-V* curves for parallel connected modules under the same solar irradiance in a PV field

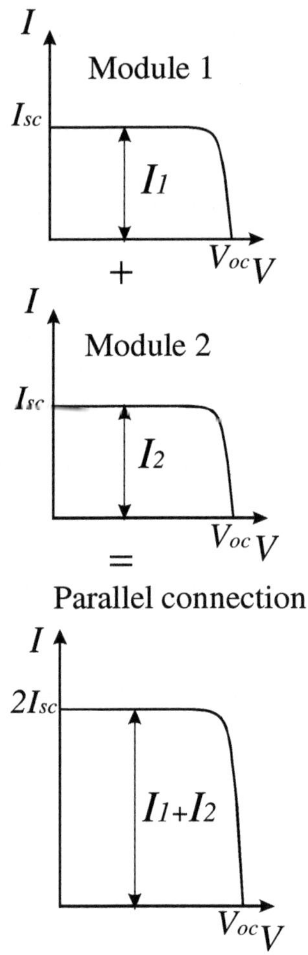

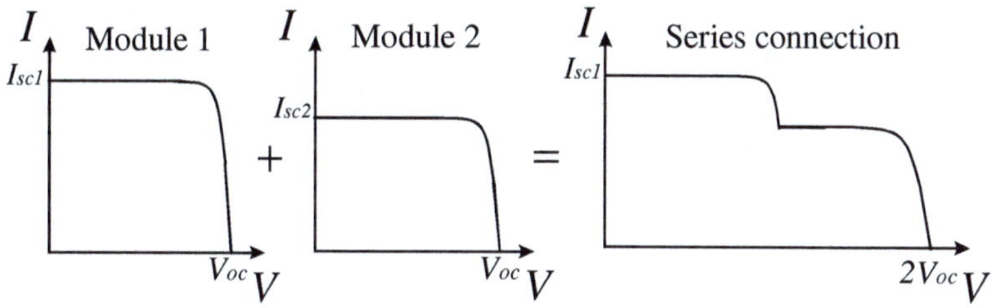

Fig. 3.16 Composition of *I–V* curves for two series connected modules under different solar irradiance in a PV field

Fig. 3.17 Comparison between *P–V* curves under uniform and nonuniform irradiance

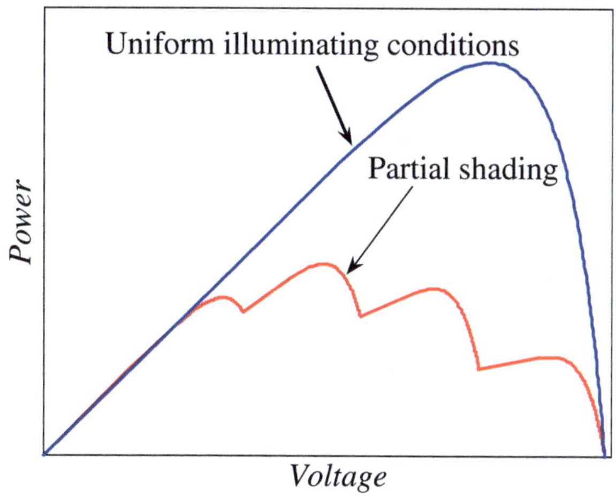

The process can be obviously extended to the case of PV field formed of an arbitrary number of series/parallel connected modules.

In the case of PV modules under different solar irradiance, the shape of the overall *I–V* curve shows steps that correspond to the occurring of local maxima in the power-voltage (*P–V*) curve, as will be explained in the following. Figure 3.16 illustrates the composition process when two modules are series connected and subject to different solar irradiance. The series connection is the most frequent case encountered when dealing with PV fields, since a high output voltage is often necessary to supply the electrical load.

When more modules are series connected, and they operate under different solar irradiance, the definition of the whole PV field characteristics is obtained by summing the single contributions.

As a matter of fact, the PV field performance is dependent on the behavior of the individual solar cells and this could be critical, especially in nonideal operating conditions.

An usual nonideal operating condition that strongly affects PV field during electric generation is the partial shading, i.e., the non-uniform distribution of the solar irradiance on the PV field surface. This problem is mainly due to clouds passing over a portion of the PV array and is particularly serious for large installations.

Partial shading leads to a reduction of the available output power. Moreover, the occurrence of multiple local maxima in the *P–V* characteristic, due to non-uniform solar irradiance, can cause a failure of the MPPT control with a relevant power loss due to the lack of the real MPP detection. Figure 3.17 illustrates the qualitative difference between *P–V* characteristics of PV field under uniform illuminating conditions and under partial shading.

It is evident that the correct determination of solar irradiance distribution over a PV array is crucial both for the optimization of the MPPT strategy and for a correct prediction of the PV system energy capability.

Usually, manufacturers give the characteristic parameters referred to a photovoltaic module supposed under uniform solar irradiance and temperature; hence, it is useful to obtain the model of the modules and to build the whole characteristic of the PV field on the basis of their electrical connections.

Depending on the design and the dimensions of the considered PV field, the best approach could be to separate its model into single modeling units (the modules), as solar irradiance and temperature are unlikely to change across the dimensions of a single unit, while they could possibly change across the dimensions of the whole field. In this way each module can be characterized on the basis of its own temperature and solar irradiance, so partial shading situations and other nonideal operating modes can be investigated, as well as the case of uniformly illuminated PV field.

To be more precise, once the module *I–V* characteristics are available, a composition process is necessary to characterize the whole PV field. This process is based on:

- the sum of module voltages, the currents being equal, for series connected modules;
- the sum of currents, the voltages being equal, for parallel connected modules.

In order to avoid possible mismatch in series or parallel connection, in a real installation, *bypass* and *block* diodes are provided.

The bypass diodes are used to protect series connected modules from problems due to possible current mismatch.

The block diodes are used to avoid that, under some operating or fault conditions, a string, i.e., the series connection of modules, could be interested by a reverse current generated by other strings.

The installation and the usefulness of such diodes are described by Fig. 3.18.

As for the *bypass* diode, it is parallel connected to each module in the PV field, as shown in Fig. 3.18a. When a module is completely shaded, it operates in the point $I \approx 0$, $V \approx 0$ and does not allow the current coming form other modules to flow. The *bypass* diode, in this case, give an alternative way to the current. When the module is illuminated, the *bypass* diode is inverse polarized and behaves as an open circuit.

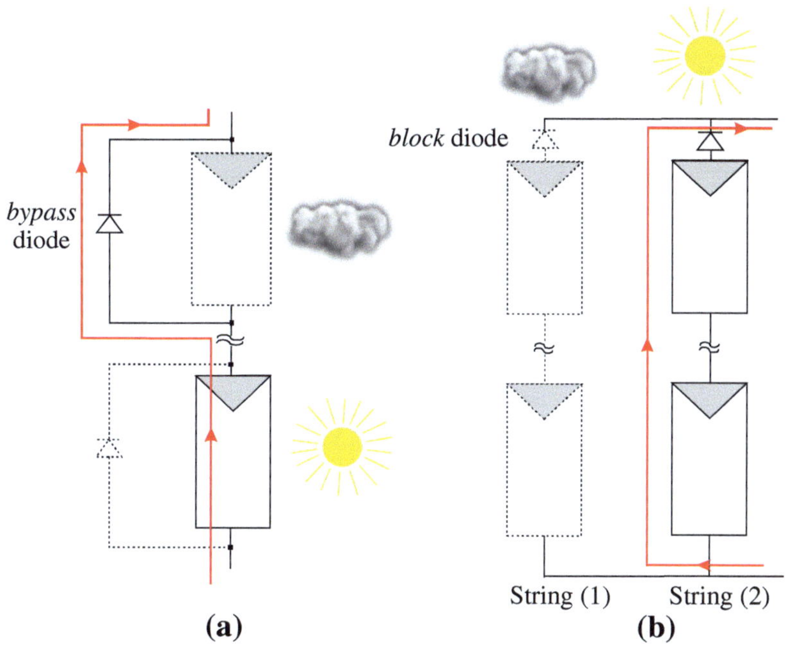

Fig. 3.18 Installation and operation of *bypass* diode (**a**) and *block* diode (**b**)

As for the *block* diode, it is series connected to a PV string, as shown in Fig. 3.18b. With reference to Fig. 3.18b, when the string (1) is partially shaded, its voltage is lower than the voltage of string (2). The *block* diode is inverse polarized and prevents string (1) from imposing its low voltage to the parallel of the two strings. On the other hand, the *block* diode causes a power loss equal to the product of the threshold voltage of the diode and the direct current ($V_{threshold} \cdot I$) and must be able to undertake a reverse voltage equal to the open circuit voltage of the string and a forward current equal to the maximum current delivered by the string.

Several configurations, in terms of modules connection and conditioning power electronics, have been proposed in the last decade in the field of PV grid connected plants, starting from centralized architectures to more and more decentralized technologies. Some examples of modules configurations used in real PV plant, in particular, the centralized technology, the string and multi-string technology, and the AC module technology are represented in Fig. 3.19. In the schemes of Fig. 3.19 the *bypass* and *block* diodes are not sketched, for the sake of simplicity.

Figure 3.19 represents the evolution of PV plants from the past (left) to the future (right). In particular, in Fig. 3.19a, a centralized inverter configuration is drawn. This kind of plant is the oldest realization to inject power coming from PV modules into the grid. The source is realized by series connection of PV modules to obtain a high voltage that supplies directly the inverter stage. This power stage was a line commutated inverter with thyristors as power switch. Among disadvantages

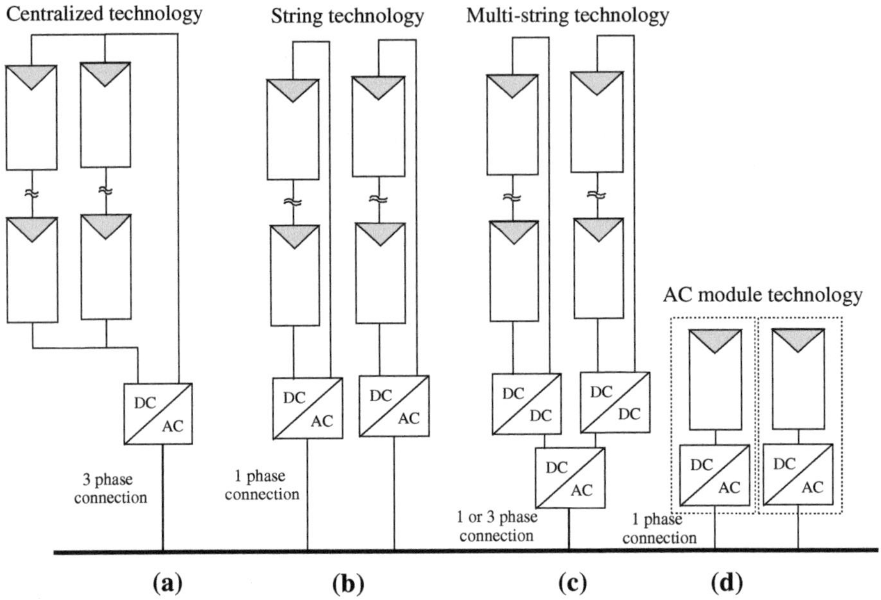

Fig. 3.19 Examples of module configurations in PV grid connected plants: centralized technology (**a**), string technology (**b**), multi-string technology (**c**), AC module technology (**d**)

there are: the high voltage of DC cables at the inverter input, the need of blocking diodes at the end of the strings, and the poor quality of power injected to the grid.

Figure 3.19b shows a more recent system in which each string is connected to a dedicated single-phase inverter. In this case there is no need of blocking diode, the input inverter voltage can be reduced by adopting a step-up transformer before the connection to the grid. In this configuration each inverter can perform a maximum power point tracking (MPPT) on its string and, compared to the centralized inverter, a greater efficiency is achieved and the system exhibits a better modularity.

Figure 3.19c shows a multi-string technology realization. Each string is connected to a DC/DC converter that performs the MPPT and raises the input voltage of the inverter. The system can be extended by adding new strings with its own DC/DC stage if the inverter is correspondingly designed.

Finally, in Fig. 3.19d the AC module technology is shown. In this case each module is connected to a DC/AC stage that boosts the module voltage, performs the MPPT, and the grid connection; no by-pass diodes are necessary.

The evolution, during the time, from (a) to (d) plants has been made possible thanks to the evolution of power devices as MOSFETs and IGBTs that have been exhibiting high blocking voltages, high conduction currents with fast commutation times.

3.6 Conclusions

An overview of mathematical models of PV sources and the corresponding derivation of equivalent circuits have been given in this chapter. Both the mathematical model and the circuit model are the basis for the PV source behavior reproduction.

Starting from a nearly ideal behavior of a PV cell, the different nonideality elements, related to the operation of a real PV device, have been included in the model. Furthermore, from a generalized PV model, some simplifications, valid when dealing with electric power generation, have been introduced.

The extension of PV cell model to the case of PV modules or fields has been presented, and finally some issues related to the use of the models under non-uniform solar irradiance have been outlined.

It should be observed that the advantageous use of the previously described models is possible only if a suitable extraction of their parameters is done. For this reason, the parameters identification is a crucial stage in the setup of a precise and reliable PV model.

The methods for the extraction of the PV model parameters will be the object of following two chapters, the former dedicated to the case of static models and the latter focused on the case of dynamic models.

Bibliography

Castañer L, Silvestre S (2002) Modelling photovoltaic systems using Pspice. Wiley, Chicester

Chan SH, Phang JCH (1987) Analytical methods for the extraction of solar-cell single-and double-diode model parameters from I-V characteristics. IEEE Trans Electron Devices ED-34(2): 286–293

Di Piazza MC, Vitale G (2010) Photovoltaic field emulation including dynamic and partial shadow conditions. Appl Energ (87)3: 814–823

Gow JA, Manning CD (1999) Development of a photovoltaic array model for use in power-electronics simulation studies. IEE Proc Elect Power Appl 146(21):193–200

Gradella Villalva M, Gasoli JR, Filho ER (2009) Comprehensive approach to modeling and simulation of photovoltaic array. IEEE Trans Power Electron 24(5):1198–1208

Luque A, Hegedus S (eds) (2003) Handbook of photovoltaic science and engineering. Wiley, Hoboken

Messenger RA, Ventre J (2005) Photovoltaic systems engineering, 2nd edn. CRC Press, Boca Raton

Sera D, Teodorescu R, Rodriguez P (2007) PV panel model based on datasheet values. IEEE Intern Symp Ind Electron 2392–2396

Kjaer SB, Pedersen JK, Blaabjerg F (2005) Review of single-phase grid-connected inverters for photovoltaic modules. IEEE Trans Ind Appl 41(5):1292–1306

Friesen G, Ossenbrink HA (1997) Capacitance effects in high-efficiency cells. Solar Energ Mater Solar Cells 48: 77–83

Jain A, Kapoor A (2005) A new approach to study organic solar cell using Lambert W-function. Solar Energ Mater Solar Cell 86:197–205

Chapter 4
Parameter Identification for Photovoltaic Source Models

4.1 Introduction

A suitable exploitation of the PV models described in the previous chapter, is possible, provided that the values of the model parameters are accurately determined.

Once the model parameters are identified and properly related to the environmental parameters (solar irradiance and temperature), the developed PV models can be used, for example, in the field of power electronics to simulate a PV source in conjunction with a power conversion system. Hence, the electrical behavior of the PV source itself and the operation of power conversion systems used within PV installations can be studied and optimized. For this reason, the parameter identification is a crucial stage in the accomplishment of a reliable simulation model for a PV source.

Due to the analytical formulations of the PV source model, its parameter identification is a nontrivial endeavor. Several approaches have been proposed in the technical literature. Among them, a rough classification can be done in analytical and numerical approaches, according to the used extraction method.

On the other hand, it is possible to distinguish methods where the parameter extraction is performed from experimental data, methods where only data provided by manufacturers are used, and methods where only a reduced amount of experimental information is employed. In addition, the variation of the model parameters with T and G has to be defined, so that the obtained simulation model is able to cover any possible operating condition.

In this chapter, starting from the most common formulations of the PV circuit static model, an overview of the existing techniques for parameter identification is presented. Furthermore, the possibility to simplify the parameter extraction problem, correlating the different quantities involved in the modeling by linear regressions, is discussed. The use of some mapping techniques to determine PV source characteristics is described hereinafter, for the sake of completeness. Examples of Matlab/Simulink® simulation of PV electrical characteristics are finally given to show the potential of an accurate PV modeling and identification.

M. C. Di Piazza and G. Vitale, *Photovoltaic Sources*, Green Energy and Technology,
DOI: 10.1007/978-1-4471-4378-9_4, © Springer-Verlag London 2013

Table 4.1 Sets of parameters in five-parameter PV model formulations

Circuit model scheme	Sets of parameters
Double diode (Eq. 4.1)	$I_{\text{ph}}, I_{s1}, I_{s2}, R_s, R_p$
Single diode (Eq. 3.11)	$I_{\text{ph}}, I_s, A_q, R_s, R_p$

4.2 Five-Parameter PV Model Formulation

One of the most common formulations of the PV source mathematical model, when representing the source through a double-diode scheme, is given in Eq. (4.1).

$$I = I_{\text{ph}} - I_{s1}\left(e^{q(V+IR_s)/kT} - 1\right) - I_{s2}\left(e^{q(V+IR_s)/2kT} - 1\right) - \frac{(V + IR_s)}{R_p} \tag{4.1}$$

Equation (4.1) is a double exponential representation of the PV source coming from (3.9) where the nonideality of the diffusion diode is neglected assuming $A_0 = 1$. With this formulation of the PV model, it is evident that five parameters are needed to define the source behavior univocally. These parameters are: the photo-generated current I_{ph}, the dark saturation current due to recombination in the quasineutral region $I_{s1,}$ the dark saturation current due to recombination in the space charge region, $I_{s2,}$ the series resistance R_s, and the shunt resistance R_p. For this reason, the analytical model in Eq. (4.1) is referred to as the five-parameter model formulation.

Another possible five-parameter model formulation is obtained starting from the single-diode PV circuit scheme as described by Eq. (3.11). This is an exponential equation rewritten here for clarity:

$$I = I_{\text{ph}} - I_s\left(e^{q(V+IR_s)/A_q kT} - 1\right) - \frac{(V + IR_s)}{R_p}$$

As in the case of Eq. (4.1), five parameters have to be determined to solve Eq. (3.11) and hence to yield the electrical characteristic of the PV source. In this case the five parameters are: the photo-generated current I_{ph}, the dark saturation current due to recombination in the quasineutral region I_s, the ideality factor of the diode A_q, the series resistance R_s, and the shunt resistance R_p.

Table 4.1 summarizes the sets of parameters to be identified in the described five-parameter PV model formulations.

The basic modeling process for a PV source is then schematized in Fig. 4.1, for the case of PV five-parameter models, corresponding to Eqs. (4.1) and (3.11). It should be noted that this process has a general validity and can be applied in any case, independently of the model formulation, and of the number of model parameters to be identified.

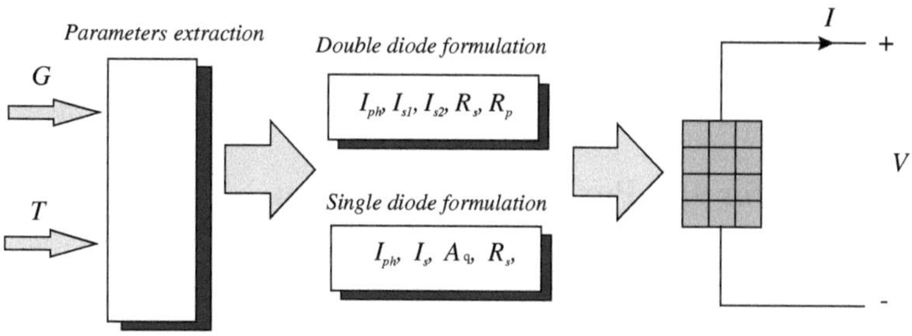

Fig. 4.1 Modeling process for the five-parameter models

Table 4.2 Sets of parameters in four-parameter PV model formulations

Circuit model scheme	Sets of parameters
Single diode (Eq. 3.12)	I_{ph}, I_s, A_q, R_s
Single diode (Eq. 3.13)	I_0, R_s, K_1, K_2

4.3 Four-Parameter PV Model Formulation

The technical literature exhibits plenty of PV model formulations that need the knowledge of four parameters to be identified. They are generally models based on the single-diode electrical scheme where either shunt or series resistance is neglected to obtain a simplification. The single-diode-based model formulation where R_p is neglected, due to its high value, will be considered hereafter. The equivalent circuit, in this case, is that shown in Fig. 3.8.

The mathematical model is expressed by Eqs. (3.12) or (3.13).

With reference to Eq. (3.12), rewritten here for clarity,

$$I = I_{ph} - I_s\left(e^{q(V+IR_s)/A_q kT} - 1\right)$$

the four parameters to be identified are: the photo-generated current I_{ph}, the dark saturation current I_s, the ideality factor of the diode A_q, and the series resistance R_s.

With reference to Eq. (3.13), rewritten here for clarity,

$$I = I_0 - e^{[(V+IR_s)K_1+K_2]}$$

the four parameters to be identified are: I_0, R_s, K_1, and K_2, where

$$I_{ph} = I_0$$
$$\frac{q}{A_q kT} = K_1$$
$$I_s = e^{K_2}$$

as previously shown.

The sets of parameters to be identified in the described four-parameter PV model formulations are summarized in Table 4.2.

4.4 Methods for the Parameter Extraction

It has been clearly stated that the *I–V* curve of a PV source depends on the internal characteristics of the device (e,g., R_p and R_s) and on external influences such as irradiance levels and temperature. Therefore, the extraction of PV model parameters implies the dependence of parameters on environmental conditions to be accounted for.

Several methods have been proposed for the identification of PV model parameters, as anticipated in the Introduction. A first classification of them can be done on the basis of the chosen extraction technique. The following solutions have been proposed:

- Analytical solutions.
- Numerical solutions.
- Heuristic methods-based solutions.

Another important aspect is related to the sets of data used for the parameter extraction. By this point of view, it is possible to distinguish extraction methods based on:

- Experimental characteristics.
- Data given by the manufacturer of the PV source (usually *remarkable points* under standard conditions).
- Experimental remarkable points.

The third method allows to limit the amount of measurements needed for the parameter identification with respect to a method based on experimental data only. This last method is particularly useful when an acquisition system is available within the arrangement of a running PV plant. In such a case, usually, the captured data are MPPs.

Figure 4.2 summarizes the PV model extraction techniques available in the technical literature. A detailed explanation of the different approaches is given in the following sections.

4.4.1 Analytical Solution

A fully analytical solution for the extraction of single- and double-diode model parameters has been proposed by Chan and Phang (1987).

The solution is first presented with reference to the double-diode model described by Eq. (4.1).

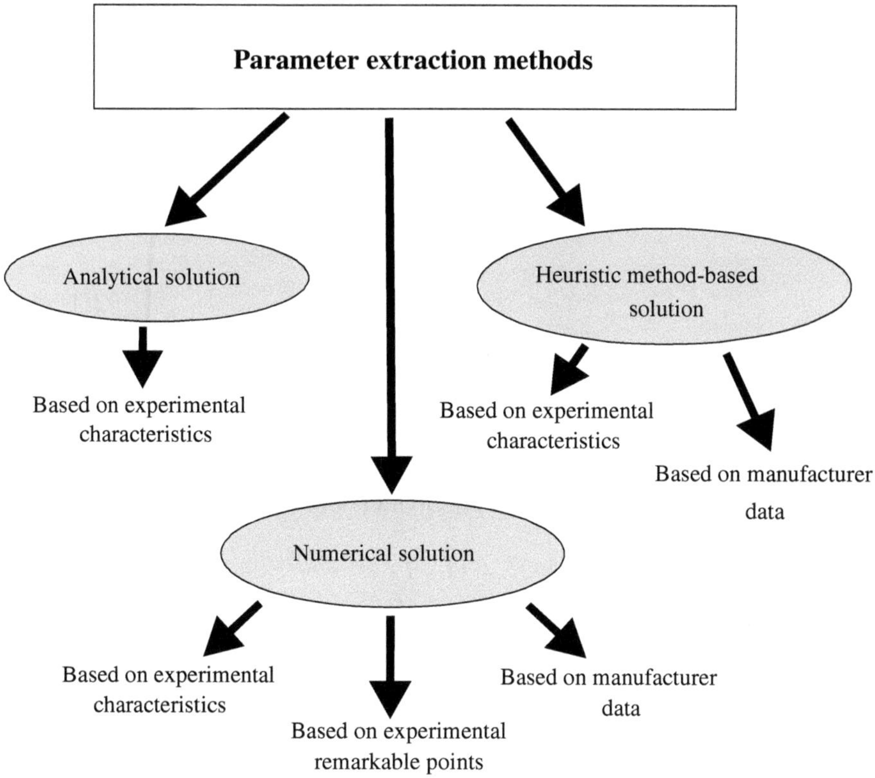

Fig. 4.2 Classification of the main PV model parameter extraction techniques in the technical literature

The unknown values of the model parameters, I_{ph}, I_{s1}, I_{s2}, R_s, and R_p, at given temperature and irradiance conditions, can be calculated on the basis of the values of the open circuit voltage V_{oc}, the short-circuit current I_{sc}, the maximum power point (V_{MP}, I_{MP}), and the values R_{so} and R_{po}, measured from the experimental I–V characteristic. In particular, R_{so} and R_{po} are given by the relationship between Eqs. (4.2) and (4.3) and their evaluation from the I–V curve is represented in Fig. 4.3.

$$R_{so} = -\left(\frac{dV}{dI}\right)_{V=V_{oc}} \tag{4.2}$$

$$R_{po} = -\left(\frac{dV}{dI}\right)_{I=I_{sc}} \tag{4.3}$$

Starting from these input parameters, the PV source model parameter values can be identified by the solution of a system of nonlinear equations that will be defined in the following.

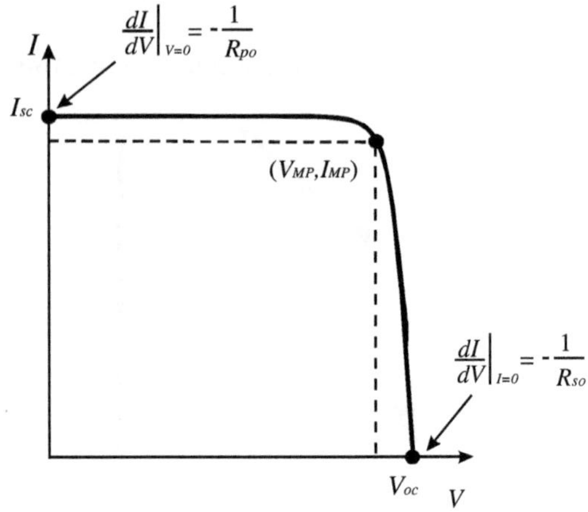

Fig. 4.3 Input parameters for model parameter extraction from the *I–V* curve

Equation (4.1), rewritten in the short-circuit point, $V = 0$ and $I = I_{sc}$, becomes:

$$I_{sc} = I_{ph} - I_{s1}\left(e^{q(I_{sc}R_s)/kT} - 1\right) - I_{s2}\left(e^{q(I_{sc}R_s)/2kT} - 1\right) - \frac{(I_{sc}R_s)}{R_p} \tag{4.4}$$

If written at the open circuit point $V = V_{oc}$ and $I = 0$, Eq. (4.1) becomes:

$$I_{ph} = I_{s1}\left(e^{q(V_{oc})/kT} - 1\right) + I_{s2}\left(e^{q(V_{oc})/2kT} - 1\right) - \frac{V_{oc}}{R_p} \tag{4.5}$$

Combining the previous equations to eliminate I_{ph}, the following relationship is obtained:

$$\begin{aligned} I_{sc} =& I_{s1}\left(e^{q(V_{oc})/kT} - e^{q(I_{sc}R_s)/kT}\right) - I_{s2}\left(e^{q(V_{oc})/2kT} - e^{q(I_{sc}R_s)/2kT}\right) \\ &+ \frac{(V_{oc} - I_{sc}R_s)}{R_p} \end{aligned} \tag{4.6}$$

Differentiating Eq. (4.1) with respect to I, yields:

$$\begin{aligned} -\frac{dV}{dI}\left(\frac{qI_{s1}}{kT}e^{q(V+IR_s)/kT} + \frac{qI_{s2}}{2kT}e^{q(V+IR_s)/2kT} + \frac{1}{R_p}\right) = \\ 1 + \frac{qI_{s1}R_s}{kT}e^{q(V+IR_s)/kT} + \frac{qI_{s2}R_s}{2kT}e^{q(V+IR_s)/2kT} + \frac{R_s}{R_p} \end{aligned} \tag{4.7}$$

Equation (4.7), rewritten at the short-circuit point, becomes:

$$\left(R_{po} - R_s\right)\left(\frac{qI_{s1}}{kT}e^{q(I_{sc}R_s)/kT} + \frac{qI_{s2}}{2kT}e^{q(I_{sc}R_s)/2kT} + \frac{1}{R_p}\right) - 1 = 0 \tag{4.8}$$

If written at the open circuit point, Eq. (4.7) becomes:

$$(R_{so} - R_s)\left(\frac{qI_{s1}}{kT}e^{q(V_{oc})/kT} + \frac{qI_{s2}}{2kT}e^{q(V_{oc})/2kT} + \frac{1}{R_p}\right) - 1 = 0 \tag{4.9}$$

Substituting the maximum power point voltage and current values (V_{MP}, I_{MP}) into Eq. (4.1) and using the expression of I_{ph} given in Eq. (4.5), the following equation is obtained:

$$\begin{aligned} I_{MP}\left(1 + \frac{R_s}{R_p}\right) &= I_{s1}\left(e^{q(V_{oc})/kT} - e^{q(V_{MP}+I_{MP}R_s)/kT}\right) \\ &+ I_{s2}\left(e^{q(V_{oc})/2kT} - e^{q(V_{MP}I_{MP}R_s)/2kT}\right) + \frac{(V_{oc} - V_{MP})}{R_p} \end{aligned} \tag{4.10}$$

Equations (4.6, 4.8– 4.10) form a system of nonlinear equations, with R_s, I_{s1}, I_{s2} and R_p as unknown variables, whose solution leads to the PV model parameter extraction. For achieving such a scope, many authors have followed numerical/ iterative approaches, which will be shown in the following.

On the other hand, for typical parameters of a silicon solar cell under AM1 illumination, the following approximations can be introduced:

$$e^{q(V_{oc})/ikT} \gg e^{q(I_{sc}R_s)/ikT} \quad i = 1, 2 \tag{4.11}$$

$$R_p, R_{po} \gg R_s \tag{4.12}$$

$$\frac{qI_{s1}}{kT}e^{q(I_{sc}R_s)/kT}, \frac{qI_{s2}}{2kT}e^{q(I_{sc}R_s)/2kT} \ll \frac{1}{R_{po}} \tag{4.13}$$

Therefore, taking into account these approximations, Eqs. (4.6, 4.8– 4.10) can be simplified and rewritten as:

$$I_{s1}e^{q(V_{oc})/kT} + I_{s2}e^{q(V_{oc})/2kT} - I_{sc} + \frac{V_{oc}}{R_p} = 0 \tag{4.14}$$

$$R_p = R_{po} \tag{4.15}$$

$$(R_{so} - R_s)\left(\frac{qI_{s1}}{kT}e^{q(V_{oc})/kT} + \frac{qI_{s2}}{2kT}e^{q(V_{oc})/2kT}\right) - 1 = 0 \tag{4.16}$$

$$-I_{s2}\left(e^{q(V_{MP}+I_{MP}R_s)/2kT}\right) + \frac{(V_{oc} - V_{MP})}{R_p} - I_{MP} = 0 \tag{4.17}$$

Finally, eliminating I_{s1} and I_{s2} from Eqs. (4.14), (4.16), and (4.17) and using Eq. (4.15), the following equation in R_s is obtained:

$$\begin{aligned} &I_{sc} - I_{MP} - \frac{V_{MP}}{R_{po}} - \left(\frac{V_{oc}}{R_{po}} - I_{sc} + \frac{2kT}{q(R_{so} - R_s)}\right)\left(e^{q(V_{MP}-V_{oc})/kT}e^{q(I_{MP}R_s)/kT}\right) \\ &- 2\left(-\frac{V_{oc}}{R_{po}} + I_{sc} + \frac{kT}{q(R_{so} - R_s)}\right)\left(e^{q(V_{MP}-V_{oc})/2kT}e^{q(I_{MP}R_s)/2kT}\right) = 0 \end{aligned} \tag{4.18}$$

Equation (4.18) can be expressed as a second order or third order polynomial, in this way either a *quadratic* or *cubic* analytical solution is achievable, as illustrated below.

The *quadratic* solution of Eq. (4.18) is based on the following approximation:

$$e^{\xi R_s} = 1 + \xi R_s \tag{4.19}$$

for $\xi = q(I_{MP})/kT$ and $q(I_{MP})/2kT$.

Introducing the new variables in Eqs. (4.20–4.23), the quadratic equation in (4.24) is obtained.

$$\alpha = I_{sc} - \frac{V_{oc}}{R_{po}} \tag{4.20}$$

$$\beta = I_{sc} - I_{MP} - \frac{V_{MP}}{R_{po}} \tag{4.21}$$

$$\gamma = e^{\frac{q(V_{MP} - V_{oc})}{2kT}} \tag{4.22}$$

$$\delta = \frac{qI_{MP}}{kT} \tag{4.23}$$

$$aR_s^2 + bR_s + c = 0 \tag{4.24}$$

The coefficients of Eq. (4.24) are defined by:

$$a = \alpha\gamma\delta(1 - \gamma) \tag{4.25}$$

$$b = \alpha\gamma(2 - \gamma) + \alpha\gamma\delta R_{so}(\gamma - 1) - \beta + \gamma\delta\frac{kT}{q}(1 - 2\gamma) \tag{4.26}$$

$$c = \alpha\gamma R_{so}(\gamma - 2) + \beta R_{so} + 2\gamma\frac{kT}{q}(1 - \gamma). \tag{4.27}$$

The solution in R_s is then obtained by simply solving the quadratic Eq. (4.24) according to:

$$R_s = \frac{-b \pm \sqrt{b^2 - 4ac}}{2a} \tag{4.28}$$

The *cubic* solution leads to more accurate results. It is based on the following approximation:

$$e^{\xi R_s} = 1 + \xi R_s + \frac{\xi^2 R_s^2}{2} \tag{4.29}$$

for $\xi = q(I_{MP})/kT$ and $q(I_{MP})/2kT$.

The equation in R_s to be solved (cubic in this case) is:

$$a_1 R_s^3 + b_1 R_s^2 + c_1 R_s + d_1 = 0 \tag{4.30}$$

where

$$a_1 = \alpha\gamma\delta^2(1 - 2\gamma) \tag{4.31}$$

$$b_1 = 4\alpha\gamma\delta(1 - \gamma) + \alpha\gamma\delta^2 R_{so}(2\gamma - 1) + \gamma\delta^2 \frac{kT}{q}(1 - 4\gamma) \tag{4.32}$$

$$c_1 = 4\alpha\gamma(2 - \gamma) + 4\alpha\gamma\delta R_{so}(\gamma - 1) + 4\gamma\delta \frac{kT}{q}(1 - 2\gamma) - 4\beta \tag{4.33}$$

$$d_1 = 4\alpha\gamma R_{so}(\gamma - 2) + 8\gamma \frac{kT}{q}(1 - \gamma) + 4\beta R_{so} \tag{4.34}$$

Finally, once the equation in R_s (regardless of quadratic or cubic) is solved, the other parameters are found. In particular, from Eq. (4.14) and Eq. (4.16) I_{s1} and I_{s2} are obtained according to:

$$I_{s1} = \left(-I_{sc} + \frac{V_{oc}}{R_p} + \frac{2kT}{q(R_{so} - R_s)}\right) e^{-q(V_{oc})/kT} \tag{4.35}$$

$$I_{s2} = 2\left(I_{sc} - \frac{V_{oc}}{R_{po}} - \frac{kT}{q(R_{so} - R_s)}\right) e^{-q(V_{oc})/2kT} \tag{4.36}$$

From Eq. (4.8) R_p is obtained as:

$$R_p = \left(\frac{1}{R_{po} - R_s} - \frac{qI_{s1}}{kT} e^{q(I_{sc}R_s)/kT} - \frac{qI_{s2}}{2kT} e^{q(I_{sc}R_s)/2kT}\right)^{-1} \tag{4.37}$$

Then, once I_{s1}, I_{s2}, and R_p are known, Eq. (4.5) is used to calculate I_{ph}.

4.4.1.1 Identification from Experimental Characteristics

The technical literature exhibits the use of analytical solution in association with experimental data, in general *I–V* characteristics, for the PV source model parameter identification.

Thanks to this approach, the double-diode PV model parameters have been determined for several silicon cells of different quality with errors ranging from 0.001 to 19 % with respect to some given reference values.

The reference values of the PV model parameters have been obtained by a more accurate method using an iterative solution of Eqs. (4.6, 4.8–4.10), using the Newton–Raphson method.

Table 4.3 Comparison of the PV double-diode model parameters of two silicon cells obtained by analytical quadratic and cubic solution and by iterative solution

	Cell CN1-500-50			Cell TL1-900-50		
PV model parameters	Quadratic solution (Error)	Cubic solution (Error)	Iterative solution	Quadratic solution (Error)	Cubic solution (Error %)	Iterative solution
I_{ph} [A]	0.534 (−0.052 %)	0.534 (−0.052 %)	0.534	0.905 (−0.21 %)	0.905 (−0.2 %)	0.907
I_{s1} [nA]	1.879 (−0.48 %)	1.88 (−0.14 %)	1.88	2.05 (−16.35 %)	2.38 (−3.4 %)	2.46
I_{s2} [μA]	8.99 (1.22 %)	8.9 (0.21 %)	8.88	33.89 (19.69 %)	29.35 (3.67 %)	28.31
R_s [mΩ]	7.3 (−4.25 %)	7.39 (−3.07 %)	7.63	28.47 (−8.62 %)	30.52 (−2.05 %)	31.17
R_p [Ω]	21.87 (0.005 %)	21.87 (−0.001 %)	21.87	19.97 (0.25 %)	19.92 (0.047 %)	19.92

Table 4.3 shows, as an example, the results obtained from analytical, both quadratic, and cubic, solutions in comparison with those obtained by the iterative method.

4.4.2 Numerical Solution

The great majority of scientific contributions on the problem of the PV model parameter identification are based on numerical solutions, usually employing iterative methods.

As a matter of fact, the analytical expressions of the PV source models are nonlinear and implicit; then, their analytical solution is quite complex and the approach to look forward numerical solutions by curve fitting procedures is often preferred since it leads to more accurate results.

When using such a numerical approach, the choice of the initial values of the unknown model parameters is a crucial issue. In fact, an inappropriate selection of the initial values of model parameters will result in unacceptable parameter values or in nonconvergence of the algorithm. A typical flowchart describing the procedure for the PV model parameter identification by a numerical solution is sketched in Fig. 4.4, for the case of single-diode five-parameter model. The procedure for the parameter calculation is often similar to that described in Sect. 4.4.1; on the contrary, the solution of the equations is performed by iterative methods. In particular, in Fig. 4.4, R_s, R_p, and A_q have to be obtained on the basis of a set of initial values. Once these parameters are obtained, I_{ph} and I_s can be calculated as well.

In the following sections, some examples of numerical-solution-based parameter identification procedures are described. In particular, the examples of identification using: experimental characteristics only, data given by the manufacturer only, and a combination of both experimental and datasheets information are detailed.

4.4.2.1 Identification from Experimental Characteristics

Numerous contributions are present in the technical literature showing PV model parameter identification based on numerical methods and experimental data.

An interesting example is offered by Gow and Manning (1999) with regard to the double-diode five-parameter PV model formulation.

The identification procedure utilizes the Levenberg–Marquardt method to obtain a solution of the exponential model equation.

The solution is obtained in an iterative way by the Newton–Raphson method. In addition, the Levenberg–Marquardt algorithm requires the solution of the first partial derivatives of the function to be fitted with respect to the function parameters that vary during iterations.

The first step is the appropriate selection of the initial set of parameters to be identified. This is done by different approaches. Among these, there are some

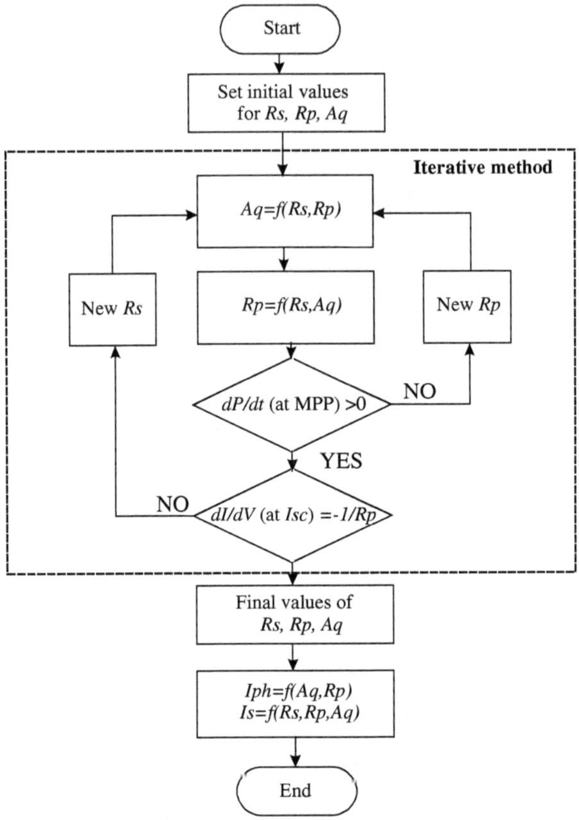

Fig. 4.4 Typical flowchart for the PV model parameter identification (iterative solution for PV five-parameter models)

so-called static solutions based essentially on suitable arrangements of the equation of the double-diode model and giving the initial values for R_s and R_p, indicated as R_{si} and R_{pi}.

When the use of such static solutions fails to deliver acceptable initial conditions, it occurs, for example, when negative values of some initial parameter values are obtained, it is possible to define the initial conditions by iterative methods. In particular, an iterative method where the approximated expressions of the saturation currents and the photocurrent are included within the iterative loop is proposed. This method is based on the initial condition equations and loop equations, expressed in Eqs. (4.38), and (4.39), where the subscript i stands for the initial parameter values.

Initial condition equations:

$$\begin{aligned} R_{\mathrm{si}} &= \left.\frac{\mathrm{d}V}{\mathrm{d}I}\right|_{I=0}, \quad R_{\mathrm{pi}} = \left.\frac{\mathrm{d}V}{\mathrm{d}I}\right|_{V=0} \\ I_{\mathrm{phi}} &= I_{\mathrm{sc}} \\ I_{s1i} &= \frac{I_{\mathrm{sc}}}{2e^{q(V_{\mathrm{oc}})/kT}-1}, \quad I_{s2i} = \frac{I_{\mathrm{sc}}}{2e^{q(V_{\mathrm{oc}})/2kT}-1} \end{aligned} \tag{4.38}$$

Iterative loop equations:

$$\begin{aligned}
X_{1v} &= \frac{q(I_{s1})}{kT} e^{q(V_{oc})/kT}, \quad X_{2v} = \frac{q(I_{s2})}{2kT} e^{q(V_{oc})/2kT} \\
R_{si} &= -\left[\left.\frac{dV}{dI}\right|_{I=0} + \frac{1}{X_{1v} + X_{2v} + 1/R_{pi}} \right] \\
X_{1\kappa} &= \frac{q(I_{s1})}{kT} e^{q(I_{sc}R_{si})/kT}, \quad X_{2\kappa} = \frac{q(I_{s2})}{2kT} e^{q(I_{sc}R_{si})/2kT} \\
R_{pi} &= -\frac{1}{\frac{1}{dV/dI|_{V=0} + R_{si}} + X_{1\kappa} + X_{2\kappa}}
\end{aligned} \tag{4.39}$$

Once acceptable values are achieved for the initial PV model set of parameters, the Levenberg–Marquardt algorithm for the curve fitting is called, leading to the model parameter extraction.

The goodness of the parameter identification is assessed comparing the experimental *I–V* curves for given solar irradiance and temperature with those back calculated from the identified double exponential model parameters.

A very good agreement among experimental and model-based I–V curves is observed by using the explained approach.

4.4.2.2 Identification from Datasheet Information

The PV model parameter identification from the knowledge of the sole information contained in the datasheets given by the manufacturer of the PV source is an advantageous method since the complication of setting up a measurement system and performing the measurements is avoided. On the other hand, a reliable extraction of the parameters by this approach is possible once the dependency of the model parameters on temperature and solar irradiance is defined very accurately. As a matter of fact, the information in datasheets refers to standard test conditions (STC), i.e., T = 25 °C and G = 1000 W/m^2, therefore, the parameter identified from datasheets is valid only under STC.

The model parameters, expressed with their dependency on *T* and *G*, can be inserted in the general expression of the mathematical PV model. In this way, a *I–V* relationship of the PV source which takes into account the irradiance and temperature conditions is obtained. This relationship represents the full-characteristic model where the contemporary dependence on temperature and irradiance is managed by using the principle of superposition.

An interesting example of PV model parameter identification is given in Sera et al. (2007) where the equivalent circuit and the basic equation of the PV source i.e., the single-diode five-parameter model, in STC, are presented as well as the parameters extraction from the datasheets values for the PV panel BP-MSX120.

The extraction procedure follows some steps in a similar way as in the analytical procedure, presented in Sect. 4.4.1; on the other hand, the dependence of the model parameters on solar irradiance and temperature is described in depth. Moreover, a novel formulation of the dependence of the dark saturation current on temperature is proposed to obtain a better correlation with the datasheet information.

For completeness, the procedure followed to extract the PV model parameters under STC is reported here. A comprehensive survey of the different proposed formulations of dependence of the PV model parameters and remarkable points on T and G will be then presented in Sect. 4.5.

The parameter extraction is obtained starting from four equations. Three equations derived from Eq. (3.11) (single-diode five-parameter PV model formulation) defined in the following remarkable points:

- short-circuit point;
- open circuit point;
- maximum power point.

An additional equation is obtained considering that the voltage derivative of the power is null at the MPP.

$$\left.\frac{\mathrm{d}P}{\mathrm{d}V}\right|_{\substack{V = V_{\mathrm{MP}} \\ I = I_{\mathrm{MP}}}} = 0 \tag{4.40}$$

In order to identify five parameters, a fifth condition is required. It is the condition expressed by Eq. (4.3) on the shunt resistor. The obtained system of equations is rewritten here for clarity.

$$I_{\mathrm{sc}} = I_{\mathrm{ph}} - I_s\left(e^{q(I_{\mathrm{sc}}R_s)/A_q kT} - 1\right) - \frac{(I_{\mathrm{sc}}R_s)}{R_p}$$

$$0 = I_{\mathrm{ph}} - I_s\left(e^{q(V_{\mathrm{oc}})/A_q kT} - 1\right) - \frac{(V_{\mathrm{oc}})}{R_p}$$

$$I_{\mathrm{MP}} = I_{\mathrm{ph}} - I_s\left(e^{q(V_{\mathrm{MP}}+I_{\mathrm{MP}}R_s)/A_q kT} - 1\right) - \frac{(V_{\mathrm{MP}} + I_{\mathrm{MP}}R_s)}{R_p}$$

$$\left.\frac{\mathrm{d}P}{\mathrm{d}V}\right|_{\substack{V = V_{\mathrm{MP}} \\ I = I_{\mathrm{MP}}}} = 0$$

$$R_{\mathrm{po}} = -\left(\frac{\mathrm{d}V}{\mathrm{d}I}\right)_{I=I_{\mathrm{sc}}}$$

Starting from this set of equations, the following expressions for the photo current and for the saturation current are obtained:

$$I_{\mathrm{ph}} = I_s\left(e^{q(V_{\mathrm{oc}})/A_q kT} - 1\right) - \frac{V_{\mathrm{oc}}}{R_p} \tag{4.41}$$

$$I_s = \left(I_{sc} - \frac{V_{oc} - I_{sc}R_s}{R_p}\right)e^{q(V_{oc})/A_q kT} \tag{4.42}$$

Substituting Eqs. (4.41) and (4.42) in the expression of the PV model evaluated at the MPP, the following equation is obtained:

$$I_{MP} = I_{sc} - \frac{V_{MP} + I_{MP}R_s - I_{sc}R_s}{R_p} - \left(I_{sc} - \frac{V_{oc} - I_{sc}R_s}{R_p}\right)e^{q(V_{MP}+I_{MP}R_s-V_{oc})/A_q kT} \tag{4.43}$$

There are three unknowns in Eq. (4.43), i.e., R_s, R_p, and A. The other two necessary relationships for extracting their values are obtained by exploiting the expression of the derivative of power with voltage at the MPP and the condition expressed by Eq. (4.40).

The following equations are deduced:

$$\left.\frac{dP}{dV}\right|_{I=I_{MP}} = I_{MP} + V_{MP}\frac{-\frac{q(I_{sc}R_p - V_{oc} + I_{sc}R_s)e^{q(V_{MP}+I_{MP}R_s-V_{oc})/A_q kT}}{A_q kTR_p} - \frac{1}{R_p}}{1 + R_s\frac{q(I_{sc}R_p - V_{oc} + I_{sc}R_s)e^{q(V_{MP}+I_{MP}R_s-V_{oc})/A_q kT}}{A_q kTR_p} + \frac{R_s}{R_p}} \tag{4.44}$$

$$\left.-\frac{1}{R_p}\right|_{I=I_{sc}} = \frac{-\frac{q(I_{sc}R_p - V_{oc} + I_{sc}R_s)e^{q(I_{sc}R_s-V_{oc})/A_q kT}}{A_q kTR_p} - \frac{1}{R_p}}{1 + R_s\frac{q(I_{sc}R_p - V_{oc} + I_{sc}R_s)e^{q(I_{sc}R_s-V_{oc})/A_q kT}}{A_q kTR_p} + \frac{R_s}{R_p}} \tag{4.45}$$

The values of the three unknown parameters can be now determined by solving the system of Eqs. (4.43–4.45) by a numerical approach, according to the philosophy of the flowchart in Fig. 4.4.

Although the results obtained by implementing the calculated parameters are in very good agreement with the PV characteristics reported in the datasheets, a direct comparison with experimental data from the PV panel is not available.

It should be considered that the actual electrical behavior of a PV source can differ from its theoretical behavior declared by the manufacturer. Therefore, the simplification of the parameters extraction, tied to the possibility to do without the experimental data, is achieved but a nonperfect matching of the obtained PV characteristics with the real one could occur.

4.4.2.3 Identification from Experimental Remarkable Points

To summarize the prerogatives of the PV parameter identification methods described thus far, it is possible to state that:

- the methods based only on experimental characteristics are precise, but involve the knowledge of plenty of I–V curves in different environmental conditions;
- the methods based only on the data given by the manufacturers do not need any measurement, but can be affected by some uncertainty.

As an alternative, some authors have proposed methods for the PV parameter identification which uses the measurements of only some remarkable points of the PV electrical characteristics.

This approach leads to a reduction of the measurement effort and allows an accurate reconstruction of the PV electrical characteristics as well.

An application of this method has been proposed for the mathematical model of the PV source based on the single-diode model in which the shunt resistance has been neglected, i.e., the model described by Eq. (3.13).

The four parameters K_1, K_2, R_s, and I_0 can be obtained by the knowledge of the short-circuit current (I_{sc}), the open circuit voltage (V_{oc}), and the maximum power point current and voltage (I_{mp} and V_{mp}), using the relations in Eq. (4.46).

$$\begin{cases} K_1 = \dfrac{\frac{I_{mp}}{I_0 - I_{mp}} + \log\left[1 - \frac{I_{mp}}{I_0}\right]}{2V_{mp} - V_{oc}} \\ R_s = \dfrac{V_{mp} - \frac{I_{mp}}{[I_0 - I_{mp}]K_1}}{I_{mp}} \\ K_2 = \log[I_0] - V_{oc}K_1 \\ I_0 = I_{sc} + e^{(I_{sc}R_s)K_1 + K_2} \end{cases} \tag{4.46}$$

The solution of the nonlinear equation system in Eq. (4.46) can be achieved with the following steps.

1. The assignation $I_0 = I_{sc}$ is done;
2. By substituting I_0 in the first equation in (4.46), K_1 is obtained;
3. By substituting I_0 and K_1 in the second and in the third equations in (4.46), R_s and K_2 are respectively obtained;
4. By substituting K_1, K_2 and R_s in the fourth equation in (4.46), a new value of I_0 is obtained.

The steps from 2 to 4 are iterated until the variation on the obtained value of I_0 is lower than a threshold. Usually, the convergence is obtained in few iterations.

As an example, for extracting the PV module model parameters of the module Solar world-Sunmodule Plus SW 160, using Eq. (4.46) and the above described steps, the convergence is obtained just in three iterations, as shown in Fig. 4.5.

As previously explained, the proposed method is based on the knowledge of: (a) maximum power point data, measured when the PV plant power converter is running, and (b) open circuit voltage and short-circuit current measured off-line. In principle, V_{oc} and I_{sc} could be deduced by the datasheets, by introducing their dependence from temperature and solar irradiance.

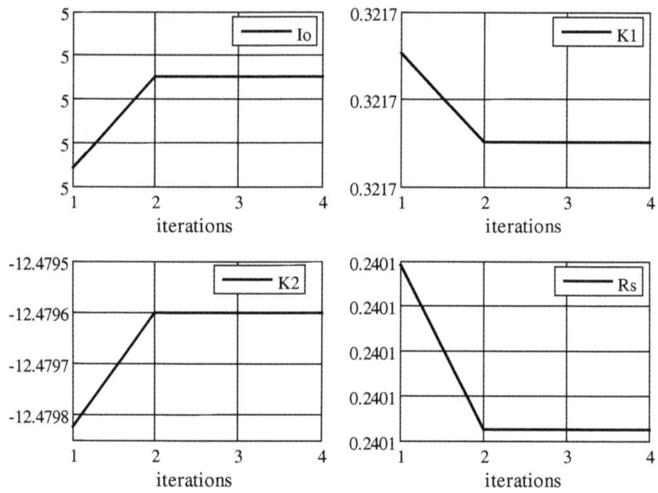

Fig. 4.5 Sunmodule Plus SW 160 PV module parameter values versus iterations

This approach allows the most accurate representation of the PV source.

The identification of the PV model parameters through the extraction of V_{oc}, I_{sc}, V_{mp}, and I_{mp} by a data logger installed in the plant has the advantage that modeling is performed on the real system and also the connection cables resistance is considered.

In the case of measurements performed at the inverter terminals, as it is usual using a data logger, for the practical case of a PV plant, the parameter identification results are presented.

The PV source is a field composed by two identical PV fields.

The electrical features of each PV field are the following:

Number of assemblies: 2
Peak power: 1450 W_p;
Open circuit voltage (V_{oc}): 228.6 V;
Short-circuit current (I_{sc}): 9.2 A;
Maximum power voltage (V_{mp}): 186 V;
Maximum power current (I_{mp}): 8 A.

Each assembly is constituted by six series connected modules.

In Fig. 4.6, a scheme of the plant under study is shown.

The PV plant is equipped with a data acquisition system that measures the following parameters: panels temperature, solar irradiance, DC voltage, and current supplied by the solar array to the inverter, AC voltage, current, and power supplied by inverter to the grid.

By performing measurements in a 5-month scale of time, during all the day and sampling each 10 min, a set of values of current and voltage, corresponding to maximum power points (MPPs), for solar irradiance and temperature ranging between 100 and 1100 W/m^2 and 293 and 323 K, respectively, have been acquired. Equations (3.13) and (3.46), which express the relations between currents and voltages of a single PV cell and of a module, can be used for describing the relations between currents and voltages of a single PV module and of an assembly, intended as the basic group of series modules connected to the inverter. In this case, N_S represents the number of series connected modules. If more assemblies are parallel connected, forming a group, N_P represents the number of assemblies.

The values of $I_{\rm mp}$ and $V_{\rm mp}$ are acquired by the data logger. In particular, only data measured under uniform conditions of solar irradiance and temperature are used. By Eq. (3.46), with $N_P = 2$ and $N_S = 6$, the data referred to a single module are obtained. In Fig. 4.7 the experimental measured MPPs of one PV module, referred to irradiance range $G = 500 \div 600$ W/m^2 and to temperature range $T = 313 \div 318$ K, are shown.

From Fig. 4.7, it is possible to observe a dispersion of experimental MPPs, which is present in all the intervals of measured data. This is due to the fact that these data do not always correspond to $V_{\rm mp}$, $I_{\rm mp}$, because there is no synchronization between the sample time of the data and the MPPT algorithm.

In other words, some couples of data might have been acquired when the algorithm is searching for the maximum after a fast variation of solar irradiance.

The estimation of MPPs is obtained by clustering the data on the basis of irradiance and temperature, then using a normal distribution for each group of data and performing a maximum likelihood estimation (MLE) of the distribution parameters. This experimental procedure allows to prescind from any analytical formulation of the parameters' dependence on T and G. As a matter of fact, this dependence is merged into the PV characteristic remarkable points experimentally determined.

The values of $V_{\rm oc}$ and $I_{\rm sc}$ are obtained by off-line measurements. The $V_{\rm oc}$ value is obtained by a voltage measurement without the inverter; $I_{\rm sc}$ is achieved by loading the plant with a low resistive load in order to approximate the short-circuit condition. As a matter of fact, it is possible since the I–V curve is nearly horizontal close to short-circuit condition.

On the basis of obtained values of $V_{\rm oc}$, $I_{\rm sc}$, $V_{\rm mp}$, and $I_{\rm mp}$, versus solar irradiance and temperature, it is possible to calculate the corresponding parameters K_1, K_2, R_s, $I_{\rm ph}$ that identify the I–V curves.

The proposed method allows a set of parameters $K_1, K_2, R_s, I_{\rm ph}$ ($=I_0$) to be obtained for each interval of temperature and irradiance in which data are clustered.

In Table 4.4, the obtained parameters values for three intervals are reported, as an example. Figure 4.8 shows the I–V characteristic for $G = 550$ W/m^2 and $T = 315$ K computed with the extracted parameters with superimposed the experimental MPPs.

It should be observed that the hypothesis to neglect R_p, simplifies the analytical calculations of either I versus V or V versus I, i.e., the inversion of the PV characteristic. When this assumption is not introduced, the calculation is more difficult due to the implicit relationship between current and voltage and can be managed by the use of the Lambert W function, as shown in Sect. 3.2.5.4.

In Fig. 4.9, the number of available experimental data is represented by means of a bar diagram. In this representation of the data, the temperature is divided into steps of 5 K and the irradiance into steps of 40 W/m^2.

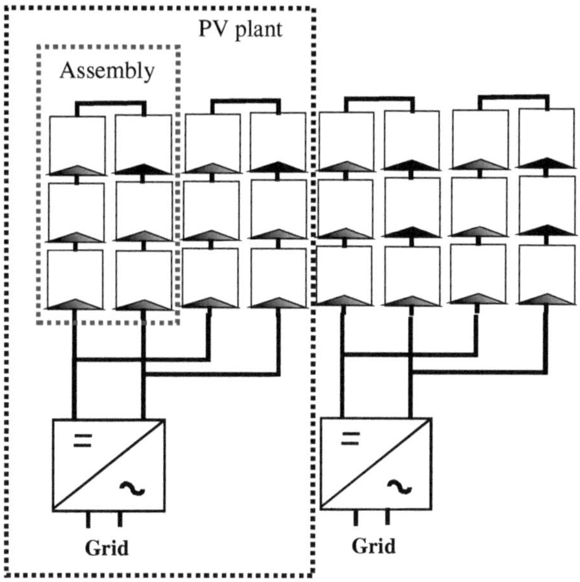

Fig. 4.6 Schematics of the PV plant

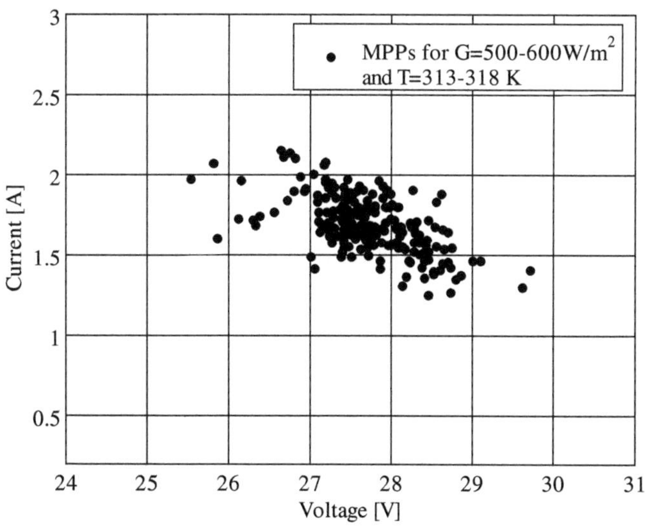

Fig. 4.7 Experimental maximum power points of one PV module for $G = 500 \div 600\,\mathrm{W/m^2}$ and $T = 313 \div 318\,\mathrm{K}$

It should be noted that there are some intervals in which no data are present, as they have not had any correspondence with the PV field operating conditions, during the period of observation. In this last case, the identification of the PV

Table 4.4 Some estimated model parameters for the single module

	$G = 725 - 775$ W/m^2, $T = 313 - 318$ K	$G = 525 - 575$ W/m^2, $T = 313 - 318$ K	$G = 225 - 275$ W/m^2, $T = 313 - 318$ K
K_1	0.2485	0.2442	0.2569
Rs [Ω]	0.309	0.851	1.7337
K_2	−8.0643	−8.25	−9.5173
I_O [A]	2.7312	1.9435	0.8682

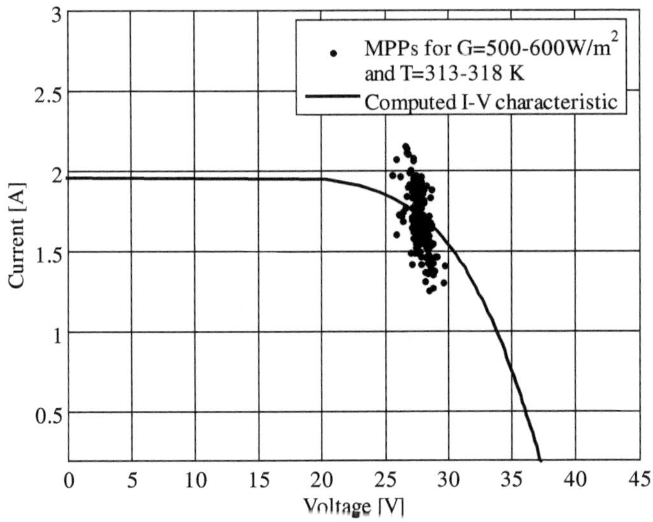

Fig. 4.8 Example of computed *I–V* characteristic with superimposed experimental MPPs

module characteristic is not possible, but this situation does not have a practical interests as it is quite unlikely.

On the other hand, the identification is more accurate where a great number of points are available, i.e., in very probable situations.

4.4.3 Heuristic Methods-Based Solution

In recent years, some authors have proposed the use of heuristic methods for the PV source model parameter identification.

The proposed methods range from genetic algorithms (GA) to Differential Evolution and examples of applications to both double- and single-diode formulations of the five-parameter PV model are given in the literature.

In the case of GAs, the core idea is to define a population of individuals where each individual is a set of parameters values and then to select the best fitted individuals, on the basis of minimizing an error function f_ε, as the base for a new population. The process is repeated until the error falls below a chosen threshold.

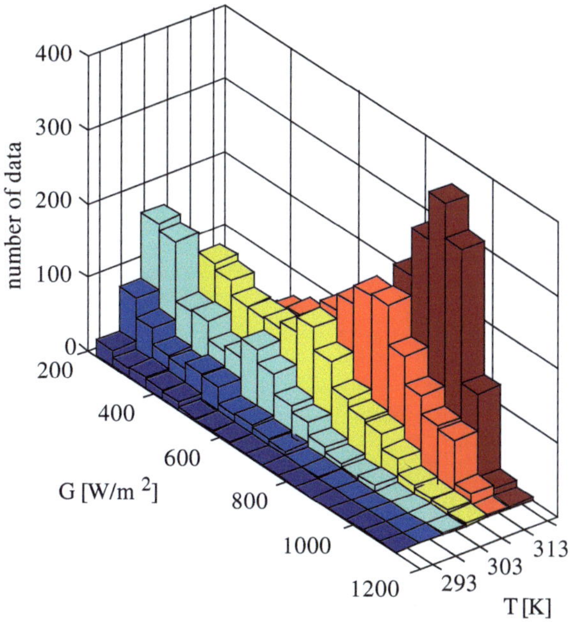

Fig. 4.9 Graphical representation of the number of experimental data

In Fig. 4.10, the flowchart describing the example of the so-called *crossing-mate* GA is shown.

Equation (4.47) defines the error function that, in this case, is tied to the difference between the measured and estimated values of the current.

$$f_{\varepsilon} = \sqrt{\sum_{i=1}^{N} \left(i_{i,\,\text{measured}} - i_{i,\,\text{estimated}}\right)^{2}} \tag{4.47}$$

The application of this algorithm for the parameters extraction uses an explicit formulation of the PV *I–V* model based on the Lambert W function. This formulation is necessary for the computation of the terms $i_{i,\ \text{estimated}}$ in Eq. (4.47).

The main advantage of this approach is that the initial values of the PV model parameters can be generated randomly without the need for particular assumption on them. Furthermore, with this approach, the use of complex equation to evaluate the model parameters is not required.

The main drawback of the described method is the quite high computation time. As a matter of fact, this may be not an issue in the research context, but it could represent a limitation when the real-time computation of the PV source behavior is desired, for example, in the case of checking out the quality of the fabricated cells in an industrial process. In these cases, the best solution could be the use of the procedure only to test statistically significant samples.

Example of heuristic method-based parameters identification based on the manufacturers' data are reported in the literature.

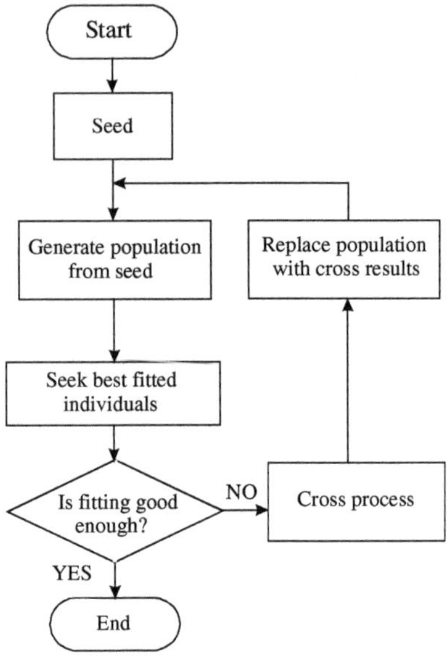

Fig. 4.10 Flowchart describing the *crossing-mate* GA

4.5 Parameters Dependence on Temperature and Solar Irradiance

In order to make the PV model completely determined, the dependence of the model parameters on temperature and solar irradiance has to be addressed.

Indeed, the model parameters are different for each type of solar cell and vary with the environmental parameters. Therefore, in general, a set of equations are to be defined, whose solution relates each PV model parameter to the actual values of temperature and solar irradiance, possibly incorporating coefficients tied to the specific PV cell.

Many contributions are present in the technical literature where analytical or empirical relationships among PV model parameters and environmental conditions, expressed by T and G, are given.

With reference to the five-parameter PV model based on the double-diode scheme, the set of Eqs. (4.48), relating the double exponential model parameters to temperature, has been proposed by Veissid and De Andrade (1991).

$$\begin{aligned} I_{\text{ph}} &= I_{\text{ph,ref}}[1 + k_0(T - 300)] \\ I_{s1} &= k_1 T^3 e^{(-qV_{\text{bi}}/kT)} \\ I_{s2} &= k_2 T^{3/2} e^{(-qV_{\text{bi}}/2kT)} \\ R_s &= R_{s,\text{ref}}[1 - k_0(T - 300)] \\ R_p &= R_{p,\text{ref}} e^{(-k_4 T)} \end{aligned} \tag{4.48}$$

In Eqs. (4.48), $I_{\text{ph, ref}}$, $R_{s,\text{ ref}}$, and $R_{p,\text{ ref}}$ are the values of the parameters at a reference temperature of 300 K, k_0 to k_4 are coefficients tied to the specific PV cell and V_{bi} is the built in voltage of the p–n junction.

The Eqs. (4.48) describe only the dependence of the model parameters on the temperature and do not take into account the effects of solar irradiance. In order to make them suitable for use in a general purpose PV model, some terms accounting for the irradiance effect have been introduced by Gow and Manning (1999). In particular, the nature of the relationships among PV model parameters and temperature and irradiance has been firstly deduced by plotting the parameters variations versus T and G; then, using some curve fitting techniques, the set of Eqs. (4.49) has been obtained.

$$\begin{aligned} I_{\text{ph}} &= k_0 G(1 + k_1 T) \\ I_{s1} &= k_2 T^3 e^{(k_3/T)} \\ I_{s2} &= k_4 T^{3/2} e^{(k_5/T)} \\ R_s &= k_6 + \frac{k_7}{G} + k_8 T \\ R_p &= k_9 e^{(k_{10} T)} \end{aligned} \tag{4.49}$$

The set of equations (4.49) is applicable to any PV cell described by the double-diode scheme. The coefficients k_0 to k_{10} are specific of the cell to be characterized.

In the case of PV models based on the single-diode scheme, regardless of the number of parameters (either four or five), plenty of formulations have been proposed to define the dependence of PV model parameters on T and G.

As for the dark saturation current I_s, some authors, such as Rauschenbach, Castañer, and Silvestre, Villalva et al (2009), etc., neglect its dependence on solar irradiance, whereas others establish a relationship between I_s and G.

According to Castañer and Silvestre, I_s is not dependent on G but they consider the dark saturation current strongly dependent on T. In particular, the expression of Eq. (4.50) is given for the dark saturation current density J_s, where a nonlinear dependence on temperature is exhibited:

$$J_s = B T^{\text{XTI}} e^{(-qV_{\text{bi}}/kT)} \tag{4.50}$$

where B and XTI are constants independent on temperature.

Several authors propose the following expression for the dark saturation current versus temperature:

$$I_s = I_{s(\text{stc})} \left(\frac{T_{(\text{stc})}}{T} \right)^3 e^{\left[\frac{qV_{\text{bi}}}{kT} \left(\frac{1}{T_{(\text{stc})}} - \frac{1}{T} \right) \right]} \tag{4.51}$$

where the subscript (stc) stands for reference values under STC, and $I_{s(\text{stc})}$ is expressed as:

$$I_{s(\mathrm{stc})} = \frac{I_{sc(\mathrm{stc})}}{e^{\left[q\left(V_{\mathrm{oc(stc)}}\right)/A_q kT_{(\mathrm{stc})}\right]} - 1} \tag{4.52}$$

Villava et al. (2009) substituted Eq. (4.51) with Eq. (4.53), where the current and voltage coefficients, K_I and K_V, respectively, are included. In this way, the net effect of the temperature on I_s is related to the predominant phenomenon given by the linear variation of the open circuit voltage with T, according to the voltage/temperature coefficient.

$$I_s = \frac{I_{\mathrm{sc(stc)}} + K_I\left(T - T_{(\mathrm{stc})}\right)}{e^{\left[q\left(V_{\mathrm{oc(stc)}} + K_V\left(T - T_{(\mathrm{stc})}\right)\right)/A_q kT\right]} - 1} \tag{4.53}$$

A similar approach is proposed by Sera et al. (2007) where the effect of temperature on dark saturation current is included by updating, with their corresponding temperature coefficients, each parameter of Eq. (4.54), representing the single-diode five-parameter model equation, written at the short-circuit point.

$$I_s = \left[I_{\mathrm{sc}}(T) - \frac{V_{\mathrm{oc}}(T) - I_{\mathrm{sc}}(T)R_s}{R_p}\right] e^{-\frac{qV_{\mathrm{oc}}}{A_q kT}} \tag{4.54}$$

It should be noted that, in this last case, no dependence of R_s and R_p on temperature has been considered.

In Xiao et al. (2004), proposing a modeling technique for a four-parameter single-diode model, the saturation current has been considered dependent both on temperature and solar irradiance. Then, the following expression for I_s has been proposed.

$$I_s = \frac{I_{\mathrm{ph}}(G,T)}{e^{\frac{qV_{\mathrm{oc}}(T)}{A_q kT}} - 1} \tag{4.55}$$

where I_{ph} is the photo-generated current, i.e., another model parameter whose dependence on environmental parameters should be defined.

As for the photo-generated current, the most common expression describing its relationship with T and G is given in Eq. (4.56).

$$I_{\mathrm{ph}} = I_{\mathrm{sc(stc)}}\left[1 + K_I\left(T - T_{(\mathrm{stc})}\right)\right]\frac{G}{G_{(\mathrm{stc})}} \tag{4.56}$$

With regard to the series and shunt resistors, it should be observed that, most of the authors consider these parameters independent on environmental conditions.

Anyway, in some cases, some relationships between the parasitic resistances and T and G have been deduced, mostly on the basis of empirical laws.

In Teixeira Da Costa et al. (2010), a dependence of R_s and R_p on temperature is established on the basis of the following equations:

$$R_s = R_{s(\mathrm{stc})}\left(\frac{T}{T_{(\mathrm{stc})}}\right)^{\delta}$$
$$R_p = R_{p(\mathrm{stc})}\left(\frac{T}{T_{(\mathrm{stc})}}\right)^{\varphi} \tag{4.57}$$

where δ and φ represent the coefficients of the temperature exponent of the standard series resistance and of the standard shunt resistance, respectively. With these positions, the parameters that need to be estimated are $R_{s(\mathrm{stc})}$ and $R_{p(\mathrm{stc})}$, since they are not provided by the manufacturers.

According to Eikelboom and Reinders (1997), R_s and R_p are strongly affected by the solar irradiance. As a matter of fact, they have observed, by an experimental investigation for G varying from 100 to 1,000 W/m^2 on several multi-crystalline silicon PV modules, that the shunt resistance decreases and the series resistance increases with increasing irradiance.

A similar study has been done by Bätzner et al. (2001) with reference to the performance of some HVE CdTe/CdS solar cells. They have observed an increase of both R_s and R_p with increasing irradiance and they have indicated in the photoconductivity of the cell the main cause for the high value of R_p at low irradiance.

Finally, as far as the diode quality factor A_q is concerned, many authors consider this parameter constant when defining their models with values comprised between 1 and 1.5.

The parameter A_q expresses the degree of ideality of the diode and is related, as previously explained in Sect. 3.2.5.1, to the current transport mechanism.

On the other hand, it is totally empirical; therefore, it is often given an arbitrary initial value then modified to adjust the model.

The dependence of the diode quality factor on temperature is assessed on an empirical base by some authors (Xiao et al. (2004) for example) that have observed, in particular, a nearly linear decrease of A with increasing temperature.

In other contributions, the dependence of A_q on solar irradiance has been experimentally determined for some silicon PV cells, obtaining a decrease from about 1.6 to 1.1 in the range between 2 and 600 W/m^2.

It is evident that, in all the cases when the model parameters dependence on environmental parameters is deduced by empirical observation, the natural way to implement this dependence is the use of look up tables.

4.6 Identification of PV Model Parameters by Linear Regression Methods

An approach to find out V_{MP} and I_{MP}, from the data to identify the PV model parameters consists in a simple clustering of measured maximum power point data on the basis of defined intervals of solar irradiance and temperature. Then, the estimation of MPP for each cluster is obtained using a normal distribution for the experimental data and performing a MLE of the distribution parameters, as described in Sect. 4.4.2.3.

In Fig. 4.7, an example of experimental measured MPPs, referred to irradiance range $G = 500 \div 600\,\mathrm{W/m^2}$ and to temperature range $T = 313 \div 318\,\mathrm{K}$ is reported.

From Fig. 4.7 it is possible to observe a dispersion of experimental MPPs, which is present in all the intervals of measured data. Therefore, the problem consists on the correct choice of irradiance and temperature step intervals.

In order to overcome this problem and simplify the parameter identification process, the formulation of relationships that give the V_{MP} and I_{MP} values versus solar irradiance and temperature could be helpful.

This issue has been handled by introducing linear regression methods to relate the different parameters involved.

The application of such a method illustrated, hereinafter, is referred to the plant described in Sect. 4.4.2.3.

The starting problem is a typical three variables problem.

Anyway, the observation of a 5-month scale measured data, allows to state that the dispersion diagram of T versus G exhibits a linear trend. It means that a linear relation between temperature and solar irradiance can be found by a least squares regression (LSR). In particular:

$$T = a_0 + a_1 G \tag{4.58}$$

where

$$\begin{aligned} a_0 &= \frac{\left(\sum_{i=1}^{N} T_i\right)\left(\sum_{i=1}^{N} G_i^2\right) - \left(\sum_{i=1}^{N} G_i\right)\left(\sum_{i=1}^{N} G_i T_i\right)}{N\sum_{i=1}^{N} G_i^2 - \left(\sum_{i=1}^{N} G_i\right)^2} \\ a_1 &= \frac{N\sum_{i=1}^{N} G_i T_i - \left(\sum_{i=1}^{N} G_i\right)\left(\sum_{i=1}^{N} T_i\right)}{N\sum_{i=1}^{N} G_i^2 - \left(\sum_{i=1}^{N} G_i\right)^2} \end{aligned} \tag{4.59}$$

being $(T_1, T_2, \ldots, T_N)$ and $(G_1, G_2, \ldots, G_N)$ the two sets of available temperature and irradiance data.

In the studied case, $a_0 = 26.37$ and $a_1 = 0.023$. The standard error value, s_{ls}, that measures the spread of data around the regression curve and is defined by Eq. (4.60), is 4.4636.

$$s_{\mathrm{ls}} = \sqrt{\frac{\sum_{i=1}^{N} T_i^2 - a_0 \sum_{i=1}^{N} T_i - a_1 \sum_{i=1}^{N} G_i T_i}{N}} \tag{4.60}$$

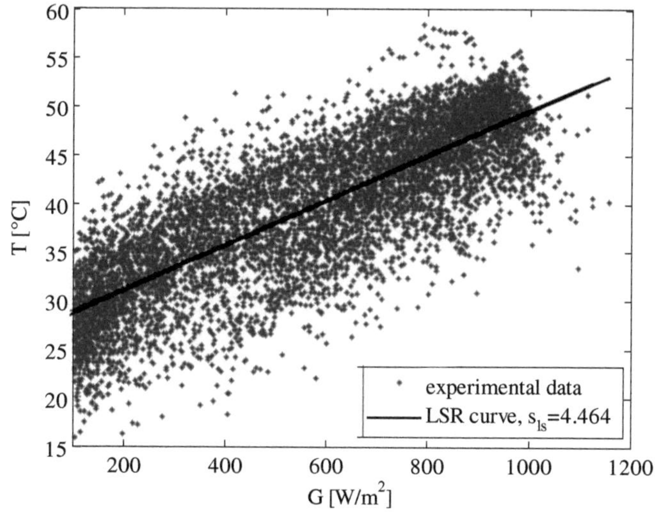

Fig. 4.11 Dispersion diagram of T versus G and LSR line

In Fig. 4.11 the dispersion diagram of T versus G with superimposed the line defined by Eq. (4.58) is shown.

Thanks to this relation between T and G, the above mentioned problem is reduced to a two variables problem. However, it is well known that, in practice, different temperatures can correspond to the same irradiance, according to other weather conditions as wind speed and environment temperature.

Using the linear regression, only one temperature is obtained for each irradiance value.

The obtained temperature is nearly the more probable, according to experimental data. This is demonstrated using an algorithm that gives the temperature value corresponding to the greater group of experimental data for the given values of solar irradiance.

Figure 4.12 shows that the temperature trend, obtained by the described algorithm, fits the linear regression curve.

By observing the dispersion diagram of V_{MP} and I_{MP} versus G, it can be noted that, in both diagrams, data are distributed in a straight way except for low irradiance where a high data dispersion is observed.

In such cases, the use of the linear LSR is not accurate enough. On the contrary, with the use of a linear robust LSR, low irradiance data are considered as outliers and filtered.

To be more precise, the linear robust LSR uses an iteratively re-weighted least squares algorithm, with the weights at each iteration calculated by applying the bisquare function to the residuals from the previous iteration. This algorithm gives lower weight to points that do not fit well.

In Fig. 4.13 the dispersion diagram of I_{MP} versus G is reported with superimposed both the lines obtained with linear LSR and with the robust linear LSR.

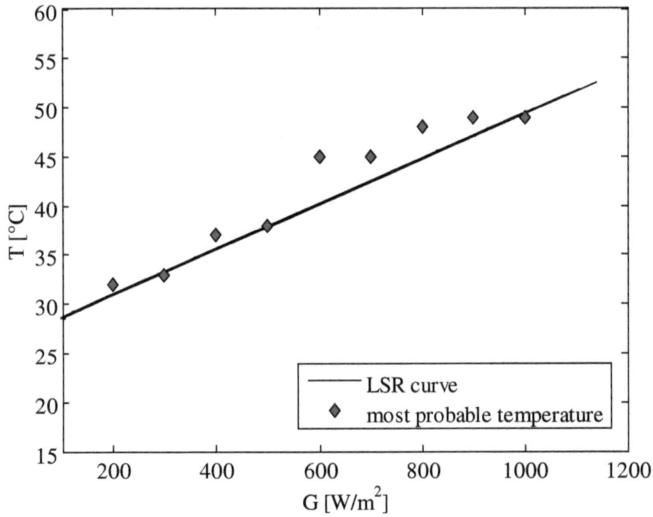

Fig. 4.12 T versus G regression line and temperature trend obtained by experimental data

In Fig. 4.14, the dispersion diagram of V_{MP} versus G is reported with superimposed both the lines obtained with the linear LSR and with the robust linear LSR.

The obtained equations for I_{MP} and V_{MP}, with LSR, obtained substituting I_{MP} and V_{MP} to T in Eq. (4.58), respectively, are:

$$\begin{cases} I_{mp} = -0.155 + 0.0066\,G \\ V_{mp} = 169.170 - 0.0069\,G \end{cases} \tag{4.61}$$

The standard errors, s_{ls}, obtained substituting I_{MP} and V_{MP} to T in (4.60), respectively, are 0.5206 and 7.0204.

With the robust LSR, the equations for I_{MP} and V_{MP}, obtained substituting I_{MP} and V_{MP} to T in (4.59), respectively, become:

$$\begin{cases} I_{mp} = -0.470 + 0.0071\,G \\ V_{mp} = 172.440 - 0.0115\,G \end{cases} \tag{4.62}$$

and the standard errors, s_r, obtained substituting I_{MP} and V_{MP} to T in (4.60), respectively, decrease becoming, 0.3176 and 5.5450.

Equations (4.58) and (4.62) are implemented in a Matlab/Simulink® model that allows to obtain continuously the photovoltaic array model parameters, according to (4.46). It should be observed that Eqs. (4.61) and (4.62) are quite accurate for $G > 100$ W/m^2.

As a result, an accurate modeling of the PV array is obtained, managing only the solar irradiance and considering the dependence of T, V_{MP}, and I_{MP} on G.

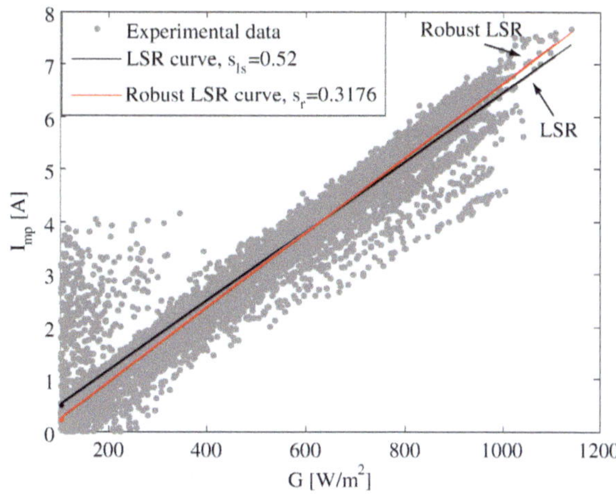

Fig. 4.13 Dispersion diagram of I_{MP} versus G with superimposed LSR and robust LSR lines

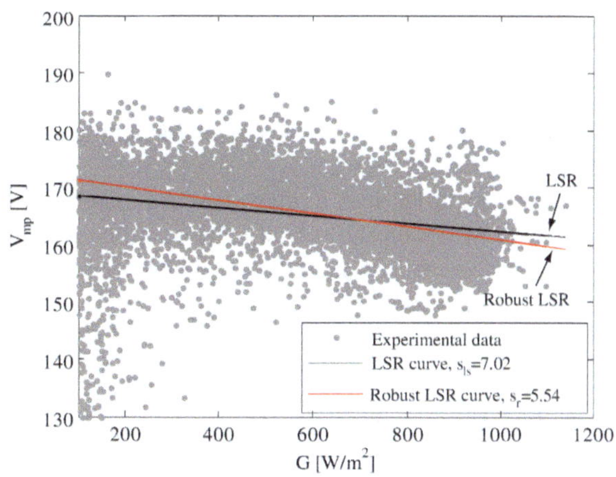

Fig. 4.14 Dispersion diagram of V_{MP} versus G with superimposed LSR and robust LSR lines

In Fig. 4.15, the *I–V* characteristics, computed for G varying in the range from 500 to 570 W/m^2, are shown, in comparison with the unique characteristic obtained when the parameter identification is performed on the basis of data clustering into defined intervals of solar irradiance and temperature (referred as step-by-step or discrete approach).

It is evident that the PV model parameters identification based on linear regressions overcomes the limitations of the step-by-step approach, giving a continuous representation of the PV source electrical behavior.

In Fig. 4.16, the *I–V* characteristics, obtained when the PV model parameters identification is based on regressions (both LSR and robust LSR), are reported.

It is possible to notice that, for low and high irradiance values, there is a deviation between the LSR based and the robust LSR-based characteristics. This is due to the great spread of data at low irradiance and to the small amount of available data at high irradiance.

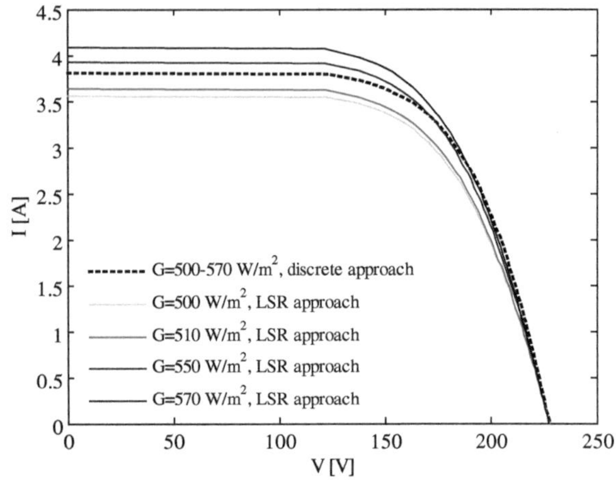

Fig. 4.15 Comparison of *I–V* characteristics obtained by linear regressions-based and discrete parameter identification approaches

On the other hand, a good matching between the LSR based and the robust LSR-based characteristics is observed in the irradiance range where experimental data are better correlated to regression lines.

In Fig. 4.16, the MPPs are reported too. In particular, the computed MPPs (both on LSR and robust LSR-based characteristics) and the experimental ones are shown.

From Fig. 4.16, it is evident that the experimental MPPs are very close to the computed MPPs on the robust LSR-based characteristics. This result is confirmed by the power values at MPPs evaluated experimentally in comparison with the computed ones and reported in Table 4.5.

4.7 PV Characteristic Representation Through Mapping Techniques

Once the PV model is completely identified, the problem of the PV electrical characteristic determination has to be faced. It has been anticipated that the relationship between current and voltage is implicit in all the five-parameter analytical formulations of the PV model. Therefore, the *I–V* characteristic determination, in this case, can be managed using numerical methods or, in alternative, introducing some mathematical tricks, such as the Lambert W function.

On the other hand, when the four-parameter formulation neglecting R_p is considered, the analytical solution of the equation $V = f(I)$ is straightforward, as shown in Sect. 3.2.5.3.

Based on the knowledge of the experimental PV characteristics, other ways to represent the *I–V* or *P–V* curves are possible. Among them, the use of a look-up-table approach and the use of a neural approach belonging to mapping techniques are described in the following subsections.

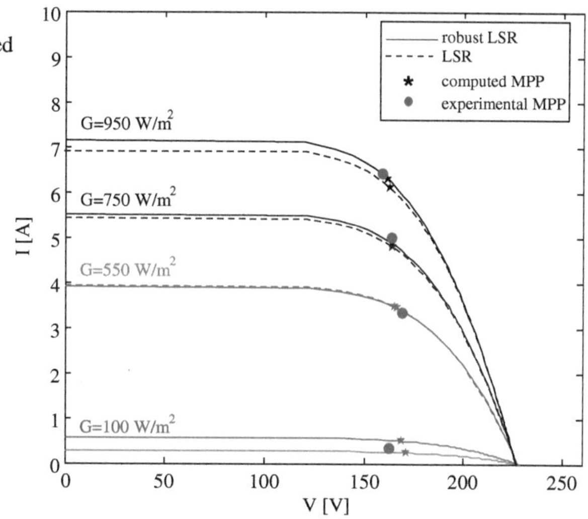

Fig. 4.16 Comparison of computed and experimental MMPs

4.7.1 Look-Up-Table Approach

The look-up-table approach is based on the acquisition of all possible *I–V* data of a real PV field and on their storage in a suitable storage unit. This method leaves out of consideration any mathematical relationship between *I* and *V*.

The success of this approach is related to the performance of the acquisition, storage, and processing system. The main advantage of this approach consists on the ease of implementation in a static memory or in a field programmable gate array (FPGA) architecture. Moreover, with the considered approach, it is possible to obtain a realistic pattern for the evolution of any voltage and current of the PV generator.

On the other hand, the look-up-table approach cannot represent the dynamic behavior of the PV source. Furthermore, the achievable accuracy is dependent on the amount of stored data that can be obtained either by an actual PV installation or by a model. In the first case, the presence of an outdoor PV plant equipped with a system for the measurements of *I–V* points under real operating conditions is needed, thus resulting in an unavoidable increase of the costs.

4.7.2 Neural Approach

An alternative way to represent the PV electrical characteristics is based on the use of suitable artificial neural networks (ANN) which can learn and reproduce the PV behavior in terms of *I–V* curves.

A neural network successfully used for this aim is the growing neural gas (GNG) network. It is an incremental network model which is able to learn the main

Table 4.5 Comparison of power at MPPs

G [W/m^2]	P_{MP} [W]–LSR	P_{MP} [W]–Robust LSR	P_{MP} [W]–Experimental
100	85.07	41.63	58.46
550	574.75	573.54	564.44
750	786.36	799.34	817.85
950	994.34	1,017.79	1,018.50

topological relations in a given set of input vector by means of simple Hebb-like learning rule.

The main idea of the GNG is to successively add new units (neurons) and connections to an initially small network in a growing structure, until a performance criterion has been met. In this approach, the network topology is generated incrementally and has a dimensionality which depends on the input data.

Figure 4.17 shows, as an example, some stages of a GNG-based simulation where the GNG network adapts to a ring-shaped distribution of data.

As in any ANN, two phases are distinguished in the operation of the GNG network: the training and the recalling.

Figure 4.18 shows the flowchart of the GNG training algorithm where n_{max} is the maximum number of neurons, n_{ep} indicates the number of epochs and n_{epmax} is the maximum number of epochs.

A possible way to represent the PV characteristics by the GNG-based approach consists of the following steps. As a first step, a set of measurements of PV voltage-current points corresponding to the different MPPs for different values of temperature and irradiance should be done. Given the set of voltage–current couples for different weather conditions, the corresponding characteristics can be deduced on the basis of a defined PV model.

The obtained set of curves, for different values of temperature and solar irradiance, is the training set of the GNG. Then the GNG is trained off-line.

The parameters of the network and the maximum number of neurons should be carefully chosen. A preprocessing of the data is also needed for placing randomly the data of the training set in the working space. In general, the GNG network has to be trained to learn a set of points in a four-dimensional space, where the dimensions of the working space are the PV current, the PV voltage, the solar irradiance, and the panels temperature.

Once the training phase is completed, the GNG network can be used on-line after having experimentally implemented only the recalling phase on a DSP.

It should be noted that the recalling phase is very simple and not computationally demanding. When a linear regression is used to relate T and G, only the solar irradiance should be known to retrieve the PV voltage for any value of PV current.

To give an example, if A GNG network with a maximum number of 400 neurons is used with an algorithm for the off-line training developed in Matlab/Simulink® environment, the results shown in Fig. 4.19 are obtained.

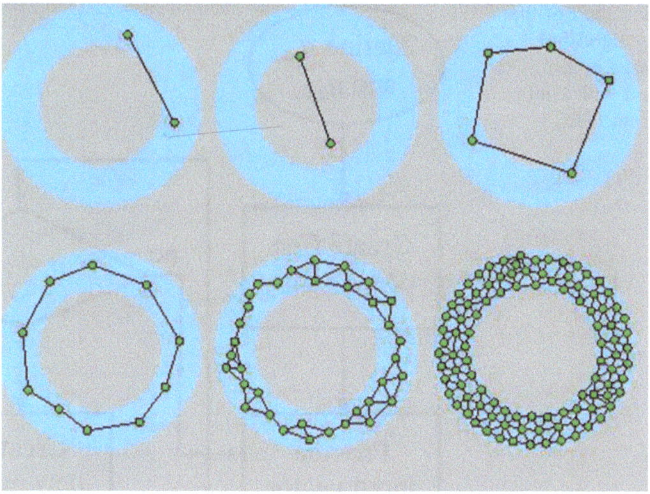

Fig. 4.17 Example of GNG reproduction of a ring-shaped data distribution

It can be observed that the GNG network correctly learns the data. As a matter of fact, the created neurons (circles) lie and correctly cover the characteristics of the PV source (curves) for all values of solar irradiance/temperature.

As for the computational requirement of this approach, the adoption of a 400 neuron network just implies that a 400 $\times$ 3 matrix of constant values has to be stored in the memory of the microprocessor. Modifying the number of neurons implies only storing a slightly different matrix. Afterwards, as anticipated, just the recalling phase has to be implemented on-line, whose most requiring task is the computation and sorting of two-dimensional vectors, where the variables are the PV voltage and current, closest to the applied input vector. This implies that the computational demand required by the GNG for this kind of application is significantly lower than that of other classic ANNs.

4.8 Examples of Matlab/Simulink® Simulation of PV Electrical Characteristics

In this Section, some examples of simulation models of PV sources developed in Matlab/Simulink® environment are presented and described in detail.

The schemes are related to the static model of the PV source obtained on the basis of the single-diode formulation where the shunt resistance is neglected. In particular, the Simulink® implementations of both the PV models obtained by a parameters identification according to the *discrete* approach and the *regression* approach, presented in Sects. 4.4.2.3 and 4.6, respectively, are described. Moreover, the simulation results obtained for both load and solar irradiance step transitions are shown.

Fig. 4.18 Flowchart of the GNG training algorithm [From Di Piazza et al (2010b)]. Used with kind permission from IEEE

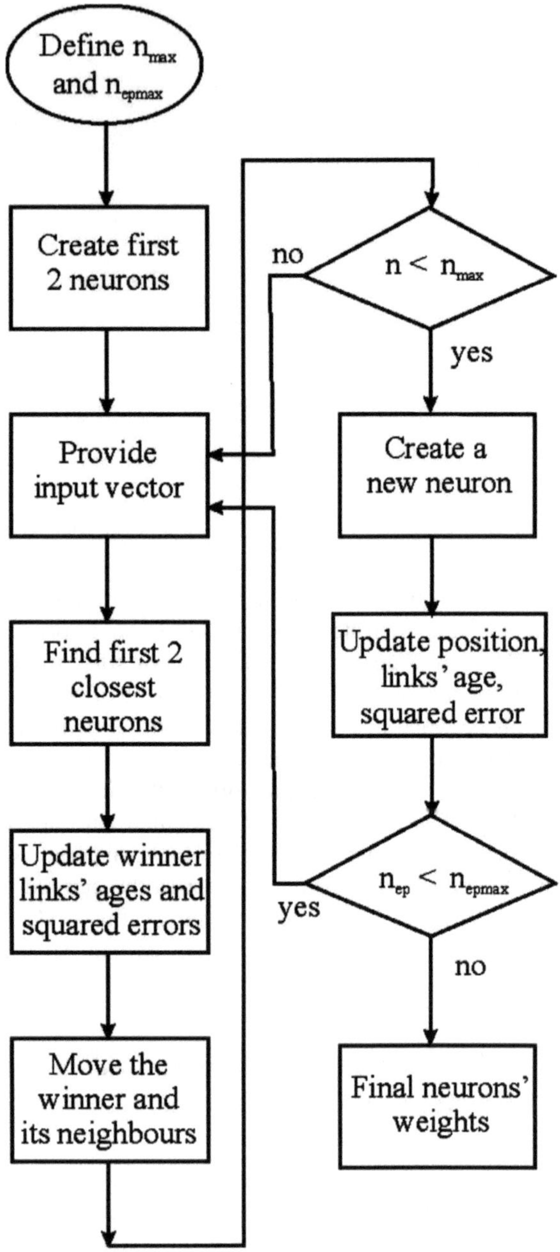

Finally, a Matlab® script, used to carry out the electrical characteristics of a PV assembly under nonuniform irradiance conditions, is given. Also in this last case, the used model for each PV module is the single-diode model, neglecting R_p, while the model parameters are obtained following the discrete approach.

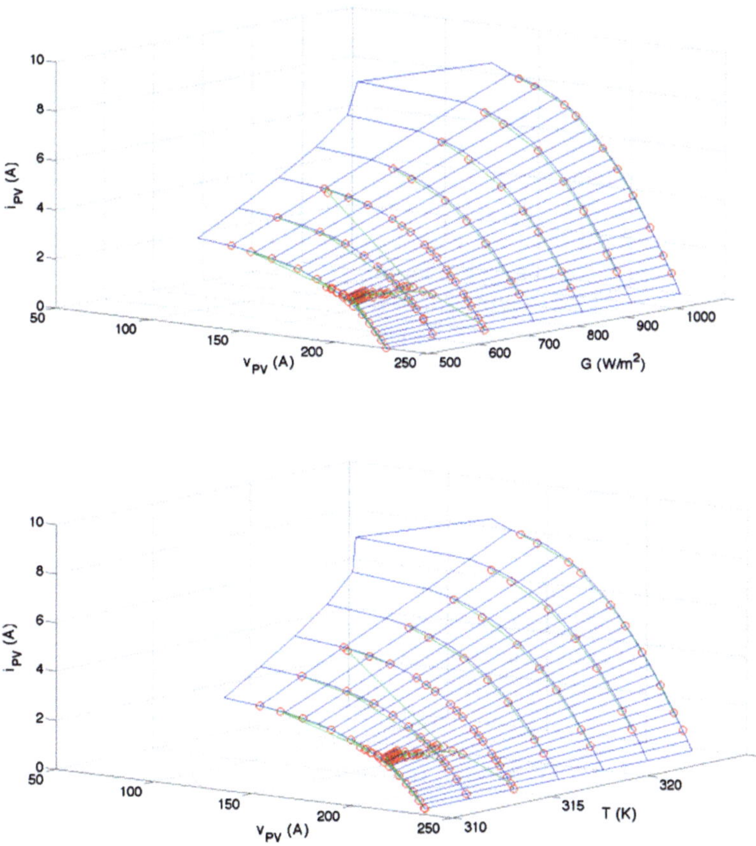

Fig. 4.19 Results of the training of GNG in the 3D subspaces: PV voltage versus PV current and irradiance (*upper*); PV voltage versus PV current and temperature(*lower*) (From Di Piazza et al. (2010b)). Used with kind permission from IEEE

4.8.1 PV Field Array Model Identified by the Discrete Approach

The scheme in Fig. 4.20 shows the PV array model as a "black box" where only the inputs and the output are highlighted.

As previously anticipated, the model is based on the single-diode formulation, neglecting R_p. Moreover, the model parameters are identified on the basis of the discrete approach, described in Sect. 4.4.2.3. This means that the experimental data (MPPs) are clustered into defined intervals of temperature and irradiance, according to the representation given in Fig. 4.9.

In this simulation model, the approximation $I_{sc} = 1.15 \times I_{MP}$ has been introduced. This approximation gives a good estimate of the short-circuit current, simplifying the model without jeopardizing the accuracy of the simulation results.

Going inside the block with I_{sc}, I, G and T as inputs and V as output, the scheme shown in Fig. 4.21 is found. This scheme describes the relationship $V = f(I)$, i.e., the inverse function of (3.13), taking into account the effects of temperature and solar irradiance, that are processed in the light grey block.

Expanding the "Parameters extraction" block, the scheme in Fig. 4.22 is obtained. Here, the temperature and solar irradiance are allocated within their own intervals and on this basis, the correct values of the three parameters K_1, K_2, and R_s are extracted from three coefficients matrices whose elements have been previously determined, according to Eq. (4.46).

As an example, a load step transition is simulated by the described model under the following condition: $T = 322$ K, $G = 1{,}000$ W/m^2. The simulation result is shown in Fig. 4.23. The initial and final operating points on the I–V electrical characteristic related to the step load transition are shown in Fig. 4.24. Furthermore, a solar irradiance step transition is simulated by the described model, going from $G = 1{,}000$ to 600 W/m^2, with a constant temperature of 322 K and being the initial and final operating points the MPPs. The simulation result is shown in Fig. 4.25. The initial and final operating points on the I–V electrical characteristics related to the step irradiance transition are shown in Fig. 4.26.

It should be noted that the described tests refer to a static PV model. When a dynamic model is used, the time domain waveforms can vary, remaining starting, and final operating points the same.

4.8.2 PV Field Array Model Identified by the Regression Approach

The scheme in Fig. 4.27 shows the PV array model as a "black box" where only the inputs and the output are highlighted. In this case, the model is based on the single-diode formulation, neglecting R_p, while the model parameters are identified on the basis of the regression approach, described in Sect. 4.6. In particular, the robust regression method has been used to correlate the solar irradiance with temperature, MPPs current and MPPs voltage. Also in this case the short-circuit current has been obtained by: $I_{sc} = 1.15 \times I_{MP}$.

Expanding the block with I and G as input and V as output, it is possible to observe how I_{MP} (I_{sc}), V_{MP}, and T are determined from G on the basis of the robust regression relationships. This is illustrated in the scheme of Fig. 4.28.

In Fig. 4.29 the expansion of the block "$V = f(I)$" is given. The obtained scheme implements the dependence of V_{oc} on temperature and describes, on a whole, the relationship $V = f(I)$, i.e., the inverse function of (3.13).

Finally, the "Parameters extraction" block calculates the model parameters according to (4.46); its Simulink® scheme is shown in Fig. 4.30.

As an example, a load step transition is simulated by the described model under a solar irradiance $G = 1{,}000$ W/m^2. The simulation result is shown in Fig. 4.31.

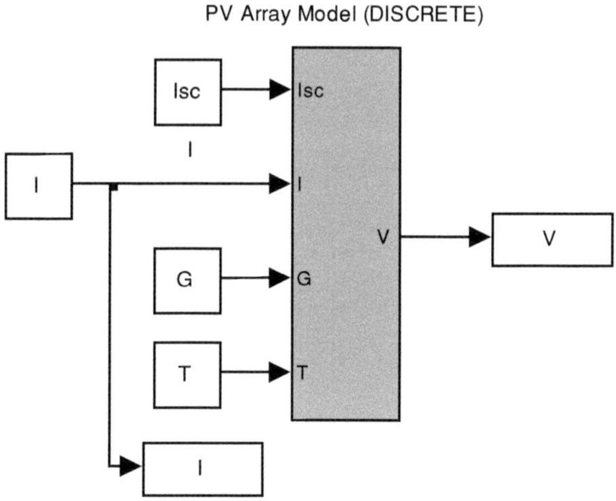

Fig. 4.20 Simulink® scheme of the PV array model identified by the discrete approach: overview

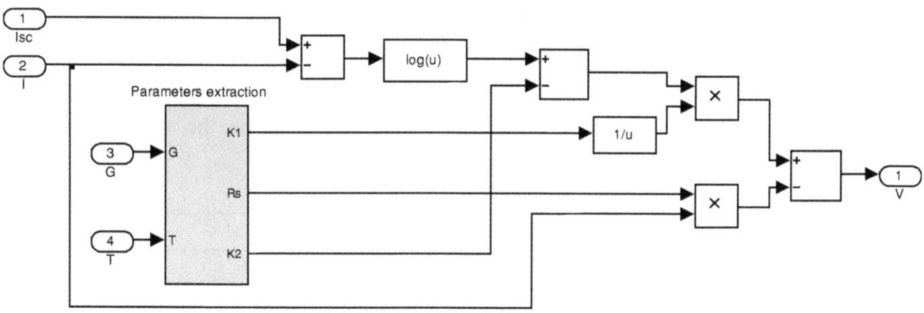

Fig. 4.21 Simulink® scheme of the PV array model identified by the discrete approach: implementation of $V = f(I)$

The initial and final operating points on the *I–V* electrical characteristic related to the step load transition are shown in Fig. 4.32. Furthermore, a solar irradiance step transition is simulated by the described model, going from $G = 1{,}000$ to 600 W/m^2, being the initial and final operating points the MPPs. The simulation result is shown in Fig. 4.33. The initial and final operating points on the *I–V* electrical characteristics related to the step irradiance transition are shown in Fig. 4.34.

It should be observed that the proposed examples are referred to the PV field described in 4.4.2.3, but obviously the implementation of model and simulation procedure is the same regardless of a PV cell, module, or field is considered.

It should be noted that the described tests refer to a static PV model. When a dynamic model is used, the time-domain waveforms can vary, remaining starting, and final operating points the same.

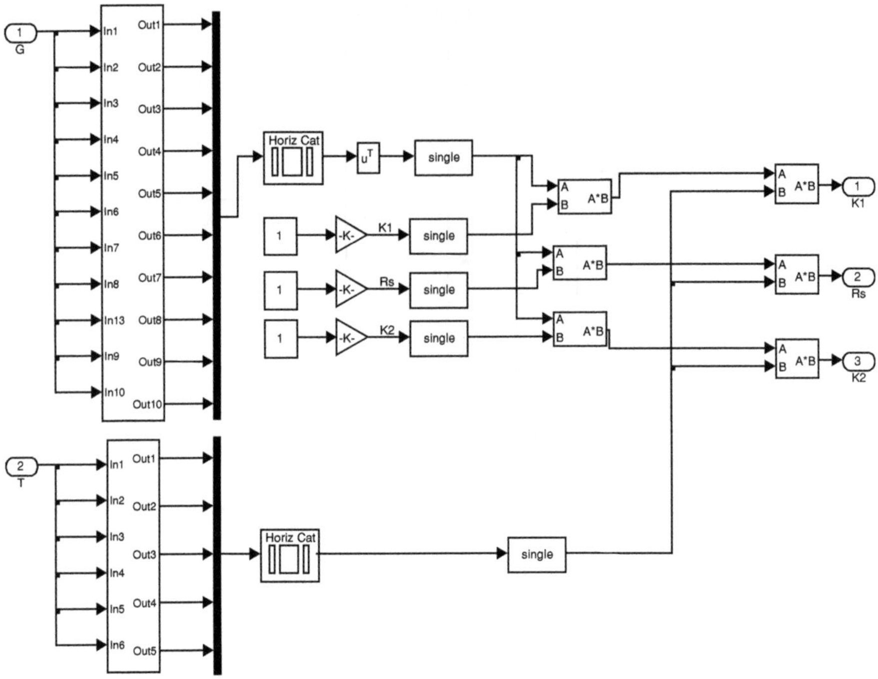

Fig. 4.22 Simulink® scheme of the PV array model identified by the discrete approach: determination of K_1, K_2 and R_s

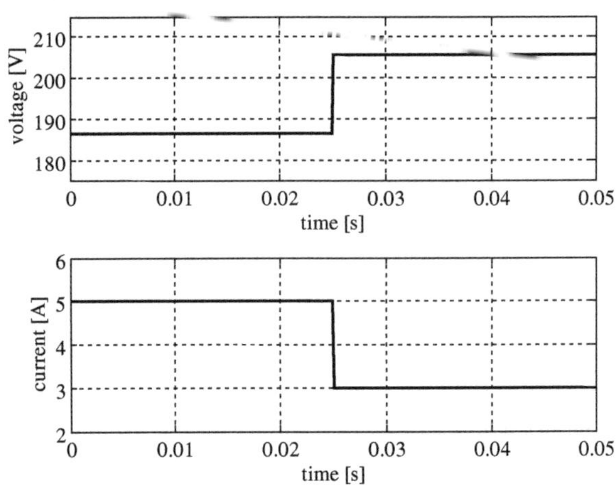

Fig. 4.23 Step load transition ($T = 322$ K; $G = 1{,}000$ W/m^2)

4.8.3 PV Source Model Under Non-Uniform Irradiance

The simulation models described until now are useful to describe the electrical behavior of a PV source under uniform irradiance conditions.

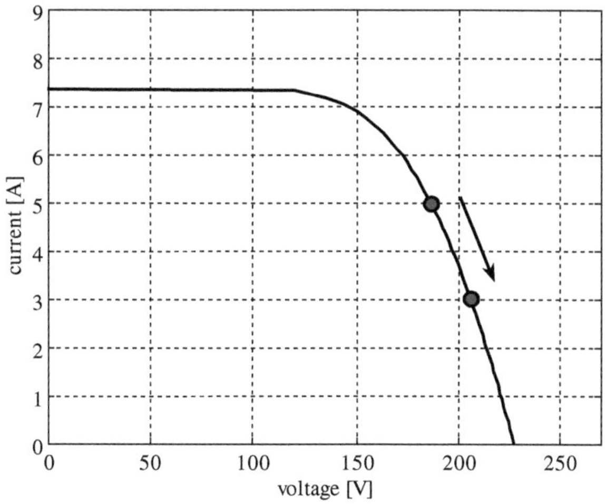

Fig. 4.24 Initial and final operating points on the *I*–*V* electrical characteristic related to the step load transition ($T = 322$ K; $G = 1{,}000$ W/m^2)

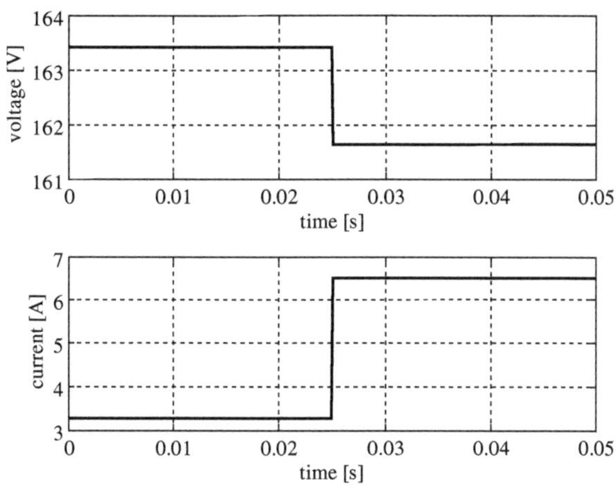

Fig. 4.25 Step irradiance transition from G = 1,000 to 600 W/m^2 ($T = 322$ K; initial and final operating conditions correspond to the MPPs)

If the analysis is focused on a PV generator subject to nonuniform irradiance, as in the case of partial shading, a different approach is needed.

At the end of this section, a Matlab® script developed to find out the electrical characteristics of PV strings under nonuniform irradiance conditions is given.

The script requires input data given in the form of a matrix with five rows and a number of columns equal to the number of series connected modules. In particular, each column contains the data in the following order: G, V_{oc}, I_{sc}, V_{mp}, I_{mp}.

The PV assembly is assumed to be formed of six modules connected in series, like in Fig. 4.6, and the features of the PV modules are those of the PV plant described in Sect. 4.4.2.3.

In Eq. (4.63) the input data matrix used for plotting the *I*–*V* curve for a PV assembly subject to three levels of solar irradiance (950, 650 and 250 W/m^2), is given.

In Fig. 4.35, the obtained *I*–*V* curve is shown.

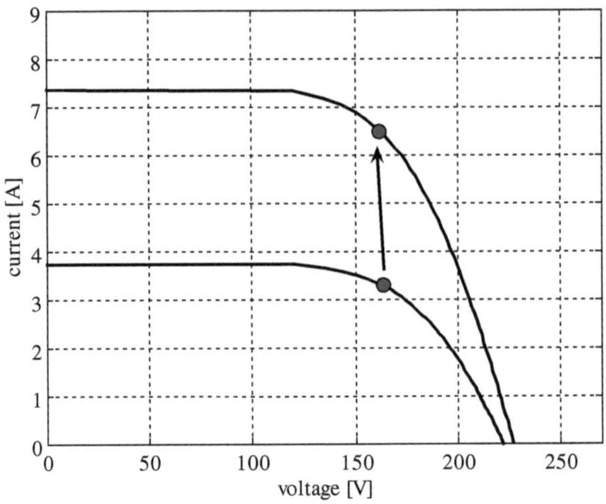

Fig. 4.26 Initial and final operating points on the I–V electrical characteristics related to the step irradiance transition from G = 1,000 to 600 W/m^2

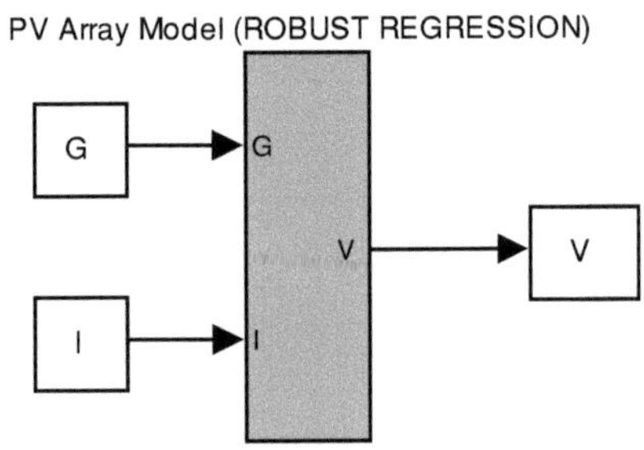

Fig. 4.27 Simulink® scheme of the PV array model identified by the robust regression approach: overview

The I–V characteristic of the assembly under a uniform solar irradiance corresponding to the highest considered level is present too in Fig. 4.35, so to highlight the different behavior of the PV source.

It should be observed that any possible partial shading situation can be simulated rearranging the input data matrix.

$$Data\_in = \begin{bmatrix} 950 & 950 & 650 & 650 & 250 & 250 \\ 36.66 & 36.66 & 36.66 & 36.66 & 36.66 & 36.66 \\ 3.37 & 3.37 & 2.37 & 2.37 & 0.86 & 0.86 \\ 27.69 & 27.69 & 27.62 & 27.62 & 27.26 & 27.26 \\ 2.93 & 2.93 & 2.06 & 2.06 & 0.75 & 0.75 \end{bmatrix} \tag{4.63}$$

MATLAB® script used for simulation

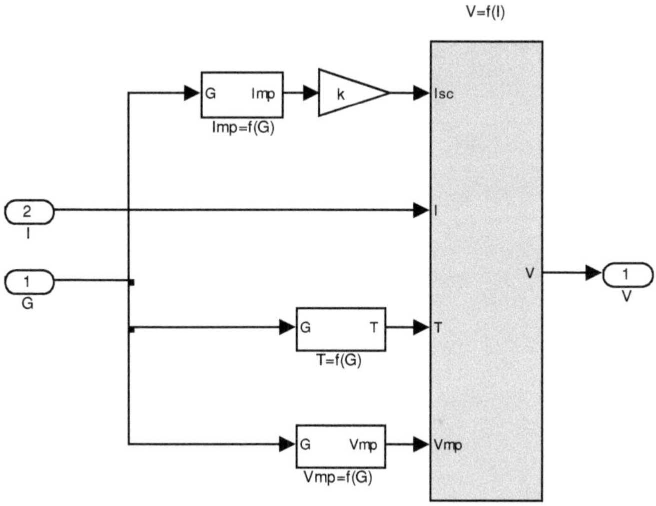

Fig. 4.28 Simulink® scheme of the PV array model identified by the robust regression approach: implementation of robust regression relationships

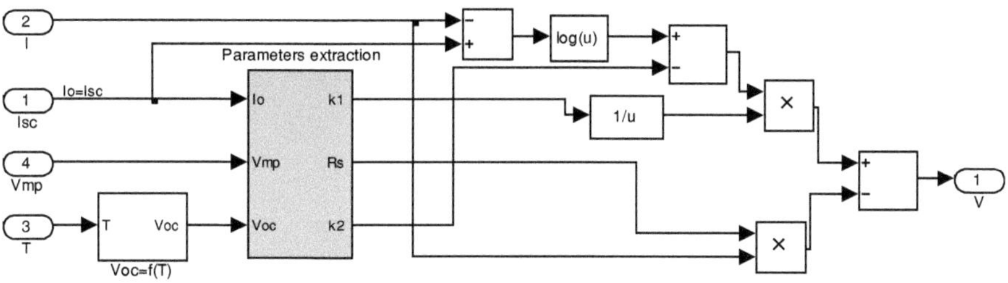

Fig. 4.29 Simulink® scheme of the PV array model identified by the robust regression approach: implementation of $V = f(I)$

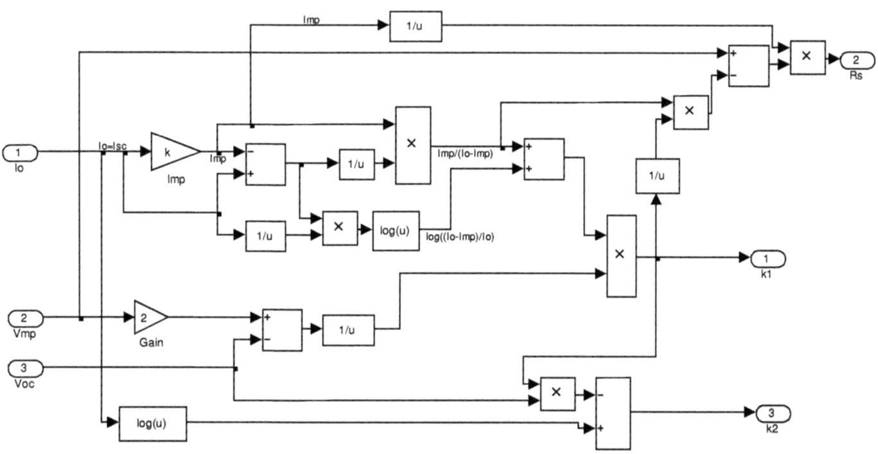

Fig. 4.30 Simulink® scheme of the PV array model identified by the robust regression approach: implementation of parameters extraction block

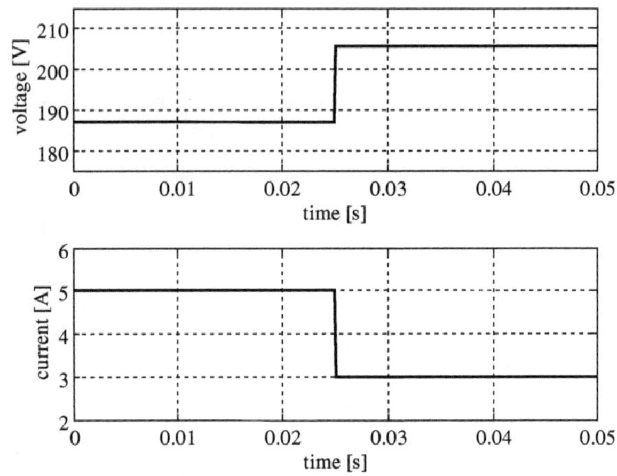

Fig. 4.31 Step load transition ($G = 1{,}000\ \mathrm{W/m^2}$)

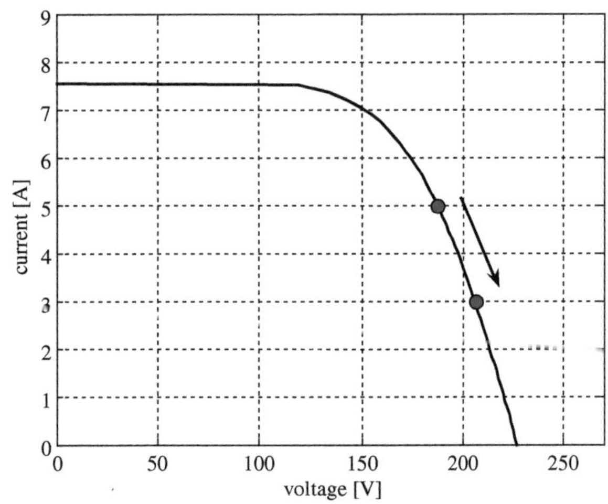

Fig. 4.32 Initial and final operating points on the I–V electrical characteristic related to the step load transition ($G = 1{,}000\ \mathrm{W/m^2}$)

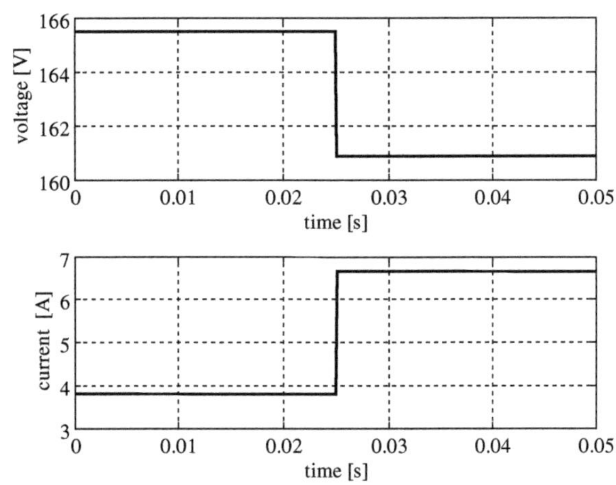

Fig. 4.33 Step irradiance transition from $G = 1{,}000$ to $600\ \mathrm{W/m^2}$ (initial and final operating conditions correspond to the MPPs)

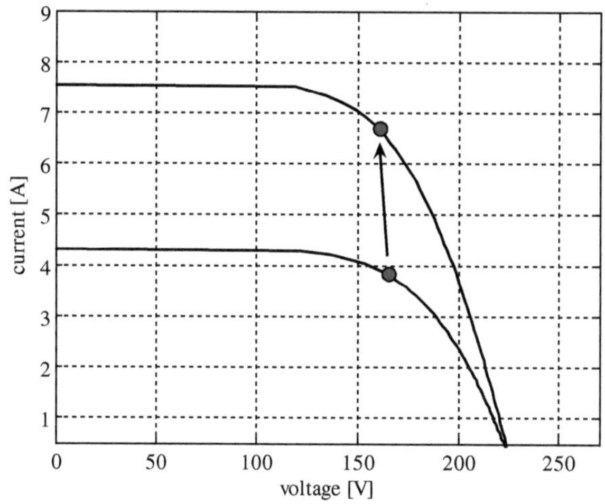

Fig. 4.34 Initial and final operating points on the *I–V* electrical characteristics related to the step irradiance transition from $G = 1{,}000$ to 600 W/m^2

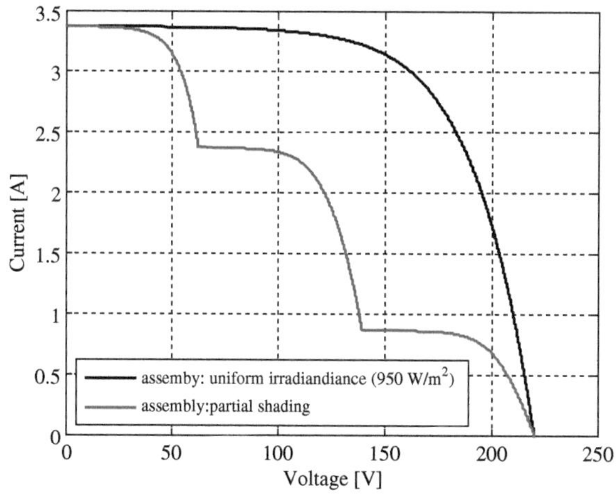

Fig. 4.35 *I–V* electrical characteristics of a PV assembly under uniform and nonuniform irradiance conditions

```
clear all
close all

% loading of the input data matrix
load Data_in;

% extraction of the number of series connected modules
D=size(Data_in);
m=D(2);

for k=1:m;
% initialization of the "remarkable points" vector
    Voc(k)=Data_in(2,k);
    Isc(k)=Data_in(3,k);
    Vmp(k)=Data_in(4,k);
    Imp(k)=Data_in(5,k);
    Iph(k)=Isc(k);

 % calculus of the PV module model parameters (in four
iterations)
        for i=1:4;
              k11(i)=((Imp(k)/(Iph(k)-Imp(k)))+log(1-
Imp(k)/Iph(k)))/((2*Vmp(k))-Voc(k));
              Rs1(i)=(Vmp(k)-(Imp(k)/((Iph(k)-
Imp(k))*k11(i))))/Imp(k);
              k21(i)=log(Iph(k))-(Voc(k)*k11(i));
              Iph1(i)=Isc(k)+exp(Isc(k)*Rs1(i)*k11(i)+k21(i));
        end

 % initializion of the "remarkable points" vector
        k1(k)=k11(4);
        Rs(k)=Rs1(4);
        k2(k)=k21(4);
        Iph(k)=Iph1(4);
end
% initializion of the current vector
I=0:5e-4:max(Iph);

% calculus of the voltage matrix
for j=1:length(I);

    for jj=1:m;

        if   I(j)<Iph(jj)
             V(j,jj)=((log(Iph(jj)-I(j)))/k1(jj))-I(j)*Rs(jj)-
k2(jj)/k1(jj);
        else
             V(j,jj)=0;
        end;
        A=0;
    end

    for kk=1:m;
        A=A+V(j,kk);
    end
VV(j)=A;
end
% plot of the I-V curve of partially shaded assembly
plot(VV,I);
hold;
xlabel('Voltage')
ylabel('Current')
title('partially shaded assembly');
```

4.9 Conclusions

In this chapter, different methods for the parameters identification of the most common static PV models have been described, including the different approaches of the scientific literature, ranging from pure analytical solutions to numerical or heuristic-based methods.

A general classification of these methods has been given considering, in addition, the used information for the parameters extraction, i.e., the rated data given by the manufacturer, the experimental characteristics directly measured at the PV source, some experimental remarkable points.

It has been found that the methods based only on the data given by the manufacturers are cost-effective since they do not need any measurement; on the other hand, they can be affected by some uncertainty because the actual electrical behavior of a PV source can differ from its theoretical behavior declared by the manufacturer. The identification methods based on experimental characteristics are more precise, but involve the knowledge of plenty of *I–V* curves in different environmental conditions. The approach based on the use of measurements of only some remarkable points of the PV electrical characteristics leads to a reduction of the measurement effort allowing, at the same time, an accurate determination of the PV electrical characteristics.

The possibility to simplify the parameters identification by introducing the use of LSRs has been shown.

Finally, the PV characteristic representation by mapping techniques, including a neural approach, has been delineated.

The PV models, suitably identified, are finally implemented in Matlab/Simulink® environment and simulation results are shown. The described simulation tests refer to a static PV model. In chapter 5, the issues related to PV source dynamic modeling, including parameters identification, and simulation, are faced up, to make the proposed study more complete.

Bibliography

Altas IH, Sharaf AM (2007) A photovoltaic array simulation model for matlab–simulink GUI environment. In: Proceedings of international conference on clean electrical and power, pp 341–345

Bätzner DL, Romeo A, Zogg H, Tiwari AN (2001) CdTe/CdS solar cell performance under low irradiance. In: Proceedings of European PV solar energy conference, paper VB1.40

Celik AN, Acikgoz N (2007) Modelling and experimental verification of the operating current of mono-crystalline photovoltaic modules using four and five parameters models. Appl Energy 84(1):1–15

Chan DSH, Phang JCH (1987) Analytical methods for the extraction of solar cell single and double diode model parameters from I-V characteristics. IEEE Trans Electron Devices ED-34(n.2):286–293

Da Costa WT, Fardin JF, Simonetti DSL, Neto L, de VBM (2010) Identification of photovoltaic model parameters by differential evolution. In: Proceedings of international conference on industrial technology, pp 931–936

Eikelboom JA, Reinders AHME (1997) Determination of the irradiation dependent efficiency of multicrystalline Si PV modules on basis of IV curve fitting and its influence on the annual performance. In: Proceedings of European PV solar energy conference, pp 293–296

Fritzke B (1995a) Incremental learning of linear local mappings. In: Proceedings of international conference on artificial neural networks, pp 217–222

Fritzke B (1995b) A growing neural gas network learns topologies, advances in neural information processing systems 7. MIT Press,

Fritzke B (1997) Growing neural gas. http://wwwold.ini.rub.de/VDM/research/gsn/JavaPaper/node19.html

Glass MC (1996) Improved solar array power point model with SPICE realization. In: Proceedings of 31st Intersoc. Energy Conversion Engineering Conference, vol 1, pp 286–291

Gow JA, Manning CD (1999) Development of a photovoltaic array model for use in power-electronics simulation studies. IEE Proc Electr Power Appl 146(21):193–200

Gradella Villalva M, Gasoli JR, Filho ER (2009) Comprehensive approach to modeling and simulation of photovoltaic array. IEEE Trans Power Electron 24(5):1198–1208

Jain A, Kapoor A (2005) A new approach to study organic solar cell using Lambert W-function. Sol Energy Mat Sol Cells 86:197–205

Kajihara A, Harakawa AT (2005) Model of photovoltaic cell circuits under partial shading. In: Proceedings of IEEE international conference on industrial technologies, pp 866–870

Kuo Y-C, Liang T-J, Chen JF (2001) Novel maximum power point tracking controller for photovoltaic energy conversion system. IEEE Trans Ind Electron 48(3):594–601

Matagne E, Chenni R, El Bachtiri R (2007) A photovoltaic cell model based on nominal data only. In: Proceedings of international conference on power engineering, energy electrical drives, pp 562–565

Moldovan N, Picos R, Garcia-Moreno E (2009) Parameter extraction of a solar cell compact model using genetic algorithms. In: Proceedings of Spanish conference on electron devices, pp 379–382

Ortiz-Rivera EI, Peng FZ (2005) Analytical model for a photovoltaic module using the electrical characteristics provided by the manufacturer data sheet. In: Proceedings of Power Electron. Spec. Conf.: 2087–2091

Petcut F-M, Dragomir T-L (2010) Solar cell parameter identification using genetic algorithms. CEAI 12(1):30–37

Phang JCH, Chan DSH, Phillips JR (1984) Accurate analytical method for the extraction of solar cell model parameters. Elect Lett 20(10):406–408

Di Piazza MC, Ragusa A, Vitale G (2009) Identification of photovoltaic array model parameters by robust linear regression methods. In: International conference on renewable energies and power quality, paper n. 268

Di Piazza MC, Vitale G (2010a) Photovoltaic field emulation including dynamic and partial shadow conditions. Appl Energy 87(3):814–823

Di Piazza MC, Pucci M, Ragusa A, Vitale G (2010b) Analytical versus neural real-time simulation of a photovoltaic generator based on a DC–DC converter. IEEE Trans Ind Appl 46(6):2501–2510

Saloux E, Teyssedou A, Sorin M (2011) Explicit model of photovoltaic panels to determine voltages and currents at the maximum power point. Solar Energy (in press). doi:10.1016/j.solener.2010.12.022

Sanchis P, Echeverria I, Ursua A, Alonso O, Gubia E, Marroyo L (2003) Electronic converter for the analysis of photovoltaic arrays and inverters. In: Proceedings of power electronics Spec. Conf, vol 4, pp 1748–1753

Sera D, Teodorescu R, Rodriguez P (2007) PV panel model based on datasheet values. In: Proceedings of IEEE international symposium on industrial electronics, pp 2392–2396

TamizhMani G, Ji L, Tang Y, Petacci L, Osterwald C (2003) Photovoltaic module thermal/wind performance: long-term monitoring and model development for energy rating, NCPV and solar program review meeting, pp 936–939

Tan YT, Kirschen DS, Jenkins N (2004) A model of PV generation suitable for stability analysis. IEEE Trans Energy Convers 19(4):748–755

Veerachary (2006) PSIM circuit-oriented simulator model for the non linear photovoltaic sources. IEEE Trans Aerosp Electron Syst 42(2):735–740

Veissid N, De Andrade AM (1991) The I–V silicon solar cell characteristic parameters temperature dependence, an experimental study using the standard deviation method. In: Proceedings of European photovoltaic solar energy conference, pp 43–47

Walker G (2001) Evaluating MPPT converter topologies using a matlab PV model. J Electr Electron Eng Aus 21(1):45–55

Xiao W, Dunford WG, Capel A (2004) A novel modeling method for photovoltaic cells. In: Proceedings of IEEE 35th Annual Power Electron. Spec. Conf, pp 1950–1956

Chapter 5
Photovoltaic Source Dynamic Modeling Issues

5.1 Introduction

In the previous chapters, static models of the PV source have been proposed with related techniques to identify the values of the model parameters. However, operating conditions imply the variation of environmental parameters and load.

As for the first, being usually their variation slow with respect to time, the static model can still be utilized assuming a quasi-static variation. On the contrary, the load can be subject to fast variations due to the customer request and the inherent switching frequency of the inverter supplied by the PV source.

For these reasons, a dynamic model represents a useful tool for the analysis of power converters, for the study of maximum power point tracking (MPPT) algorithms and, for simulating the PV system using circuit simulators or real-time emulators.

In Sect. 2.12 it has been explained that the presence of a p–n junction implies a capacitive effect due to excess of minority carriers stored in the quasi-neutral region of a diode. These effects can be modeled by a voltage-dependent capacitance, as explained in Sect. 3.4.1 where it has been demonstrated by (3.53) that the nonlinear junction capacitance influences the current supplied by the PV source with a term proportional to the series resistance and to the time derivative of the same current. This term becomes relevant for fast load current variations. In any case a precise model of the nonlinear capacitance is cumbersome to be implemented.

This last consideration suggests that it is worth trying to adopt a constant value for the parasitic capacitance; it can be obtained with a best fitting of experimental data. In such a way, all transients can be studied also in the s-domain and the model implementation becomes quite easy.

In particular, the approach followed in Sects. 5.2 and 5.3 consists, firstly, on the determination of a static model, and on its parameter identification. Then, by imposing suitable load variations, the electric transients in time domain are used to identify the values of a parallel capacitance and of a series inductance. This last component models the inductive effects of the connecting cables, and allows a

M. C. Di Piazza and G. Vitale, *Photovoltaic Sources*, Green Energy and Technology,
DOI: 10.1007/978-1-4471-4378-9_5, © Springer-Verlag London 2013

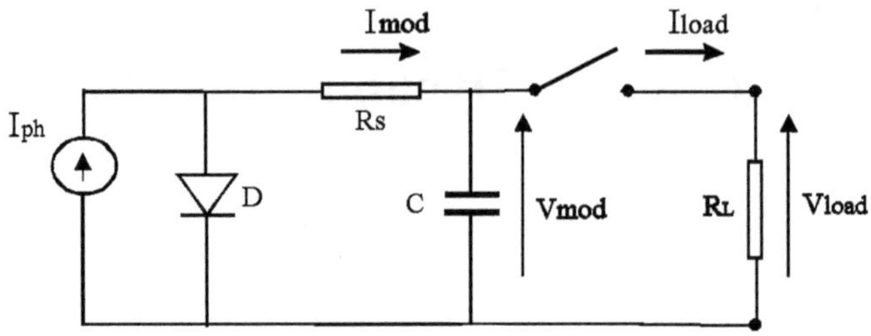

Fig. 5.1 Equivalent circuit of a PV module with its internal equivalent capacitance

better reproduction of transients. The nonlinear capacitance model is discussed in Sect. 5.4 where its effects on different load transients are shown.

5.2 Dynamic PV Model Formulation

In order to model the electric dynamic of a PV module, the simplest way is to add an equivalent capacitance parallel connected to the output of the single diode model, as proposed by King et al. (2004). The related circuit with an additional switch to impose the load variation is shown in Fig. 5.1.

The output voltage derivative of this circuit is given by:

$$\frac{\partial V_{\text{mod}}}{\partial t} = \frac{1}{C}\left[I_{\text{mod}}(V) - I_{\text{load}}\right] \tag{5.1}$$

where $V = V_{\text{mod}} = V_{\text{load}}$ when the switch is closed. Hence, by measuring I_{load} and the voltage derivative during a suitable transient and by using the model expressed by (3.12) for achieving the corresponding values of $I_{\text{mod}}(V)$, the value of C could be calculated by least square regression (LSR).

It can be observed that it is cumbersome to acquire the voltage signal and to perform a post-processing derivative, since it will be affected by noise.

Moreover, the circuit drawn in Fig. 5.1 is not sufficiently accurate to model the electrical dynamic behavior of a PV module because, although the effect of the capacitance is predominant, other effects are present which are not negligible.

Firstly, the junction capacitance effect implies the presence of a conductance $1/R_c$, whose time constant reproduces the time needed by the quantity of charge in the capacitance to vary according to the variation of the junction voltage. Moreover, a further effect given by the connection parasitic inductance is to be expected due to the cables connecting the cells and the whole source to the load. This value is higher when a series connection of PV cells is exploited to obtain a higher V_{oc}. Due to these reasons, when a resistive load is abruptly connected to a PV module, the load current is expected to have a second-order circuit response and the module voltage an exponential decay that depends on all the three resistances: R_s, R_c, and R_L. The equivalent dynamic circuit, including a switch to verify the load insertion, is given

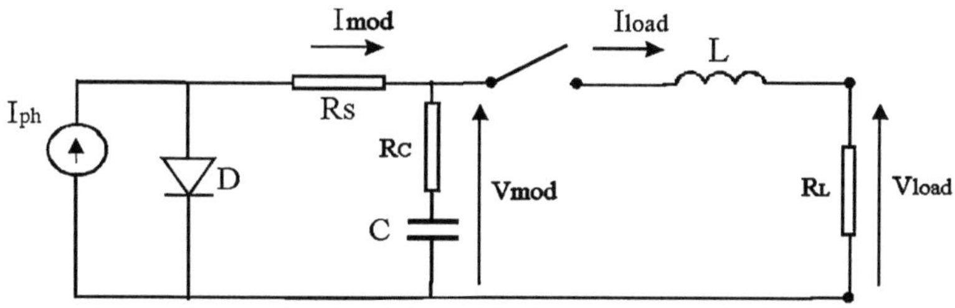

Fig. 5.2 The complete equivalent circuit of a PV module

in Fig. 5.2. On the contrary, the model sketched in Fig. 5.1 gives a first-order response for voltage and current: in particular, an exponential decay that depends on R_s and R_L is obtained.

It should be noted that the sudden load connection is a typical situation that occurs for example, when an inverter supplied by PV modules re-starts to operate after the shadowing of the modules due to the presence of clouds.

The following analysis is devoted to setup an identification method to obtain the values of C, L, R_c.

Under the hypothesis that the operating point moves in the nearly constant voltage region, a portion of the PV characteristic which exhibits just a very light deviation from the linear trend is involved. In this way, it is possible to simplify the circuit replacing the current generator and the diode with a voltage generator. Therefore, it is possible to consider the circuit represented in Fig. 5.3, which is equivalent to the real PV module in that linear zone.

The load current waveform obtained by the circuit drawn in Fig. 5.3 is expected to be composed of the sum of a constant and two exponential curves, just as in any RLC circuit. In addition, it is possible to observe that with a suitable value of R_L the effects of the inductor (front edge of the signal) and of the capacitor (trailing tail of the waveform) are clearly distinct, since the two time constants τ_L and τ_C are about an order of magnitude different. This allows to say that the initial part of the front edge lies on an inductive exponential curve and that the trailing tail, which starts a bit after the peak, lies on a capacitive exponential trend.

By choosing a time origin equal to the instant the switch is closed, the load current is given by the following equation:

$$I_{\text{load}}(t) = -A_L \times e^{-\frac{t}{\tau_L}} + A_C \times e^{-\frac{t}{\tau_C}} + I_{\text{load}}(\infty) \tag{5.2}$$

Aiming to determine coefficients A_L and A_C it is possible to rearrange (5.2) in the following form:

$$I_{\text{load}}(t) = I_1(t) + I_2(t) + K \tag{5.3}$$

where:

$$I_1(t) = A_L \times \left(1 - e^{-\frac{t}{\tau_L}}\right) \tag{5.4a}$$

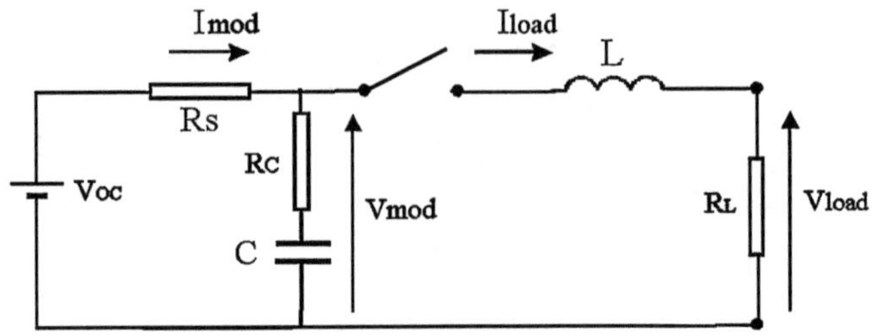

Fig. 5.3 Equivalent circuit of a PV module in the nearly constant voltage region

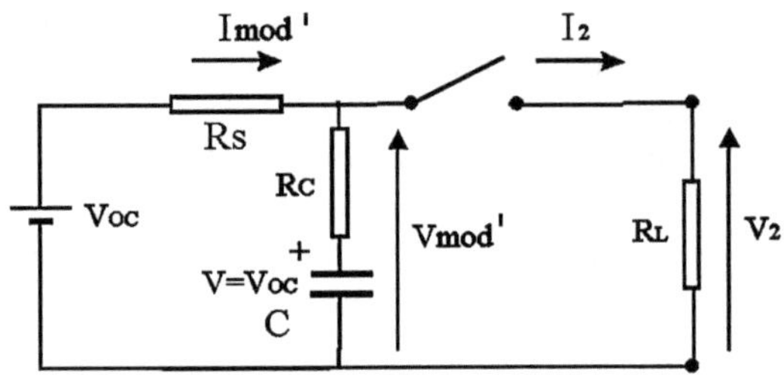

Fig. 5.4 Equivalent circuit for determining A_C and τ_C

$$I_2(t) = A_C \times e^{-\frac{t}{\tau_C}} + I_{\text{load}}(\infty) \tag{5.4b}$$

$$K = -A_L \tag{5.4c}$$

Since $I_{\text{load}}(0^-) = I_{\text{load}}(0^+) = 0$ because of the presence of the inductor, it is clearly:

$$A_C + I_{\text{load}}(\infty) - A_L = 0 \tag{5.5}$$

Equations (5.4a) and (b) represent, respectively, the inductive and capacitive exponential curves on which the initial part of the front edge and the trailing tail of the load current lie. In order to determine the coefficient A_C, it is necessary to consider the circuit without the inductor, which is shown in Fig. 5.4.

When the switch in Fig. 5.4 is closed, the module voltage decreases instantly from $V'_{\text{mod}}(0^-) = V_{\text{oc}}$ to a new value $V'_{\text{mod}}(0^+) = V'_{\text{oc}}$. Observing that the generator and the capacitor exhibit the same voltage, it is easy to determine that:

$$V'_{\text{oc}} = V_{\text{oc}} - R_{\text{CS}} \times I_2(0^+) = R_L \times I_2(0^+) \tag{5.6}$$

where:

$$R_{\text{CS}} = \left(\frac{1}{R_C} + \frac{1}{R_S}\right)^{-1} \tag{5.7}$$

and

$$I_2(0^+) = \frac{V_{\text{OC}}}{R_L + R_{\text{CS}}} \tag{5.8}$$

Hence, by combining Eqs. (5.8) and (5.4b) evaluated at $t = 0^{+}$, the following expression is obtained for the coefficient A_C:

$$A_C = I_2(0^+) - I_{\text{load}}(\infty) = \frac{V_{\text{OC}}}{R_L + R_{\text{CS}}} - I_{\text{load}}(\infty) \tag{5.9}$$

Then, from (5.5), the following expression is obtained for the coefficient A_L:

$$A_L = I_2(0^+) = \frac{V_{\text{OC}}}{R_L + R_{\text{CS}}} \tag{5.10}$$

As for the time constants of the complete equivalent model, considering that the capacitor and the inductor do not interfere with each other, it can be concluded that:

- the capacitive time constant in Eq. (5.4b), deduced by a simple analysis of the circuit in Fig. 5.4, is $\tau_C = (R_C + R_{\text{SL}}) \times C$, where $R_{\text{SL}} = \left(\frac{1}{R_S} + \frac{1}{R_L}\right)^{-1}$;
- since its response time is extremely short, the inductor in the circuit of Fig. 5.3 sees a nearly constant voltage; therefore, the inductive time constant is $\tau_L = \frac{L}{R_L + R_{\text{CS}}}$.

The previous considerations complete the description of the proposed electrical dynamic model for a PV source and they allow its parameters to be identified by means of a linear regression of the natural logarithm of suitable portions of the load current or voltage. The load current is a preferred choice since, being equal to the inductor current, it is a state variable.

What above supposed, means a current transient, caused by the abrupt insertion of a suitable load formed by the sum of an exponential curve due to the inductance contribution growing from zero value (at $t = 0$ the current is null) to a maximum value at $t = t^*$ and then an exponential decay that reaches the final value. Hence, it is possible to choose a time interval ΔT_1 before time t^* and a time interval ΔT_2 after time t^*, during which the current is originated by the parasitic inductance and by the capacitance. In formulas:

$$\begin{cases} I_{\text{load}}\big|_0^{\Delta T_1} = I_1\big|_0^{\Delta T_1} \\ I_{\text{load}}\big|_{t^*+\varepsilon}^{t^*+\varepsilon+\Delta T_2} = I_2\big|_{t^*+\varepsilon}^{t^*+\varepsilon+\Delta T_2} \end{cases} \tag{5.11}$$

where ε is a suitable nonzero coefficient.

In order to identify the capacitance value, it is possible to write Eq. (5.4b) as:

$$\ln(I_2 - I_{\text{load}}(\infty)) = \ln(I_2(0^+) - I_{\text{load}}(\infty)) - \frac{t}{\tau_C} = q_1 + m_1 \times t \tag{5.12}$$

and, thus, to apply a LSR to the quantity:

$$\ln\left[I_{\text{load}}\big|_{t^*+\varepsilon}^{t^*+\varepsilon+\Delta T_2} - I_{\text{load}}(\infty)\right] \tag{5.13}$$

The opposite of the slope m_1 will give the inverse of the capacitive time constant. The quantity e^{q_1} allows $I_2(0^+)$ to be determined:

$$I_2(0^+) = e^{q_1} + I_{\text{load}}(\infty) \tag{5.14}$$

Then, the parasitic resistance R_C can be calculated by rewriting Eqs. (5.7) and (5.8) as:

$$R_{\text{CS}} = \frac{V_{\text{OC}}}{I_2(0^+)} - R_L \tag{5.15}$$

$$R_C = \left(\frac{1}{R_{\text{CS}}} - \frac{1}{R_S}\right)^{-1} \tag{5.16}$$

Finally, the capacitance will be given by:

$$C = \frac{\tau_C}{R_C + R_{\text{SL}}} \tag{5.17}$$

Similarly, in order to identify the inductance value, it is possible to write Eq. (5.4a) as:

$$\ln(I_2(0^+) - I_1) = \ln I_2(0^+) - \frac{t}{\tau_L} = q_2 + m_2 \times t \tag{5.18}$$

and, thus, to apply a LSR to the quantity:

$$\ln\left[I_2(0^+) - I_{\text{load}}\big|_0^{\Delta T_1}\right] \tag{5.19}$$

The opposite of the slope m_2 will give the inverse of the inductive time constant, whereas the inductance will be given by:

$$L = \tau_L \times (R_L + R_{\text{CS}}) \tag{5.20}$$

Finally, the quantity e^{q_2} can be used to verify the goodness of the whole procedure, since it should be equal to $I_2(0^+)$.

It should be borne in mind that the hypothesis that the operating point moves in a quite linear portion of the PV characteristic does not affect the generality of the results. As a matter of fact, the determined equivalent circuit is able to reproduce the PV source behavior even if the operating point is changed. The advantage of the proposed approach consists on the determination of a constant capacitance value that allows transients to be correctly reproduced. Since this value can be assumed as an average value, the obtained results can be slightly different when transitions involve points located at the boundary of the I–V curve or when nonlinear capacitance effects are relevant.

Moreover, the circuit parameters achieved by the proposed methodology can be added to any static PV model (double diode with or without shunt resistor) because the fitting is performed on real data.

5.3 Parameters Identification

In this section an example of parameter identification, using the above explained approach, is proposed. The identification procedure is based on experimental data; thus, a dedicated experimental setup with the following features has been used: a measurement system for the detection of current and voltage at the PV source output, a system for the measurement of solar irradiance and temperature, and finally a programmable resistive load.

In particular, for current and voltage measurement, an Agilent MSO6104A 4 Gsamples/s oscilloscope has been used, the solar irradiance is obtained by a pyranometer Kipp & Zonen CM6b, installed with the same tilt as the modules.

Firstly, the static model of the PV source has been identified. The adopted model is the one described by (3.13) and (3.14) and the technique described in Sect. 4.4.2.3, dealing with the identification from remarkable experimental points, has been adopted. The temperature has been estimated by performing a LSR according to Eq. (4.58).

For a measured value of solar irradiance equal to $G = 655\ \mathrm{W/m^2}$, the characteristic parameters of the considered module are: $V_{oc} = 19.6$ V, $I_{sc} = 0.96$ A, $V_{MP} = 14.96$ V, $I_{MP} = 0.92$ A. By substituting these values in the set of Eq. (4.46), the four parameters of the model described by Eq. (3.13) have been found:

$$\begin{cases} I_0 = 0.96\,\mathrm{A} \\ R_s = 3.245\,\Omega \\ K_1 = 1.9207 \\ K_2 = -37.6872 \end{cases} \tag{5.21}$$

Firstly, a step load current transient has been acquired during the transition from the open circuit point to a point which is located halfway between V_{oc} and the MPP. Taking into consideration that the MPP is obtained by a load resistance equal to $R_{L,MPP} = 16.2\ \Omega$ a value of $R_{load} = 23.1\ \Omega$ has been chosen. The obtained waveform is shown in Fig. 5.5. It should be noted that this curve corresponds to a second-order system response, as expected. This test gives the peak time ($t^* = 2 \times 10^{-6}$ s) and the steady-state load current ($I_{load}(\infty) = 0.712$ A). Then, after plotting Eq. (5.13), the chosen interval for the first LSR has been fixed by means of the following parameters: $\varepsilon = 5 \times 10^{-8}$ s, $\Delta T_2 = 6 \times 10^{-6}$ s. Figure 5.6 shows the plot of Eq. (5.13) and Fig. 5.7 shows a zoom on the chosen interval. The regression line $y = q_1 + m_1 \times t$ has been superimposed.

The opposite of the inverse of the slope gives the following capacitive time constant:

$$\tau_c = 3.186\ \mu\mathrm{s} \tag{5.22}$$

Subsequent calculations, performed according to Eqs. (5.14–5.16), allow the following quantities to be determined:

$$I_2(0^+) = 0.762\ \mathrm{A} \tag{5.23}$$

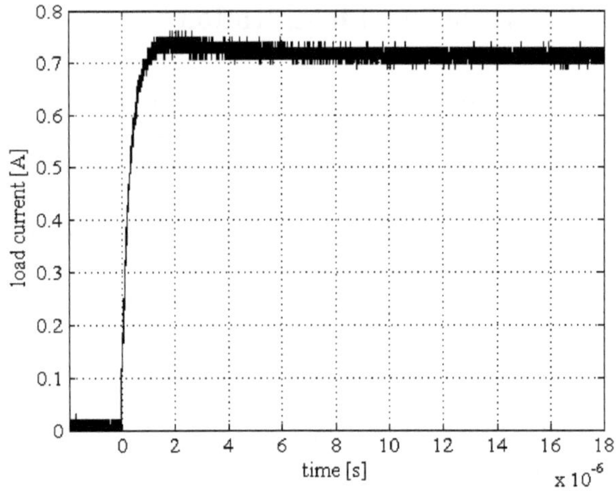

Fig. 5.5 Typical transient waveform of load current

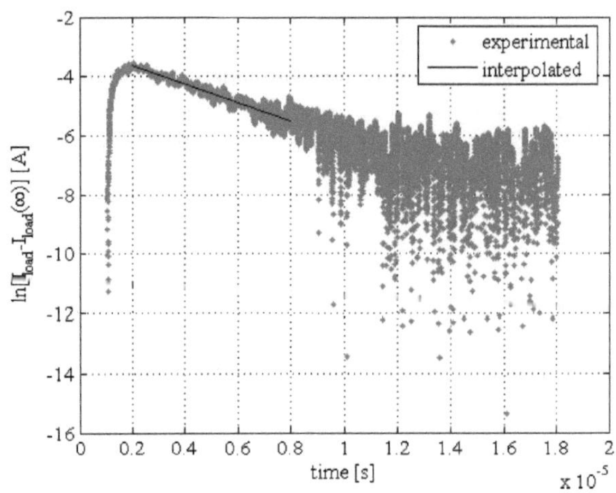

Fig. 5.6 Plot of $ln[I_{\mathrm{load}}-I_{\mathrm{load}}(\infty)]$ versus time, with the interpolating line superimposed

$$R_C = 9.56\,\Omega \tag{5.24}$$

Therefore, the capacitance value has been determined according to (5.17):

$$C = 256.92 \text{ nF} \tag{5.25}$$

Furthermore, by using Eq. (5.6) it is possible to determine that $V'_{\mathrm{oc}} = 17.6$ V.

In order to perform the second LSR, Eq. (5.19) has been plotted and the parameter $\Delta T_1 = 1 \times 10^{-6}$ s has been chosen. Figure 5.8 shows the plot of Eq. (5.19) and Fig. 5.9 shows a zoom on the chosen interval. The regression line $y = q_2 + m_2 \times t$ has been superimposed.

The opposite of the inverse of the slope gives the following inductive time constant:

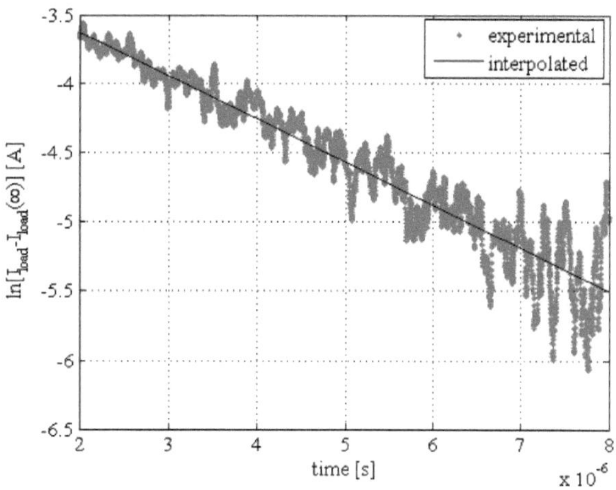

Fig. 5.7 Zoom of Fig. 5.6 which shows the chosen interval for the LSR

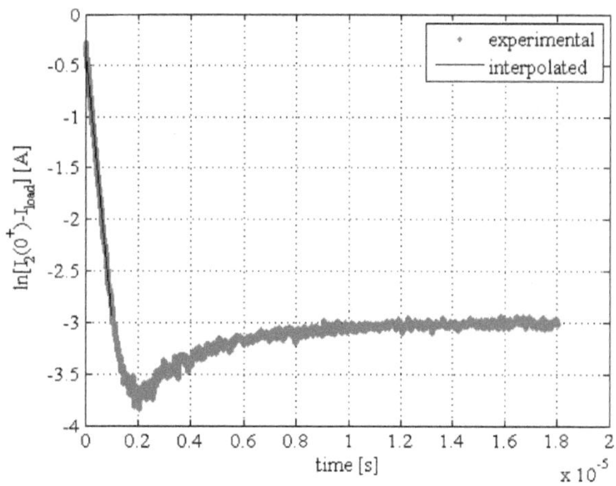

Fig. 5.8 Plot of $ln[I_2(0^+) - I_{load}]$ versus time, with the interpolating line superimposed

$$\tau_L = 0.373\ \mu\text{s} \tag{5.26}$$

Therefore, according to Eq. (5.20), the inductance value is:

$$L = 9.52\ \mu\text{H} \tag{5.27}$$

In order to verify the goodness of the whole procedure, the quantity e^{q_2} has been evaluated and compared to the value given in Eq. (5.23). The following error has been calculated:

$$e_{\%} = \frac{I_2(0^+) - e^{q_2}}{I_2(0^+)} \times 100 = \frac{0.762 - 0.746}{0.762} \times 100 = 2.1\% \tag{5.28}$$

which is acceptable.

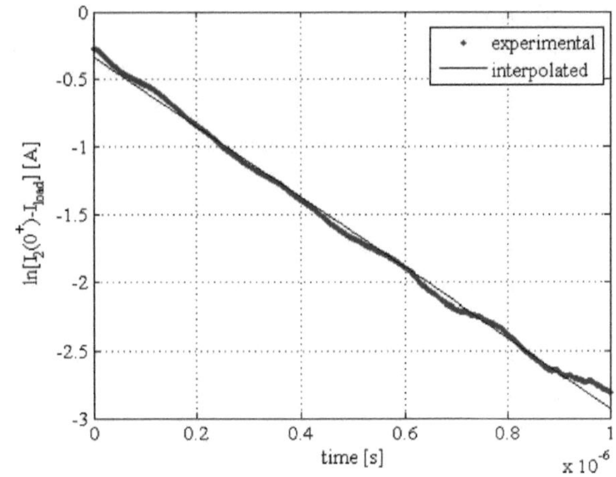

Fig. 5.9 Zoom of Fig. 5.10 which shows the interval chosen for the LSR

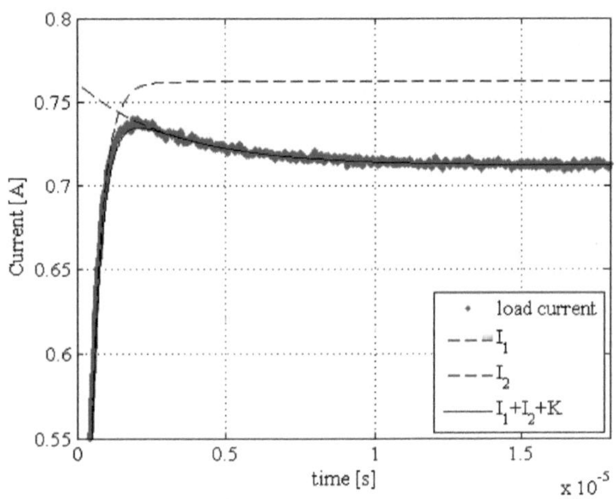

Fig. 5.10 Load current I_{load}, curves I_1 and I_2 and their sum shifted by K

Finally, Fig. 5.10 shows the load current I_{load}, the curves I_1, and I_2 on which the load current waveform lies and their sum shifted by K according to (5.3).

In order to provide an additional validation of the electrical dynamic model, a new transient has been acquired by means of the same set-up (module, oscilloscope, and pyranometer). A different load resistor has been used ($R_{load} = 21.85\ \Omega$) and the pyranometer has given a value of $G = 773$ W/m^2, at which the characteristic parameters of the modules are: $V_{oc} = 19.8$ V, $I_{sc} = 1.145$ A, $V_{MP} = 14.16$ V, $I_{MP} = 1.07$ A. Following the procedure of Sect. 4.4.2.3, the four parameters of Eq. (3.13) have been found:

$$\begin{cases} I_0 = 1.145\ \text{A} \\ R_s = 3.3905\ \Omega \\ K_1 = 1.3546 \\ K_2 = -26.6852 \end{cases} \tag{5.29}$$

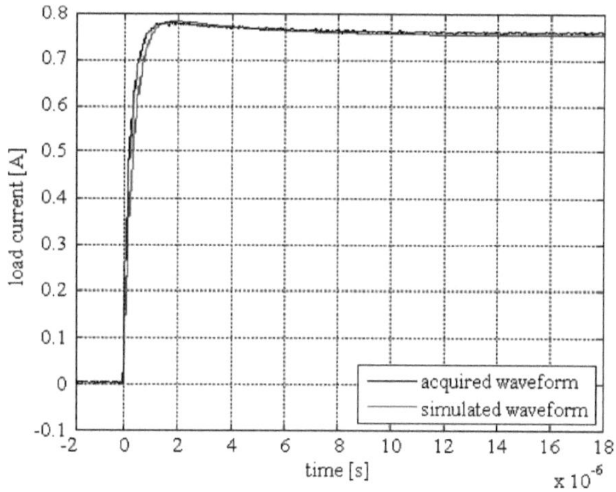

Fig. 5.11 Acquired and simulated waveforms of load current

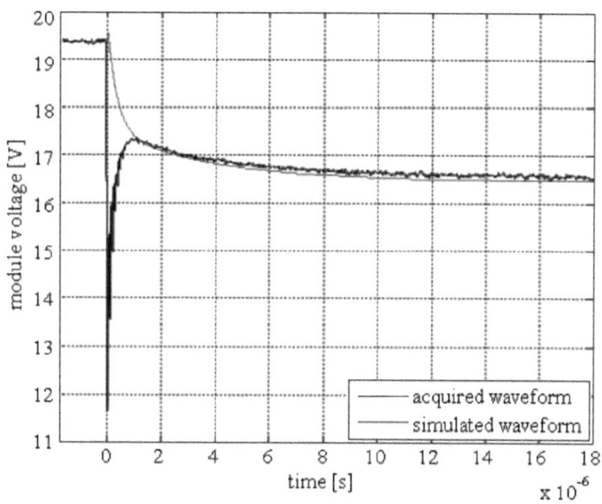

Fig. 5.12 Acquired and simulated waveforms of module voltage

Then, the model described by Eq. (3.13) with parameters (5.29) and by the circuit of Fig. 5.2 with parameters (5.24), (5.25), (5.27) has been implemented in the Simulink®-PLECS® environment and a simulation has been performed.

Figures 5.11 and 5.12 show the acquired and simulated waveforms of load current and module voltage. As it can be seen, the model is able to reproduce the electrical transient very faithfully.

The difference during the initial voltage drop in Fig. 5.12 could be eliminated by complicating the model, however, this would complicate also the procedure for the parameters identification, despite a minimal gain in precision because the spike duration is less than 1 microsecond.

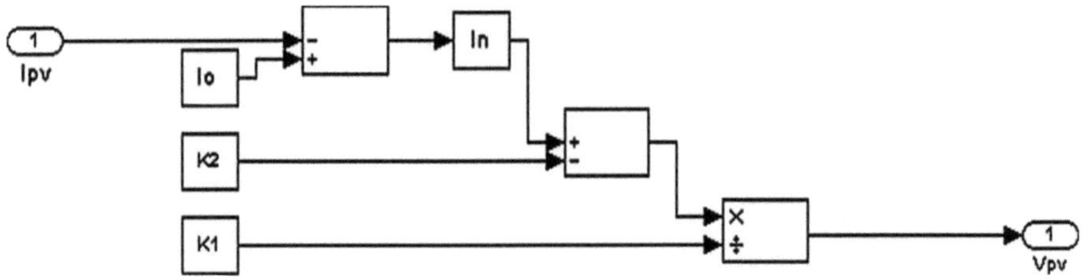

Fig. 5.13 Simulink® implementation of $V = f(I)$ model neglecting R_s

5.4 Matlab/Simulink® Simulation of PV Electrical Characteristics

In this section the Matlab/Simulink® schemes, used for simulating the PV source dynamic behavior, are described. In order to introduce this subject, an overview of the PV model implementation schemes, including static formulations, is presented.

5.4.1 $V = f(I)$ Static Model Formulation

Equation 3.13, here rewritten, can be inverted to obtain the photovoltaic source voltage having its current as input.

$$I = I_0 - e^{[(V+IR_s)K_1+K_2]}$$

The following equation is obtained.

$$V = \frac{\ln(I_0 - I) - K_2}{K_1} - IR_s \tag{5.30}$$

If the series resistance can be neglected, the model can be implemented in Simulink® as sketched in Fig. 5.13.[1]

Otherwise, if the R_s has to be included, the corresponding model is drawn in Fig. 5.14.

It should be noted that the $V = f(I)$ formulation implies the calculation of a logarithm. Normally, it is $I_0 > I$; however, this conditions could not be satisfied during transients, so the application of a very small inferior limit to $(I_0 - I)$ by a saturator block (not shown in Fig. 5.14) avoids errors.

[1] In this figure, as well as in the next Simulink diagrams, the voltage and current at the output of the PV source are indicated as V_{PV} and I_{PV} respectively.

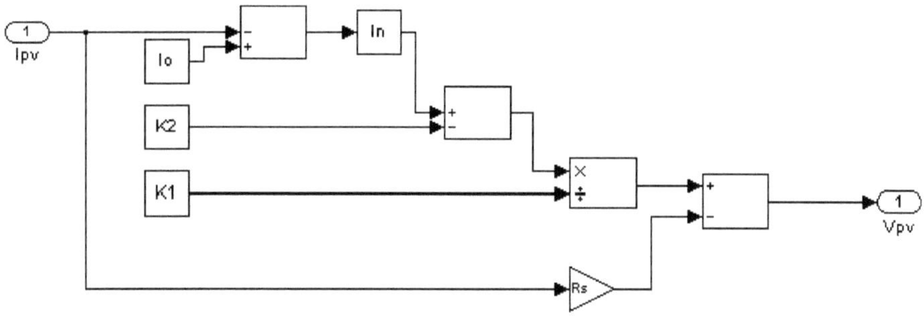

Fig. 5.14 Simulink® implementation of $V = f(I)$ model including R_s

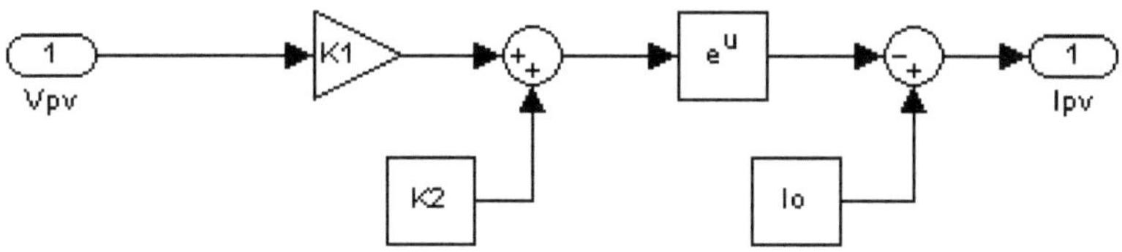

Fig. 5.15 Simulink® implementation of $I = f(V)$ model neglecting R_s

5.4.2 *I = f(V) Static Model Formulation*

In some implementations it is useful to use a $I = f(V)$ formulation. In this case, if the series resistance, R_s, can be neglected, Eq. (3.13) gives:

$$I = I_0 - e^{(VK_1+K_2)} \tag{5.31}$$

This last equation can be simply implemented as in the scheme of Fig. 5.15.

To include the series resistance R_s, a recursive model is utilized. In this case, Eq. (3.13) is used to obtain a first value of I neglecting R_s. Then it is re-introduced in the second member of (3.13) to obtain a new value of I, and so on. This model can be implemented as shown in Fig. 5.16.

Finally, the parallel resistance R_p can be included in the model. In this case, the equation to be considered is:

$$I = I_0 - e^{[(V+IR_s)K_1+K_2]} - \frac{V + IR_s}{R_p} \tag{5.32}$$

The effect of R_p consists on lessening the output current and the corresponding model is shown in Fig. 5.17.

In the following example, the above explained Simulink® models have been used to achieve the PV characteristics for different values of R_s and R_p.

The used remarkable points refer to an actual PV module and they have been measured experimentally at $G = 773\ \text{W/m}^2$. Then, by using the algorithm described in Sect. 4.4.2.3, the four parameters of the model have been determined.

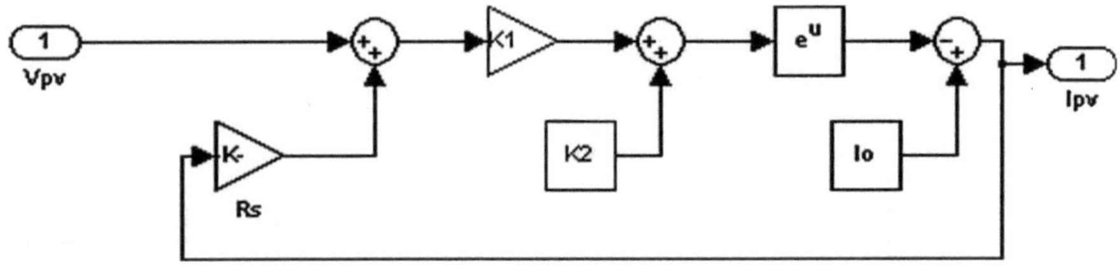

Fig. 5.16 Simulink® implementation of $I = f(V)$ model including R_s

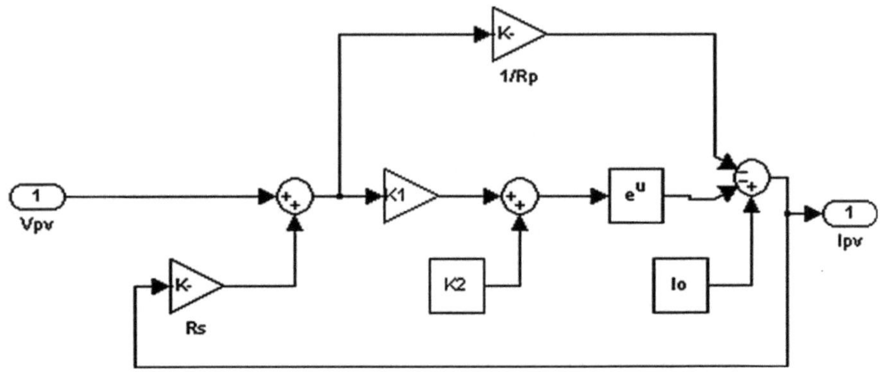

Fig. 5.17 Simulink® implementation of $I = f(V)$ model including R_s and R_p

The parasitic parallel resistance has been calculated on the basis of the slope of the I–V characteristic near the short-circuit current as shown in (4.3).

All parameters can be loaded by the following *m-file*:

```
**************************************************
    % PV module parameters measured at G=773 W/m2

    % Voc=19.8        [V]
    % Isc=1.145       [A]
    % Vmp=14.16       [V]
    % Imp=1.07        [A]

    % PV model calculated characteristic parameters

    Io=1.145
    K1=1.3546
    K2=-26.6852
    Rs=3.390

    % estimated parallel module resistance
    Rp=3000
**************************************************
```

Figure 5.18 shows the photovoltaic source characteristics calculated for four couples of R_s and R_p values. It should be noted that the effect of R_s is to lessen the maximum power points voltage and current values. The value $R_p = 3000\ \Omega$ does

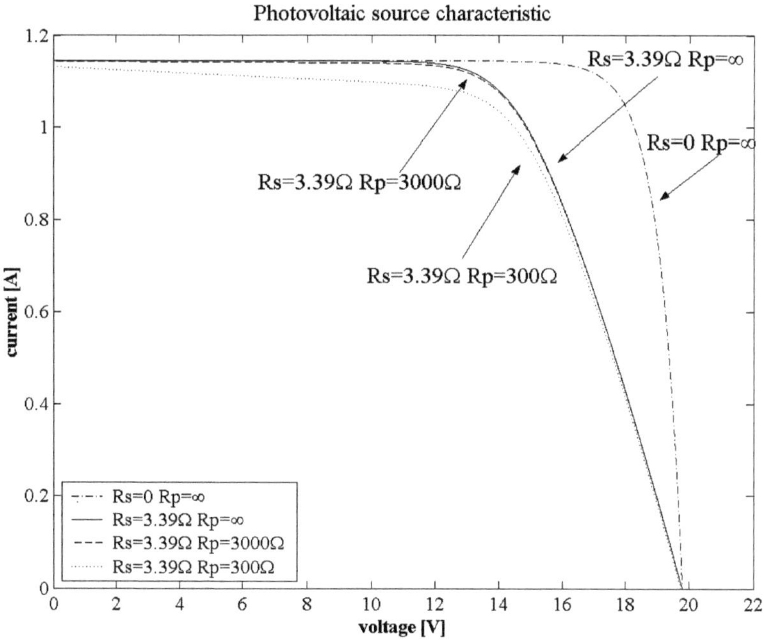

Fig. 5.18 Photovoltaic source characteristic for different values of R_s and R_p

not influence significantly the characteristic with respect to the ideal case of $R_p = \infty$. A further value of $R_p = 300\ \Omega$ has been introduced for comparison in the simulation. In this case the output current drop in the short circuit region is noticeable.

5.4.3 S-Domain Dynamic Model

Figure 5.19 shows the equivalent circuit for s-domain Simulink® implementation of a PV source dynamic model. Three main circuit blocks are present. The single diode PV model, the capacitance with its series parasitic resistance, and the parasitic inductance, series connected to the load resistance. These three blocks are connected in parallel.

The implemented model is a recursive one in which the voltage, applied at the same time to the single diode PV model and to the capacitance–resistance series, origins the current. By subtracting from the PV source current the one that flows in the capacitance–resistance series, the load current is obtained. This is finally imposed to the inductance-load block; the voltage is obtained and it is again applied to the parallel formed by the PV source and capacitance–resistance series.

In Fig. 5.20 a principle block diagram for the s-domain Simulink® implementation of the PV source dynamic model is shown.

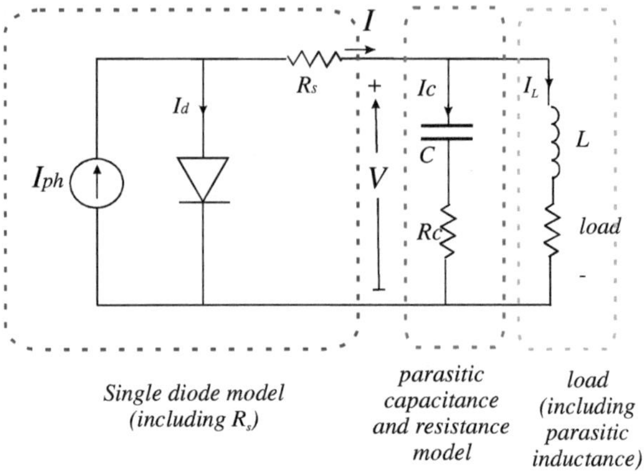

Fig. 5.19 Equivalent circuit for s-domain Simulink® implementation

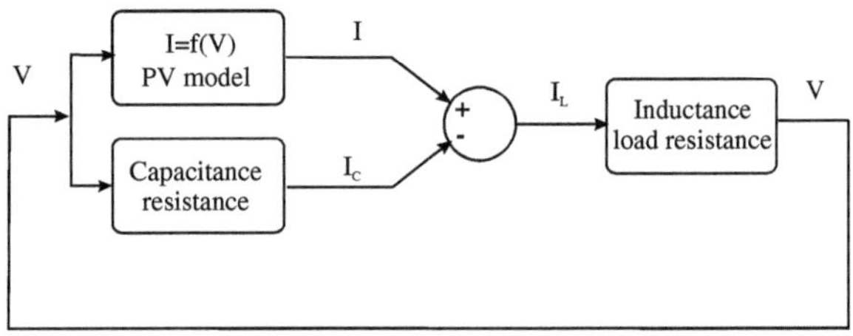

Fig. 5.20 Principle block diagram for s-domain Simulink® implementation

The circuit blocks are implemented as follows. As for the PV source, the model shown in Fig. 5.16 has been used. It represents a $I = f(V)$ formulation neglecting R_p. The block capacitance–parasitic resistance is implemented using the Simulink® transfer function block for the conductance transfer function.

$$Y_C(s) = \frac{I_C(s)}{V(s)} = \frac{sC}{sR_cC + 1} \tag{5.33}$$

Finally, the parasitic inductance and the load resistance have the load current in common; their voltage is calculated separately and then they are summed. The inductance is implemented as its impedance in the Laplace domain, sL, with an additional far pole to obtain a proper system in the transfer function. The load resistance is connected by a multiplier block to simulate a step load variation.

In the feedback loop an "initial condition" block is inserted to avoid algebraic loop errors. The final Simulink® scheme is illustrated in Fig. 5.21.

5.4.4 Nonlinear Junction Capacitance Implementation

In Sect. 2.12 an expression for the nonlinear junction capacitance has been obtained and it is given in (2.98). This expression is valid under the hypothesis $\omega\tau_p \ll 1$ and it can be rewritten as :

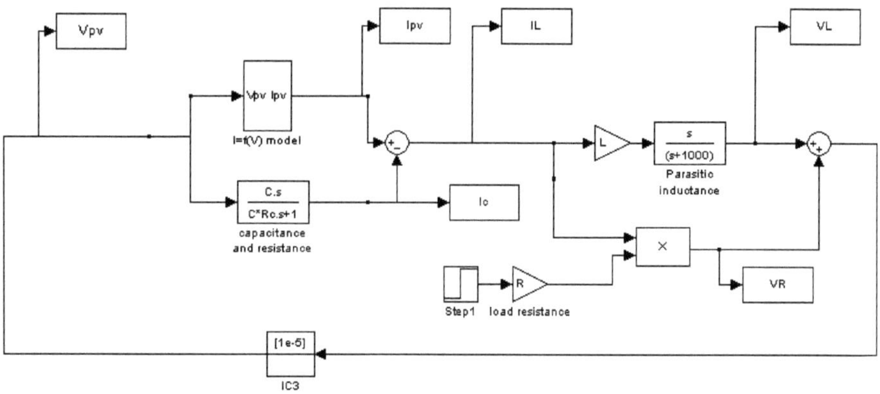

Fig. 5.21 S-domain Simulink® implementation of PV source including parasitic capacitive and inductive effects

$$C_D = ae^{bV} \tag{5.34}$$

where V is the voltage across the capacitance, a and b are two constant values, given, respectively, by:

$$\begin{cases} a = \frac{AqD_p p_{n0} \tau_p}{2L_p V_T} \\ b = {}^1\!/_{V_T} \end{cases} \tag{5.35}$$

In order to obtain the capacitance constitutive equation, it can be observed that the charge stored in the capacitance is given by the product of the same capacitance multiplied by the voltage and that the current is the time derivative of the charge.

$$I = \frac{\mathrm{d}}{\mathrm{d}t} Q = \frac{\mathrm{d}}{\mathrm{d}t} (V \times C_D) = \frac{\mathrm{d}}{\mathrm{d}t} \left(aVe^{bV}\right) = ae^{bV}(1 + bV)\frac{\mathrm{d}V}{\mathrm{d}t} \tag{5.36}$$

The current is obtained as the product of four terms, where ae^{bV}represents the junction capacitance defined by (5.34).

The nonlinear capacitance can be simulated in Matlab-Simulink®. If the junction parameters are known, the capacitance parameters can be calculated analytically. As an example, considering the following values:

$$\begin{cases} A = 10\,\mu\mathrm{m}^2 \\ q = 1.6 \times 10^{-19}\,\mathrm{C} \\ D_p = 1.3 \times 10^{-3}\ \mathrm{m/s^2} \\ V_T = 0.026\ \mathrm{V} \\ p_{n0} = 2.1 \times 10^{11}\ \mathrm{m^{-3}}\ (N_D = 10^{15}\,\mathrm{cm^{-3}}) \\ \tau_p = 10\,\mu\mathrm{s} \end{cases} \tag{5.37}$$

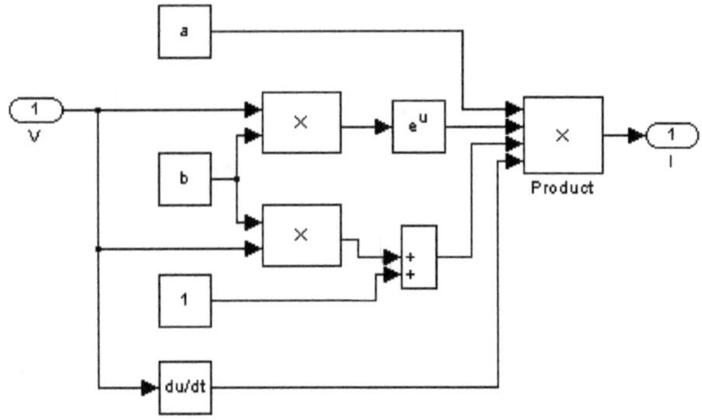

Fig. 5.22 Simulink® scheme for the simulation of a nonlinear junction capacitance

It follows that $L_p = \sqrt{\tau_p D_p} = 1.1 \times 10^{-4}$ m and finally

$$\begin{cases} a = 7.1 \times 10^{-16}\,\mathrm{F} \\ b = 38.5\,\mathrm{V}^{-1} \end{cases} \tag{5.38}$$

On the basis of Eq. (5.36), the Simulink® implementation shown in Fig. 5.22 is obtained. The output current is calculated as the product of four terms as in Eq. (5.36). The condition $\omega\tau_p \ll 1$, corresponding to a highest frequency related to the rise time by the relationship $f_h = 1/\pi\tau_r$ gives: $\tau_r \gg 2 \times 10^{-5}$s.

It should be noted that, differently from the constant capacitance C in Fig. 5.2 which represents an average value that can be determined only by experiment-based which identification procedures, the nonlinear junction capacitance can be analytically calculated by using Eqs. (5.34) and (5.35).

Figure 5.23 shows the simulated output current for the nonlinear capacitance when a sinusoidal voltage with frequency equal to 1 kHz and amplitude 0.6 V is given as input. As a comparison, the output current for a linear capacitance, whose value is equal to the constant a, is plotted as well. It should be noted that the nonlinear capacitance exalts the input voltage variation and the output current shape differs from that of the input voltage, contrarily to what happens for the linear capacitance.

Another interesting property of the nonlinear capacitance is shown in Figs. 5.24 and 5.25. In particular, both the linear and the nonlinear capacitance experience a step voltage transient from 0.1 to 0.2 V and from 0.4 to 0.5 V. The step voltage has a rise time of about 3 ms.

It should be noted that the current given by the linear capacitance remains the same; on the contrary, the current given by the nonlinear capacitance depends on the starting value too.

5.4.5 PV Model Including Nonlinear Junction Capacitance

A new dynamic model of a PV source can be obtained including the effect of the nonlinear junction capacitance. The equivalent circuit is shown in Fig. 5.26.

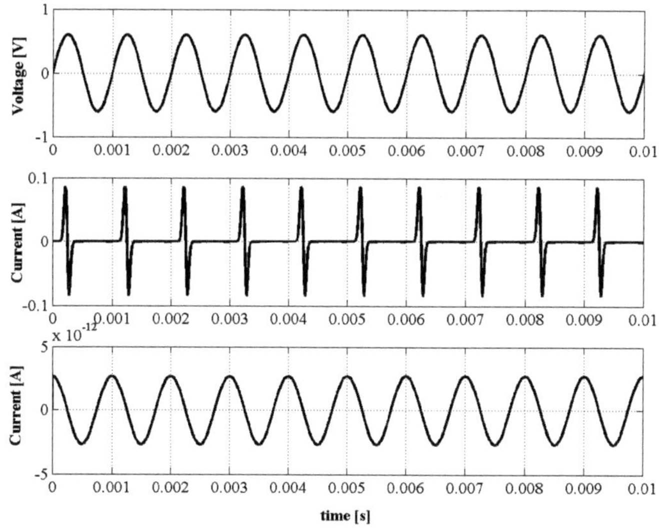

Fig. 5.23 Simulation of nonlinear and linear capacitance behavior: input voltage (*top*), output current at the nonlinear capacitance (*middle*) and output current at the linear capacitance (*bottom*)

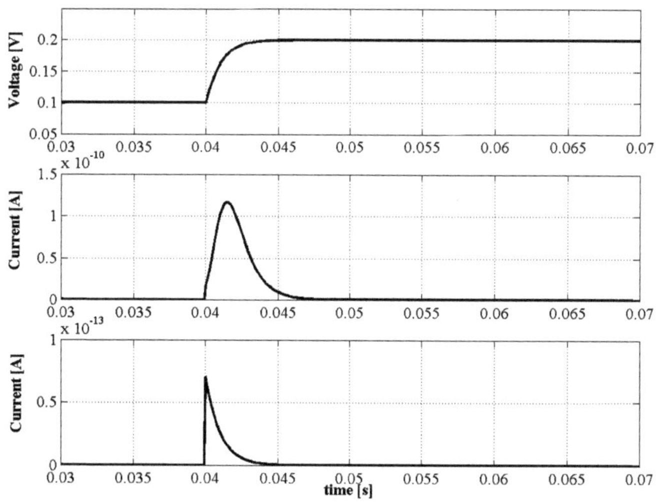

Fig. 5.24 Simulation of nonlinear and linear capacitance behavior for a step voltage from 0.1 V to 0.2 V: input voltage (*top*), output current at the nonlinear capacitance (*middle*) and output current at the linear capacitance (*bottom*)

It differs from the circuit of Fig. 5.19 because the non-linear capacitance is connected in parallel to the diode and the series resistance is included in the load resistance. However, the philosophy for the implementation remains the same. The PV source and the nonlinear capacitance have the same voltage and supply the

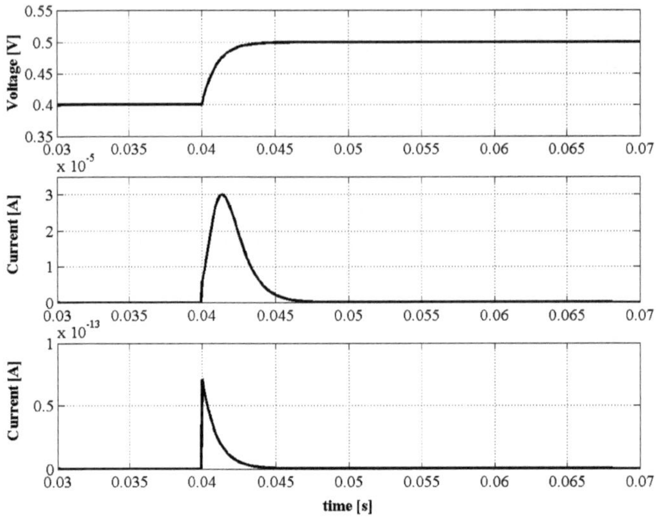

Fig. 5.25 Simulation of nonlinear and linear capacitance behavior for a step voltage from 0.5 V to 0.6 V: input voltage (*top*), output current at the nonlinear capacitance (*middle*) and output current at the linear capacitance (*bottom*)

load current to the series formed by R_s and the load resistance. If necessary, the parasitic inductance can be considered too as shown in Sect. 5.4.3.

The related block diagram is shown in Fig. 5.27.

As for the Simulink® implementation, some further considerations are necessary. The PV source model is the one shown in Fig. 5.15; the nonlinear capacitance model is the one shown in Fig. 5.22. It should be borne in mind that this model refers to a single cell. If one wants to simulate the series connection of more cells, as in a module, the input voltage has to be divided for the number of the cells connected in series. The higher is the number of series connected cells, the lower is the influence of the nonlinear capacitance on the dynamic behavior of the PV source.

The load contains a constant part given by R_s and a variable term to simulate load transitions.

Finally, the presence of a derivative block in the nonlinear capacitance model could lead to errors due to the solver failing to converge. Therefore, as a workaround, the first derivative of the capacitance can be calculated in a different way, as it will be shown hereinafter.

The Simulink® scheme is shown in Fig. 5.28a.

Finally, the parasitic inductance series connected to the load resistance can be considered as well. In this case, the corresponding Simulink® scheme is shown in Fig. 5.28b.

It should be noted that the parasitic inductance becomes relevant when a great number of cells are connected in series.

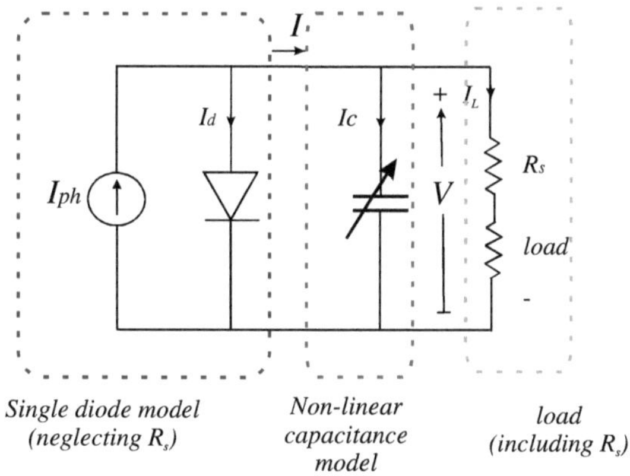

Fig. 5.26 Dynamic model of the PV source including the nonlinear junction capacitance and the load

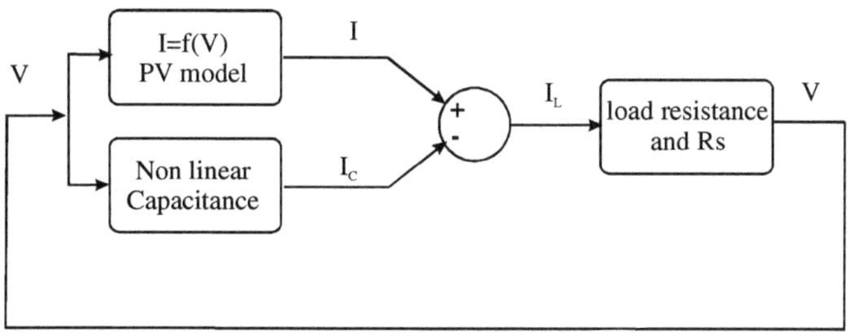

Fig. 5.27 Block diagram for Simulink® s-domain implementation of dynamic PV model

5.4.6 Circuit Implementation Using PLECS®

A complete dynamic model of a PV generator, which takes into account the voltage-dependent junction capacitance, can be implemented in the Matlab/Simulink® environment in several ways. In particular, the time-domain equations of the model can be implemented either with Simulink® basic blocks, similarly to what has been presented in Fig. 5.21 for the S-domain implementation, or with dedicated circuit-oriented toolboxes, such as SimPowerSystems™ Blockset by The Mathworks, Inc., or PLECS® by PLEXIM GmbH.

PLECS® is a very useful software tool for modeling electrical circuits within MATLAB/Simulink® environment. It allows to perform high-speed simulations of power electronic systems and offers a rich component library. The use of this toolbox allows an actual plant to be implemented, for example a power electronic circuit, as a PLECS® subsystem, and to develop the related control using standard Simulink® blocks.

In the following, the electrical circuit which models the PV source has been implemented using PLECS® Toolbox and connected to some external Simulink blocks, as shown in Fig. 5.29.

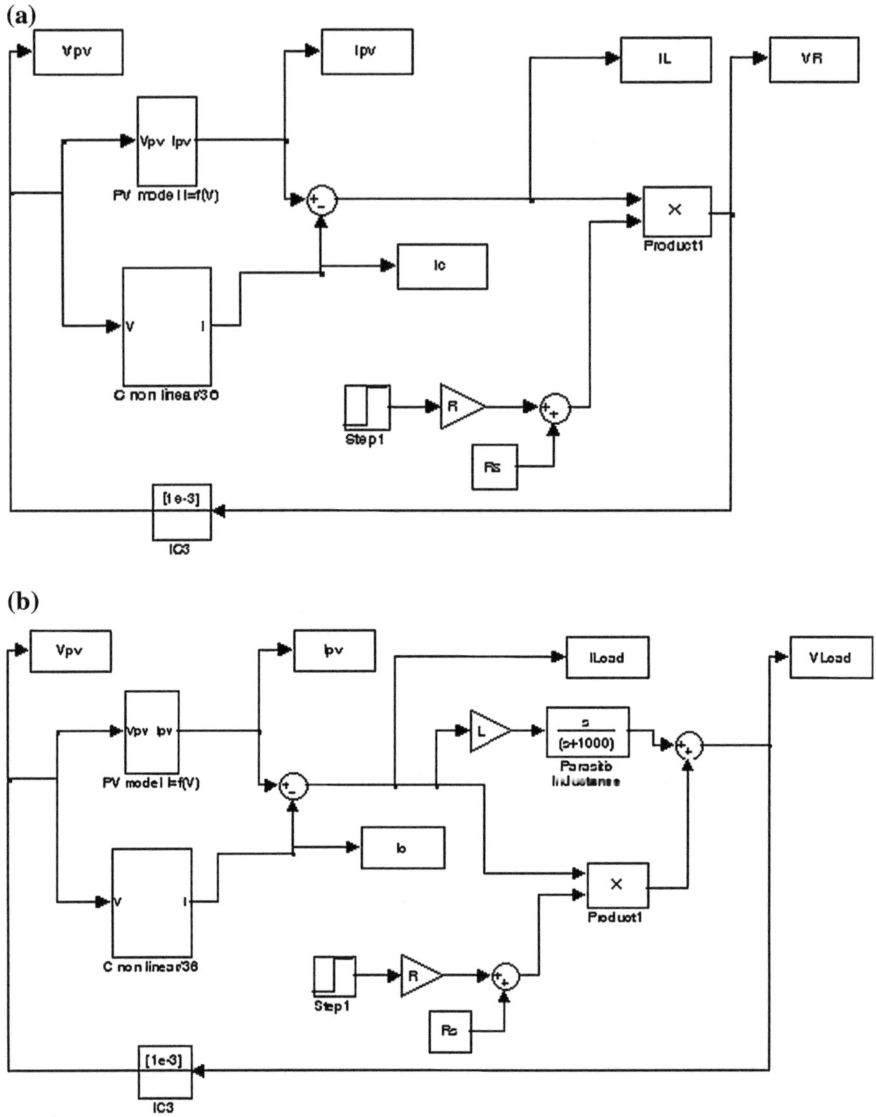

Fig. 5.28 **a** Simulink® scheme for the simulation of the PV source including the effects of the nonlinear junction capacitance. **b** Simulink® scheme for the simulation of the PV source including the effects of the nonlinear junction capacitance and the parasitic inductance

The circuit model, shown in Fig. 5.30, encompasses the following components: a voltage dependent current generator, which is driven by the "$I = f(V)$" block in the first feedback loop, according to the PV source model shown in Fig. 5.15; the shunt resistance; the nonlinear capacitance, driven by the "Calc_C_and_C_dot" block in the second feedback loop; the series resistance and the variable resistive

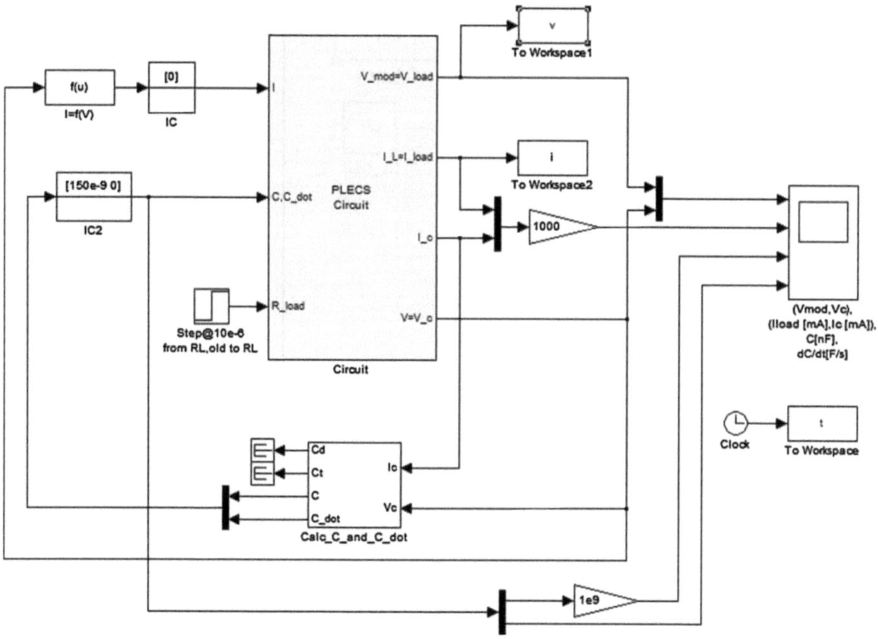

Fig. 5.29 Complete model of a PV array, which takes into account the voltage-dependent junction capacitance

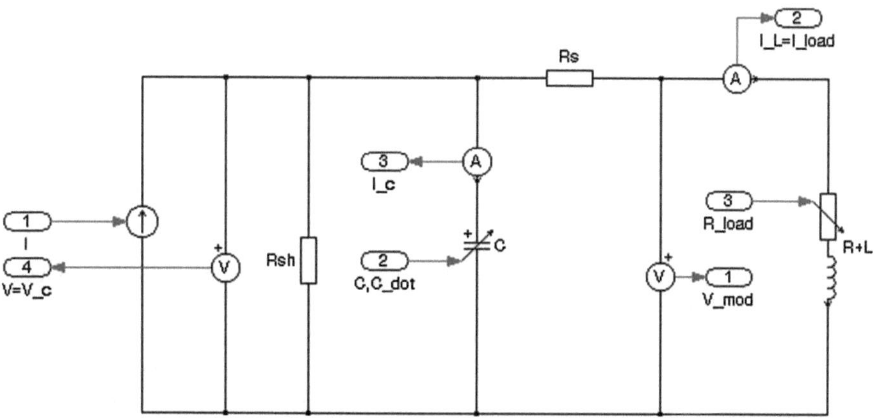

Fig. 5.30 Circuit model implemented in PLECS®

load with the parasitic inductance. Initial condition (IC) blocks have been placed in both loops to avoid algebraic loop errors.

The internal structure of the block which drives the nonlinear capacitance is shown in Fig. 5.31.

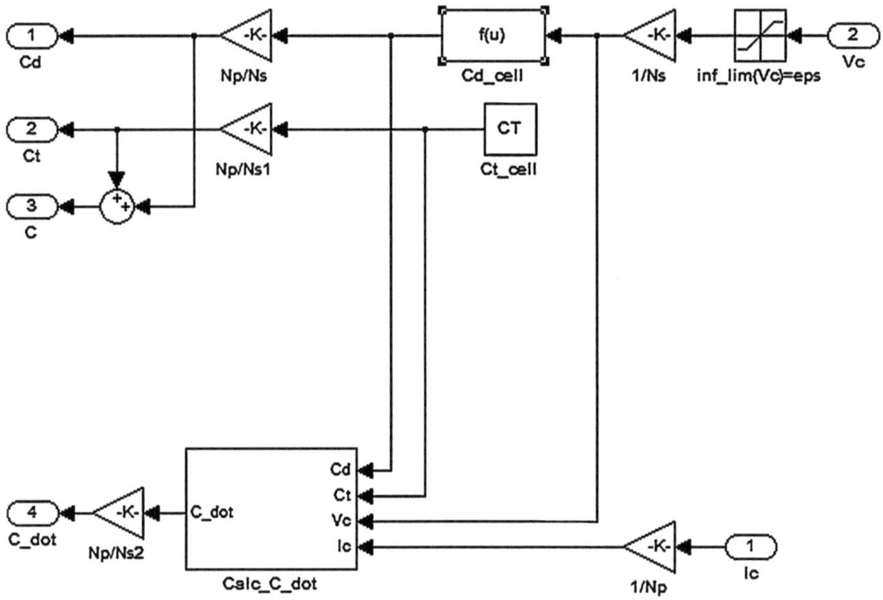

Fig. 5.31 Internal structure of the "Calc_C_and_C_dot" block

The total capacitance is the sum of the transition capacitance C_T, which is almost constant with the junction voltage, and the diffusion capacitance C_D, which is voltage dependent according to Eq. (5.34). More details about nonlinear capacitance are given in Garrigós et al. (2006) and Schwander (2002).

The transition capacitance depends on the cell area. Taking into account the results obtained in Schwander (2002) and Block et al. (2002), typical values are about 40 nF/cm^2 for silicon cells and 50 nF/cm^2 for triple junction GaAs cells.

In general, a PV module can be made up of N_p groups of cells in parallel; each group having N_s cells in series. The simulated PV module is a p-Si module with $N_p = 1$ and $N_s = 36$; the cell area is $S = 5 \times 10 = 50$ cm^2. Hence, the transition capacitance of a single cell is $C_{T,\mathrm{cell}} = 40 \times 50 = 2000$ nF. This value is then multiplied by N_p/N_s to obtain the transition capacitance of the whole PV module: $C_T = 2000 \times 1/36 = 55.5$ nF.

As for the diffusion capacitance, it can be observed that coefficient a in (5.36) has a very small value; on the contrary, the term e^{bV} has a very large value: assuming a cell voltage of 0.9 V (not uncommon during transients) and using for coefficient b the value calculated in Sect. 5.4.4, it is $e^{bV} = 1{,}117{,}644{,}789{,}978{,}933$ so an overflow error in the calculation of Eq. (5.34) can occur. As a workaround, (5.34) can be rewritten as:

$$C_D = a \times e^{bV} = 10^{-9} \times a \times 10^9 \times e^{bV} = 10^{-9} \times e^{bV+d} \qquad (5.39)$$

with $d = \log(a \times 10^9)$. Coefficient d is negative so now the exponential term is raised to a smaller quantity. Equation (5.39) is implemented in block "Cd_cell", which calculates the diffusion capacitance of a single cell, given the cell voltage. The saturation block applies a very small inferior limit to the source voltage so, when the cell voltage is negative, it is $C_D = a \cong 0$ and the total cell capacitance equals C_T.

The chosen limit, indicated by *eps*, is the so-called machine epsilon, i.e., the distance from 1.0 to the next largest floating-point number on the specific machine Matlab® is running on. The diffusion capacitance of a single cell is then multiplied by N_p/N_s to obtain the diffusion capacitance of the whole PV module.

Finally, the first derivative (C_dot) of the capacitance has to be calculated and passed to the variable capacitance component inside the circuit model. However, using Matlab®'s built-in derivative block, errors can occur or the simulation could hang due to the solver failing to converge. Therefore, as a workaround, the first derivative of the capacitance can be calculated in a different way.

Two expressions can be written for the capacitive current:

$$\begin{cases} I = \frac{dQ}{dt} = \frac{d(C \times V)}{dt} = V\frac{dC}{dt} + C\frac{dV}{dt} \\ I = \frac{dQ}{dt} = \frac{d(C \times V)}{dt} = \frac{d}{dt}(C_T \times V) + \frac{d}{dt}(C_D \times V) \end{cases} \tag{5.40a, b}$$

From (5.36) it is:

$$\frac{d}{dt}(C_D \times V) = C_D \times (1 + bV) \times \frac{dV}{dt} \tag{5.41}$$

Substituting Eq. (5.41) in Eq. (5.40b) it is possible to rewrite Eqs. (5.40a, b) as:

$$\begin{cases} \frac{dC}{dt} = \frac{I - C^{dV}/dt}{V} \\ \frac{dV}{dt} = \frac{I}{C_T + C_D \times (1+bV)} \end{cases} \tag{5.42a, b}$$

Finally, substituting Eq. (5.42b) in Eq. (5.42a) it is possible to obtain:

$$\begin{aligned} \frac{dC}{dt} = \frac{I - \frac{I \times C}{C_T + C_D \times (1+bV)}}{V} &= \frac{I}{V}\left[\frac{C_D \times (1+bV) - C_D}{C_T + C_D \times (1+bV)}\right] \\ &= \frac{I \times C_D \times b}{C_T + C_D \times (1+bV)} \end{aligned} \tag{5.43}$$

The "Calc_C_dot" block implements Eq. (5.43); then, the first derivative of the capacitance is multiplied by N_p/N_s to scale the result to the whole PV module.

The model has been simulated with the parameters expressed by Eq. (5.29) which are defined in the following *m-file*, together with other required parameters:

Table 5.1 Outline of the performed tests

Load resistance variation	Test #	$R_{L,\text{old}}$ [Ω]	R_L[Ω]	I_{old}	$V_{C,\text{old}}$
Decreasing	1	1e6	14	0	Voc
Decreasing	2	1e6	9	0	Voc
Decreasing	3	1e6	0	0	Voc
Increasing	4	0	14	Io×Rsh/(Rsh + Rs)	Iold×Rs
Increasing	5	9	30	1.1398	Iold×(Rs + $R_{L,old}$)
Increasing	6	0	1e6	Io×Rsh/(Rsh + Rs)	Iold×Rs
Increasing	7	9	1e6	1.1398	Iold×(Rs + $R_{L,old}$)

```
% ***************************************************
% PV module parameters measured at G=773 W/m2

Voc=19.8;
Isc=1.145;
Vmp=14.16;
Imp=1.07;

% PV model calculated characteristic parameters

Io=1.145;
K1=1.3546;
K2=-26.6852;
Rs=3.390;

% estimated parallel module resistance
Rsh=3000;

% capacitive parameters
Ns=36;
Np=1;
S=5*10; % cm2
CT=40e-9*S; % per cell
a=7.1e-16;
b=38.5;
a_ln=log(a*1e9);

% load parameters
L=5e-6;
RLold=1e6;
RL=0;

% initial conditions for C and L
Iold=0;
Vcold=Voc;
% ***************************************************
```

Several simulations have been performed, which are summarized in Table 5.1.

In general, the behavior of a second-order system is expected for the presence of the parasitic inductance and of the nonlinear capacitance. However, the step response can be very different depending on the capacitance value (as explained higher values of cell voltage imply higher capacitance values) and on the final value of the load resistor (high values of the load resistance imply negative real and noninteracting poles; on the contrary for small values of the load resistance poles become complex conjugates).

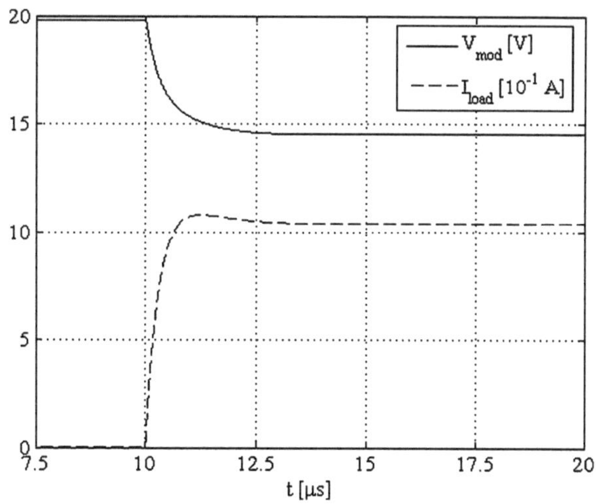

Fig. 5.32 PV output voltage and current (test #1)

Seven tests have been simulated. The starting load resistance value is denoted by $R_{L,old}$. The first three tests start from an open circuit condition (indicated as a high value load resistance). The load is step changed to a finite value (indicated as R_L) that corresponds to a point just under the MPP for test #1, just above the MPP for test #2 and to a short-circuit condition for test #3.

The couples of tests #2–#7 and #3–#6 have the starting and arriving point exchanged to show the different characteristics of the related transients.

Table 5.1 contains the expression of the load current initial value (indicated as I_{old}) and of the initial value of the voltage across the capacitor (indicated as $V_{C,old}$).

5.4.6.1 Test #1

In this test, the load is step changed from a very high value (reproducing the open circuit condition) to a value just under the MPP. Figure 5.32 shows the module voltage V_{mod} and the load current I_{load} multiplied for a scale factor of ten. The current exhibits a slightly underdamped trend in which the two poles are real and negative and where the two related time constants are recognizable. Figure 5.33 shows the capacitance value and its time derivative.

It should be noted that the capacitance value decreases and that its variation occurs during transient as expected. Finally, in Fig. 5.34, the static I–V characteristic is shown with the trajectory locus superimposed. In this case, the locus goes from the starting to the final point it exhibits higher voltage and current compared to the static ones. Although the capacitance reduction is significant, the operating zone is quite linear. Therefore, the distortion in voltage and current waveforms is not so evident.[2]

[2] The term distortion denotes a behavior different from that of linear circuit.

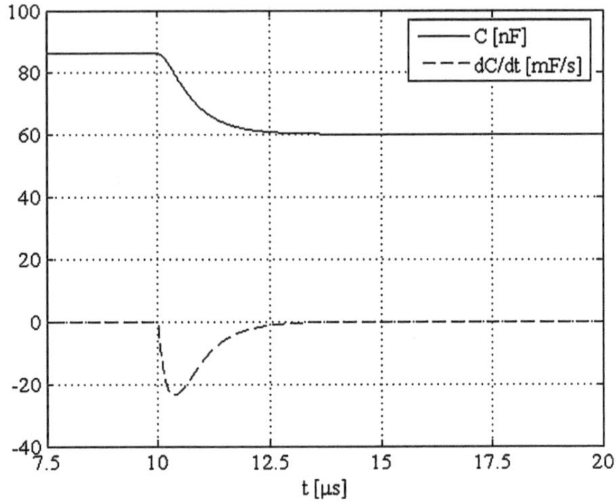

Fig. 5.33 Value of junction capacitance and its derivative (test #1)

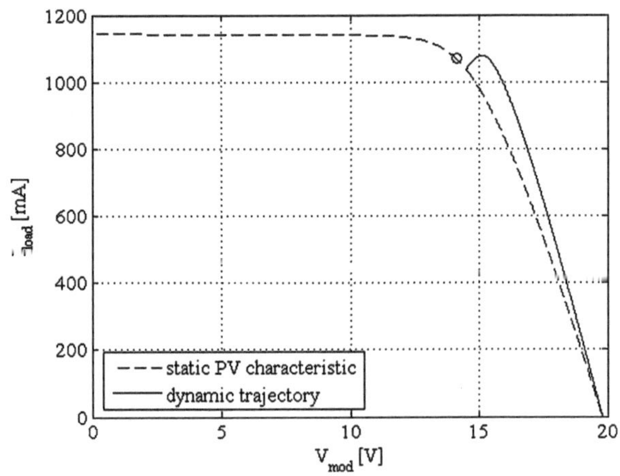

Fig. 5.34 Static PV characteristic and dynamic trajectory (test #1)

5.4.6.2 Test #2

Test #2 is similar to test #1, but the load is step changed from a very high value (reproducing the open circuit condition) to a value which corresponds to a current higher than the MPP current; as a matter of fact, a lower final value of the load resistor is adopted.

Figure 5.35 shows the module voltage V_{mod} and the load current I_{load} multiplied for a scale factor of ten. As in the previous test, the current exhibits an underdamped trend but, in this case, there is a more evident overshoot and the current rises up to a value equal to about 20 % of the final value; the oscillation continues with a barely noticeable negative amplitude; then the waveform increases again, reaching the steady-state value.

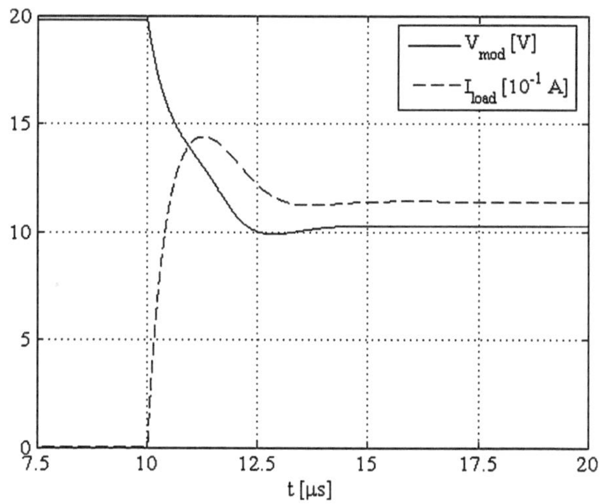

Fig. 5.35 PV output voltage and current (test #2)

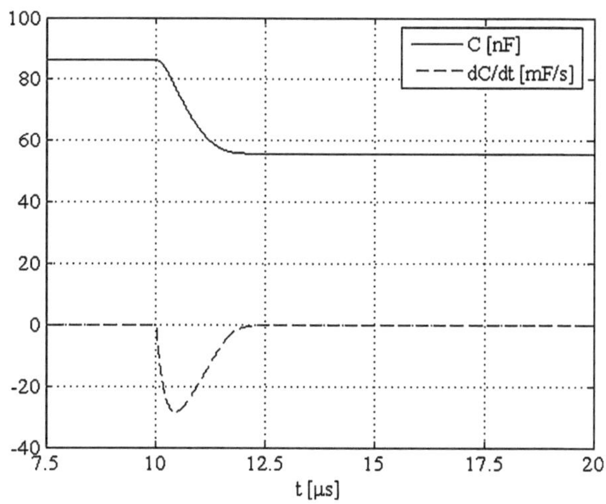

Fig. 5.36 Value of junction capacitance and its derivative (test #2)

The load resistance value, lower than the one adopted in test #1, has lessened the damping ratio. The poles are complex conjugates with a high value of the real part. Figure 5.36 shows the capacitance value and its time derivative. Also in this case the capacitance value decreases corresponding to lower values of the output voltage. Finally, in Fig. 5.37, where the static I–V characteristic is shown with the trajectory locus superimposed, the higher values of load current can be appreciated as well. Now the operating point passes the knee of the characteristic when the capacitance variation is significant. Both effects combine, causing an evident distortion of the voltage waveform; on the contrary, the load current waveform is not distorted, thanks to the series inductor. The current drained by the junction capacitor, not shown, is distorted as well.

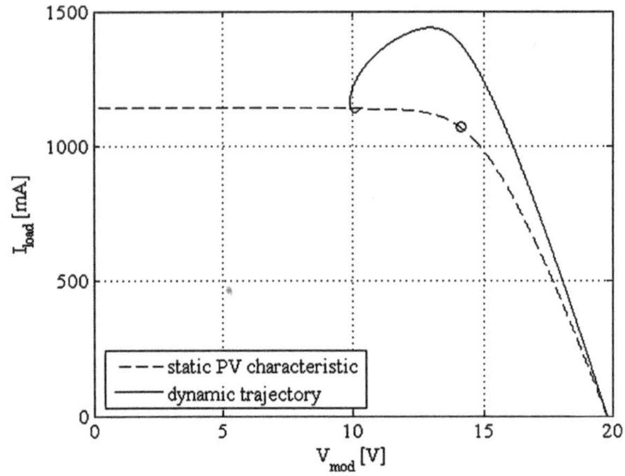

Fig. 5.37 Static PV characteristic and dynamic trajectory (test #2)

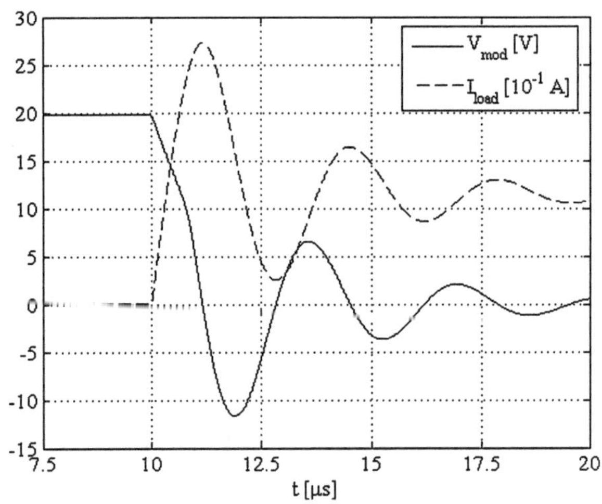

Fig. 5.38 PV output voltage and current (test #3)

5.4.6.3 Test #3

In this test, the load is step changed from a very high value (reproducing the open circuit condition) to a null value corresponding to the short-circuit condition. Figure 5.38 shows the module voltage V_{mod} and the load current I_{load} multiplied for a scale factor of ten. Both current and voltage exhibit a damped oscillatory behavior with high overshoot. In particular, the current peak rises to about 150 % of its final value. The final null value of the load resistance imposes that the poles are complex conjugates with a low value of the real part. Figure 5.39 shows the capacitance value and its time derivative.

It should be noted that the final capacitance value is almost the same as for the previous test, since in both cases the arriving point belongs to the nearly constant current region, where the cell voltage is lower than 0.35 V. Despite that, the capacitance reduction is faster, so the time derivative value during transient is

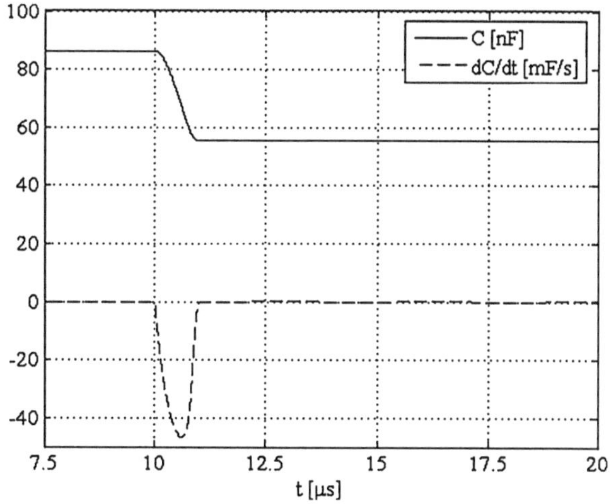

Fig. 5.39 Value of junction capacitance and its derivative (test #3)

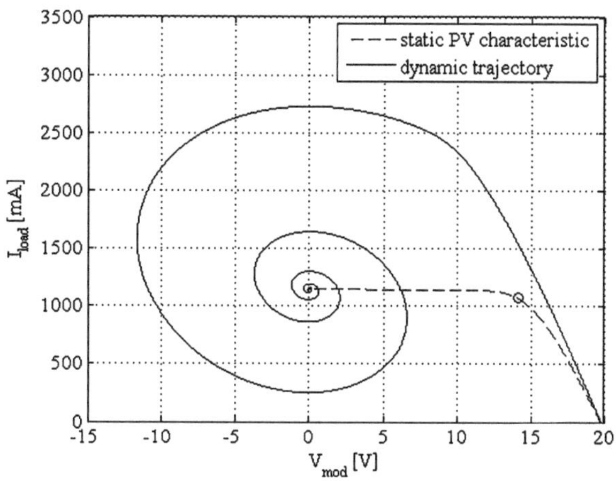

Fig. 5.40 Static PV characteristic and dynamic trajectory (test #3)

much higher than in the other two tests. In Fig. 5.40, where the static I–V characteristic is shown with the trajectory locus superimposed, the locus exhibits a particular spiral shape.

Once again, the nonlinearity of the characteristic combines with the significant capacitance variation, causing an evident distortion of the voltage waveform; on the contrary, the load current waveform is not distorted, thanks to the series inductor.

5.4.6.4 Test #4

Test #4 is performed by raising the load resistance value; in this case the current is lowered and the voltage rises. In particular, in this test the load is step changed from a null value (reproducing the short circuit condition) to a value near the MPP.

Fig. 5.41 PV output voltage and current (test #4)

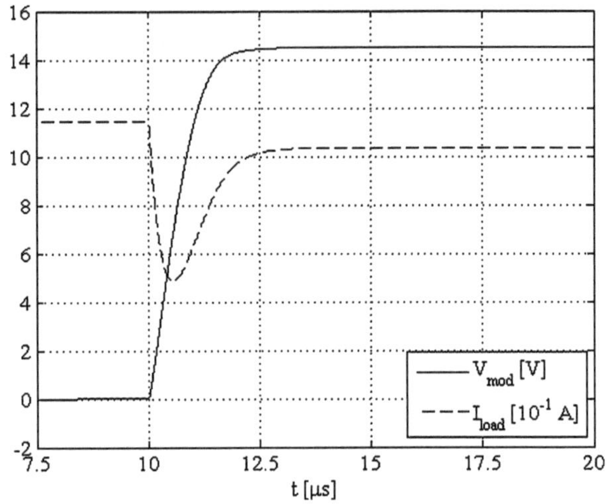

Fig. 5.42 Value of junction capacitance and its derivative (test #4)

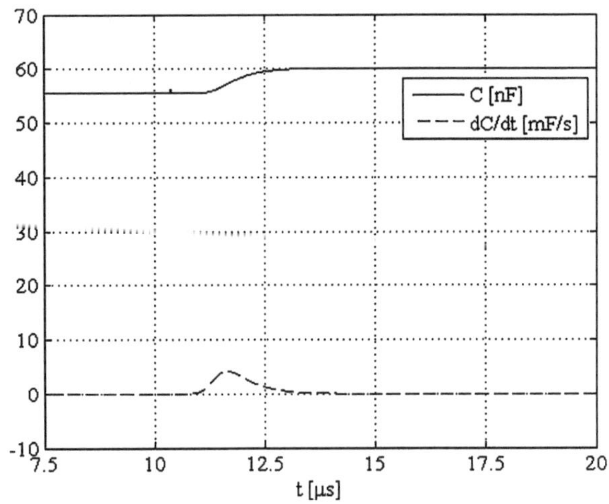

It should be noted that the final value is the same as in test #1, so it implies a similar behavior in terms of step response, as it can be noted from Fig. 5.41, where the module voltage V_{mod} and the load current I_{load} multiplied for a scale factor of ten are shown. As expected, the current exhibits an underdamped curve with two different time constants, due to the contribution of the parasitic inductance and of the capacitance, respectively.

The current has a peak whose value is much lower than the final one. Figure 5.42 shows the capacitance value and its time derivative. It can be noted that, being the voltage smaller compared with the previous test, a smaller variation of the capacitance value is to be expected. Finally, the dynamic trajectory shown in Fig. 5.43 shows that this locus is quite far from the static I–V curve. Since the

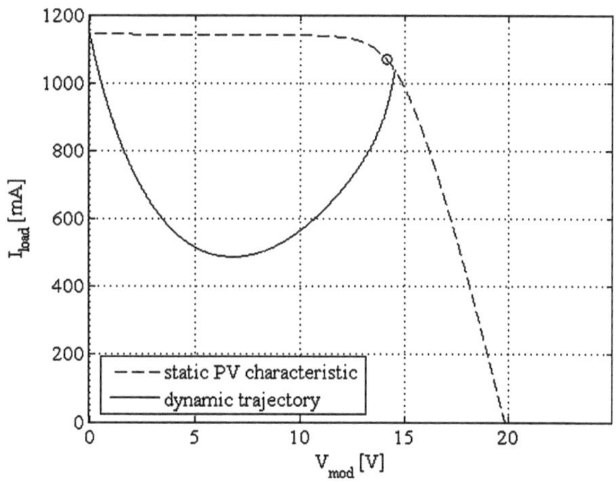

Fig. 5.43 Static PV characteristic and dynamic trajectory (simulation #4)

operating point passes the knee of the characteristic when the capacitance variation is not significant, the voltage distortion is not evident.

5.4.6.5 Test #5

Test #5 is performed by step changing the load between two finite values of the resistance load i.e., $R_{L,\mathrm{old}} = 9\ \Omega$ and $R_L = 30\ \Omega$. The corresponding points are at the left and at the right of the MPP, respectively. It should be noted that the final resistance value is higher than the one corresponding to the MPP; it implies that the poles are real and negative; moreover they are more distant compared to the previous test.

Figure 5.44 shows the module voltage V_{mod} and the load current I_{load} multiplied for a scale factor of ten. The current peak is reduced and a higher variation of the nonlinear capacitance can be appreciated in Fig. 5.45.

As in test #4, the dynamic trajectory is quite far from the static I–V curve as shown in Fig. 5.46. Since the operating point passes the knee of the characteristic when the capacitance variation is not significant, the voltage distortion is not evident.

5.4.6.6 Test #6

This test has the starting and final point exchanged compared to test #3. The presence of a high value of the final load resistance value (corresponding to an open circuit) makes the poles real, negative, and very distant. In particular, the contribution of the parasitic inductance is hardly recognizable; to this aim, the output current and voltage profiles are plotted with two different time-scale in

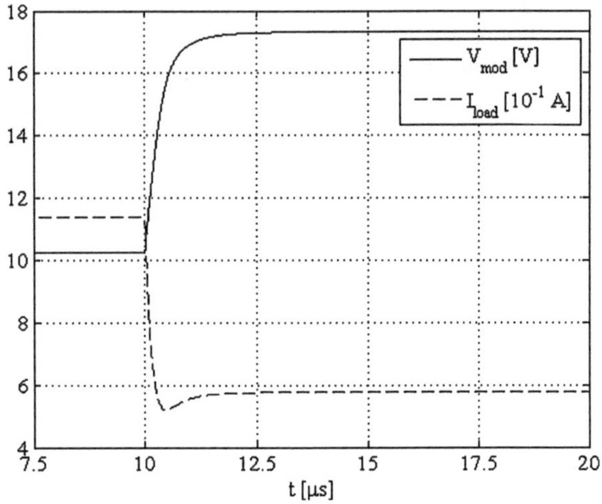

Fig. 5.44 PV output voltage and current (test #5)

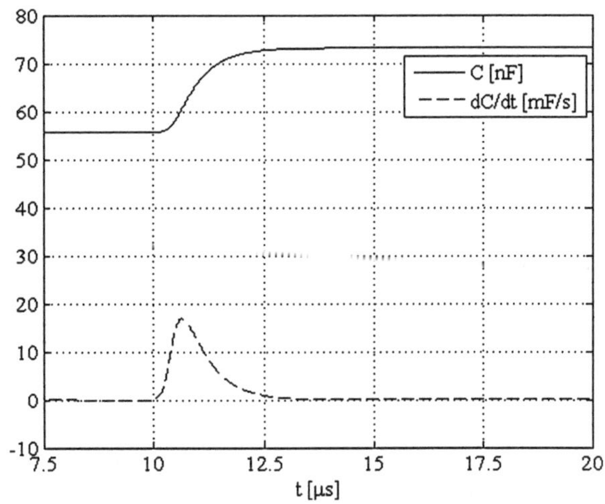

Fig. 5.45 Value of junction capacitance and its derivative (test #5)

Figs. 5.47a and b. Figure 5.47a shows the contribution of the capacitance, being it a microseconds time scale. On the contrary, the contribution of the inductance requires a time scale expressed in tens of nanoseconds to be appreciated as shown in Fig. 5.47b. When the circuit is opened, firstly the current drops to zero with the very fast inductive time constant and the module voltage rises to equal the capacitor voltage, which stays constant; then the load resistor and the inductor are definitely disconnected, so the module voltage is coincident with the capacitor voltage and they start increasing with the very slow capacitive time constant.

As for the capacitance, the final voltage near V_{oc} implies higher values compared to all the previous tests, as shown in Fig. 5.48.

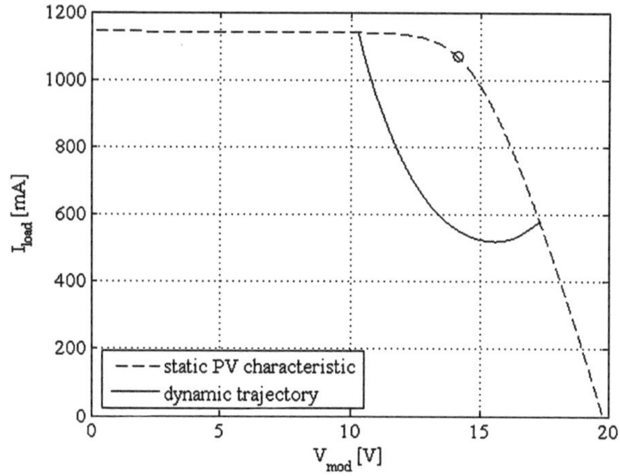

Fig. 5.46 Static PV characteristic and dynamic trajectory (test #5)

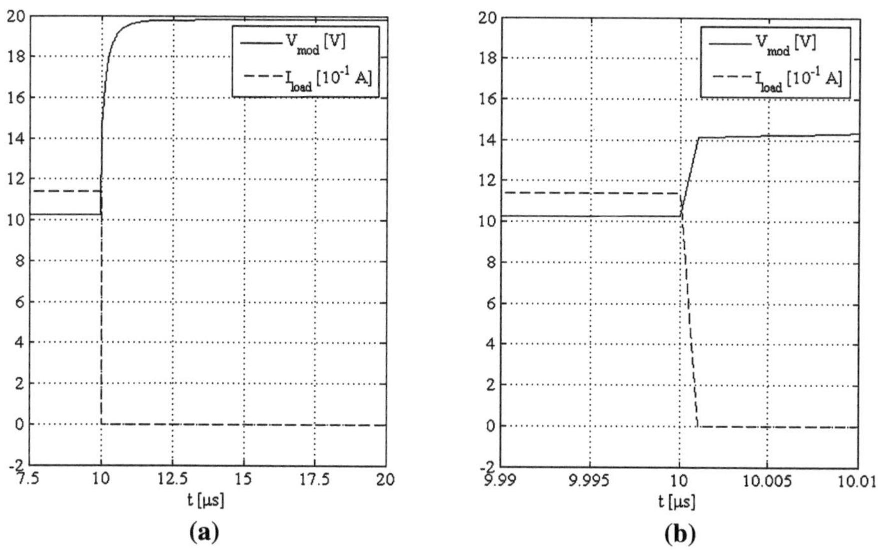

Fig. 5.47 PV output voltage and current (test #6) **a** (*left*) microseconds time scale, **b** (*right*) nanoseconds time scale

Finally, the dynamic trajectory, sketched in Fig. 5.49 exhibits two straight lines: firstly, the locus goes from the starting point to a point with null current and a low value of the voltage; then the voltage rises from this point to V_{oc}, remaining the current null. No distortion appears since the knee is passed when the current is null, i.e., the module voltage is coincident with the capacitor voltage.

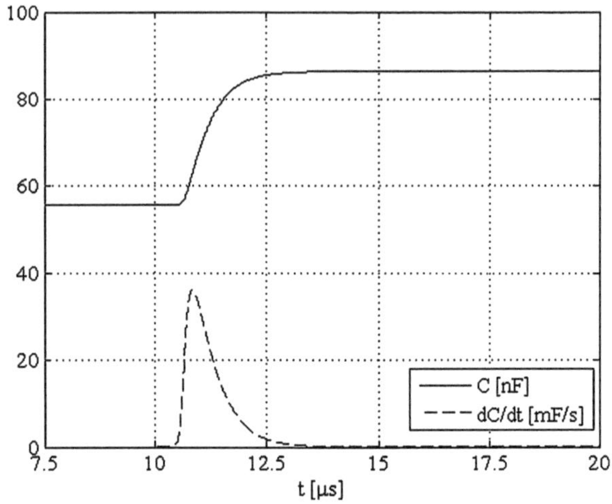

Fig. 5.48 Value of junction capacitance and its derivative (test #6)

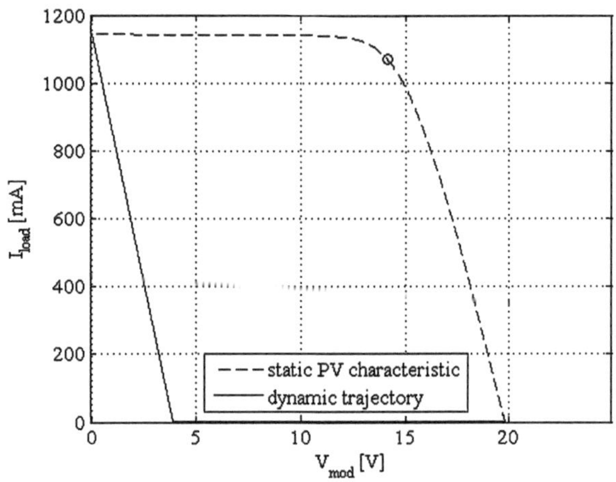

Fig. 5.49 Static PV characteristic and dynamic trajectory (test #6)

5.4.6.7 Test #7

This test has the starting and final point exchanged compared to test #2. Also, in this case, the presence of a high value of the final load resistance value (corresponding to an open circuit) makes the poles real, negative and very distant, and the contribution of the parasitic inductance is hardly recognizable. The output current and voltage profiles are plotted with two different time-scale in Figs. 5.50a and b.

As for the capacitance value, the obtained curves are almost the same of test #6 and they are shown in Fig. 5.51.

Finally, the dynamic trajectory, sketched in Fig. 5.52 exhibits two straight lines: firstly, the locus goes from the starting point to a point with null current and a low value of the voltage; then the voltage goes from this point to V_{oc} remaining

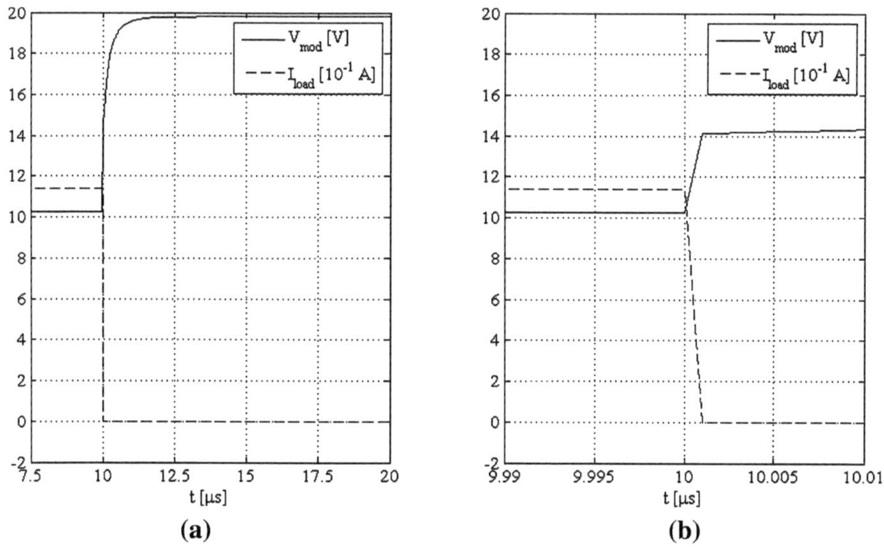

Fig. 5.50 PV output voltage and current (test #7) **a** (*left*) microseconds time scale, **b** (*right*) nanoseconds time scale

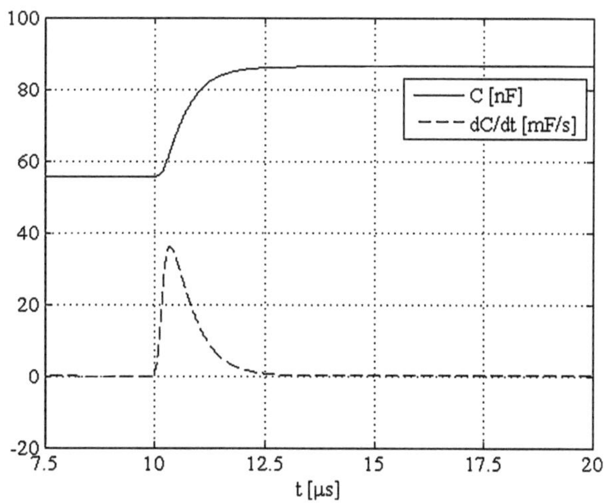

Fig. 5.51 Value of junction capacitance and its derivative (test #7)

the current null. The slope of the first line of the locus equals that of the corresponding line in Fig. 5.49. This is correct, because the inductive time constant does not change and the starting point has almost the same current value, so the two inductive transients are almost identical.

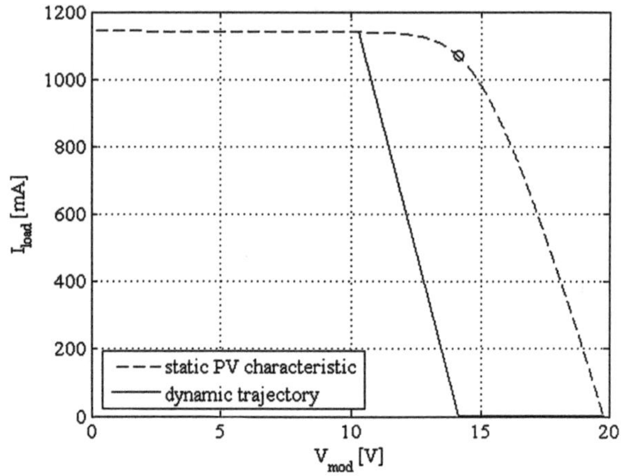

Fig. 5.52 Static PV characteristic and dynamic trajectory (test #7)

5.5 Conclusions

A model reproducing the electric dynamics of a PV module as seen at the load terminals is set up. The model considers the junction capacitive effect and the inductive contribution from cells or connecting cables. For this reason a second-order response of the circuit to step load variations is obtained.

A parameter identification method based on experimental measurements, which considers an equivalent lumped parameter model, is proposed. The related tests have to be performed with step variations of a purely resistive load to put in evidence the dynamics of the model. The nonlinear junction capacitance effects are then accounted for with a dedicated model implemented in PLECS® within MATLAB/Simulink® environment and several tests are carried out to highlight how the step response varies and how the two time constants are influenced by the inductance, by the capacitance, and by the final value of the load resistance.

Bibliography

Block R, Van den Berg E, Slootweg D (2002) Solar cell capacitance measurement. In: Proceedings 6th European space power conference, ESA SP-502

Campbell Ryan C (2007) A circuit-based photovoltaic array model for power system studies. Power Symposium. NAPS '07, 39th North American, pp 97–101

Chayavanich T, Limsakul C, Chayavanich N, Chenvidhya D, Jivacate C, Kirtikara K (2007) Voltage and frequency dependent model for PV module dynamic impedance. In: Proceedings 17th International photovoltaic science and engineering conference

Di Piazza MC, Ragusa A, Vitale G (2009) Identification of photovoltaic array model parameters by robust linear regression methods. Renew Energy Power Qual J. ISSN: 2172-038X, n. 7, http://www.icrepq.com/RE&PQJ-7.htm

Di Piazza MC, Luna M., Ragusa A, Vitale G (2010) A dynamic model of a photovoltaic generator based on experimental data. Renew Energy Power Qual J. ISSN: 2172-038X, n. 8, http://www.icrepq.com/RE&PQJ-8.htm

Garrigós A, Blanes JM, Carrasco JA, Ejea JB (2006) Influence of the parasitic solar array capacitance in the sequential switching shunt series regulator. In: Proceedings IEEE Mediterranean Electrotechnical Conference, pp 1198–1201

King DL, Boyson WE, Kratochvil JA (2004) Photovoltaic array performance model. SAND2004-3535, Unlimited Release, http://photovoltaics.sandia.gov/docs/PDF/King%20SAND.pdf

Merhej P, Dallago E, Finarelli D (2010) Effect of capacitance on the output characteristics of solar cells. In Proceedings Conference on Ph.D. Research in microelectronics and electronics (PRIME), pp 1–4

Schwander D (2002) Dynamic solar cell measurement techniques: new small signal measurement techniques. In: Proceedings 6th European space power conference, ESA SP-502

Xiao W, Dunford WG, Capel A (2004) A novel modeling method for photovoltaic cells. In: Proceedings IEEE 35th Annual power Electronics Specialists Conference, Aachen, Germany, pp 1950–1956

Xiao W, Dunford WG, Palmer PR, Capel A (2007) Regulation of photovoltaic voltage. IEEE Trans Ind Electron 54(3):1365–1374

Part II

Chapter 6
Photovoltaic Source Emulation

6.1 Introduction

Up to now, it has been amply demonstrated the possibility for solar cells, modules, or fields to be modeled theoretically. However, the obtained results require a validation through suitable experiments involving the PV source as well as various kinds of loads, such as resistive loads, DC motors, storage batteries, and inverter-connected loads with their maximum power point trackers (MPPT).

The experimental analysis of PV systems, especially when power electronic conversion circuits are present, is a very difficult challenge. Indeed, a real plant needs a wide outer surface and high costs; moreover, its produced energy is strongly dependent on uncontrollable weather conditions. Finally, the necessary investment to set up an experimental plant can be made more severe by the continuous evolution of the involved technologies. Therefore, alternative solutions to the use of an actual outdoor plant for managing problems like tests on MPPT, appropriate control for interfacing the PV generator to the grid or load, and so on, have to be taken into consideration.

For PV sources with power up to hundreds of watts, this difficulty can be overcome creating artificial environmental conditions inside the laboratory. This solution may imply the use of the so-called *solar simulators*, i.e., suitable lamps that have a spectrum reproducing the natural sunlight and can replace the Sun for repeatable and accurate indoor testing of the I–V characteristics of PV generators. The characteristics of the spectrum to be reproduced are described in Sect. 1.2. The problem related to the use of such systems is tied to the relevant power requirement and heat generation. As an example, up to about a decade ago, to get an irradiance of 1000 W/m^2 on a 50 W PV panel, an electric power of about 10 kW is needed; while for a 2 kW PV array, the required electric power is even about 400 kW. The use of novel discharge lamps allows to obtain a better efficiency, but the produced spectrum differs from that produced by the sun light. Recently the use of light emitting diode (LED) solar simulators has been proposed to mimic the spectrum and intensity of natural sunlight; this is a quite cheap and efficient

M. C. Di Piazza and G. Vitale, *Photovoltaic Sources*, Green Energy and Technology,
DOI: 10.1007/978-1-4471-4378-9_6, © Springer-Verlag London 2013

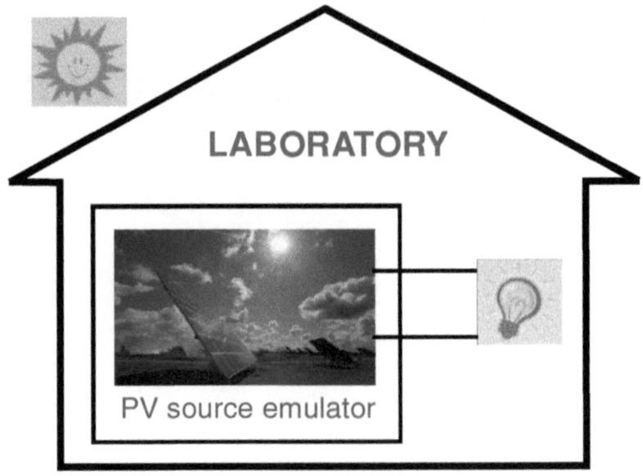

Fig. 6.1 Conceptual scheme illustrating the scope of a PV source emulator

solution in a market dominated by very expensive and high power devices. Generally, the *solar simulators* present some problems with achieving a uniform distribution of the irradiance.

For all these reasons, when higher powers are involved, the most appropriate solution is the use of other kind of devices, that will be referred, hereafter, as PV source emulators, or simply PV emulators.

An emulator is, in general, intended as a system that duplicates the functions of an equipment using a different hardware. Such a system differs form computer simulation tools that concern abstract models of the equipment to be simulated.

The possibility to setup an emulator of a PV generator, in which voltage and current behave as in the real source, is very advantageous.

The PV source emulator should show the same electrical characteristics (power, voltage, current) of a photovoltaic source, including their dependence on weather conditions, partial shadow, and dynamics, allowing the user to manage a virtual plant. Such an experimental facility would allow measurements and tests to be carried out, without the constraints of the real environmental conditions and, above all, more cheaply, since the use of an actual PV array is avoided. So, for example, the optimal choice and design of the power converter interfacing the PV generator to the utility or load and the study of all the problems related to the power electronic control would be performed more rapidly and effectively. This is a very important issue, especially in consideration of the more and more growing interest in the application of renewable energy sources in distributed generation.

Figure 6.1 synthesizes, in a perceptual way, the basic idea and scope of a PV source emulator.

Summarizing the contents of this chapter, the basic concepts of the PV source emulators are given with specific details of the possible hardware topologies and control issues. Aspects related to rated power, modularity, and dynamics are discussed, and finally some examples of commercial solutions are described as well.

6.2 PV Emulators: Concepts and Realization

The emulation of a PV generator is tied to two main tasks: the first one is the knowledge of the I–V characteristics of the generator, the latter consists in their reproduction by suitable power amplifiers, which defines the maximum power that can be generated.

In general, the I–V characteristics of a PV array can be deduced by a PV cell model, otherwise a data base can be set up by measuring I–V curves in different environmental conditions. With regard to this first task, an extensive discussion on PV source modeling and parameters identification has been given in the previous chapters.

As for the second task, a broad variety of realizations for PV emulators have been proposed in the technical literature, already as from the beginning of the 1980s.

Among the basic ideas suggested in the past for the setup a PV source emulator there are:

1. the modification of a voltage source so that its internal resistance is variable according to an exponential law with current;
2. the amplification of the current and voltage of a single cell;
3. the definition of a PV generator equivalent circuit formed of an equivalent constant current source and a diode resistor network.

All these methods have shown a limited flexibility in readily follow the influence of solar irradiance, temperature, and other parameters of the PV generator. Moreover, some of these methods are unpractical for the emulation of PV sources in the range of kW. For example, the method at point three would require the use of unacceptable number of diodes and high power current sources.

For all these reasons, the research in the field of PV emulators has been onwards oriented to the use of active power sources suitably driven to give characteristics close to reality.

Conceptually, modern PV source emulators are power electronic converters whose output voltage and current are controlled so as to reproduce the electrical behavior of PV generators.

In the following sub-sections, the key issues of PV emulators proposed in the technical literature are described both considering the chosen hardware solutions and the control issues.

6.2.1 Power Stage: A Survey of Proposed Solutions

A first rough classification of PV emulators, on the basis of their hardware structure, can be made considering the power amplifier used to realize its operation.

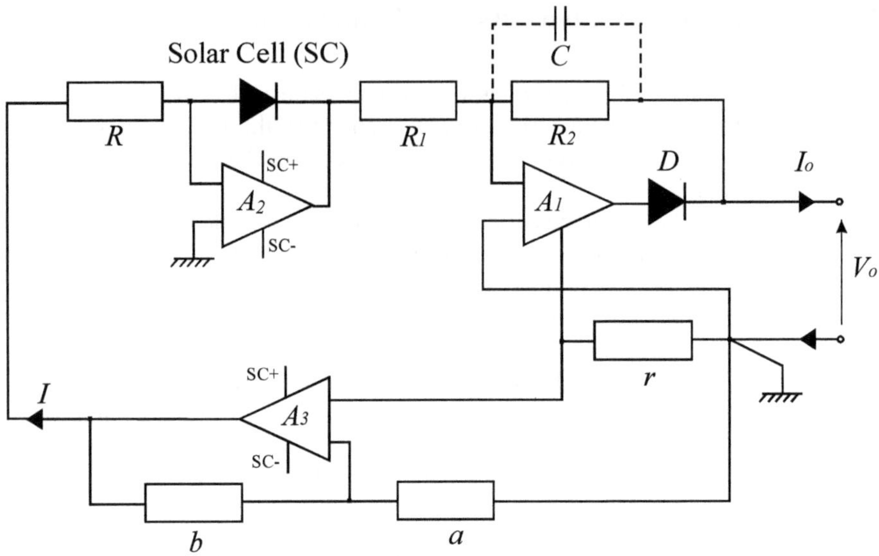

Fig. 6.2 Scheme of the PV source emulator proposed by Baert

In the technical literature realizations with operational amplifiers, series regulators, and switching converters have been proposed. Some typical examples are discussed in Sects. 6.2.1.1, 6.2.1.2, and 6.2.1.3, respectively.

6.2.1.1 Circuits Based on Operational Amplifiers

One of the earliest contributions on PV source emulators is due to Baert (1979) that proposed a circuit having the effect of virtually increasing the area of a small reference solar cell. This circuit has been intended for studies on both inverters and battery storage systems.

The circuit is based on power operational amplifiers, as shown in Fig. 6.2. In particular, the amplifier A_1 allows the output solar cell voltage to be multiplied by a factor $-R_1/R_2$ that is equivalent to consider a string of R_1/R_2 cells in series.

The output current is sensed by a low value resistance r by which it is transformed into a voltage, then it is fed back into the cell by the amplifier A_3; therefore, the new output current is the reference cell current multiplied by a factor aR/br. This situation is equivalent to consider aR/br strings in parallel. With a suitable choice of the components' values an area multiplication factor up to 10^4 can be obtained.

The capacitor C is used as a bandwidth limiter to suppress high frequency oscillations, while the diode D is used to protect the output from undesired reverse current or voltage.

A similar approach is followed by Nagayoshi (2004) that proposes a multi I–V magnifier circuit whose elementary unit is based on a DC linear amplifier.

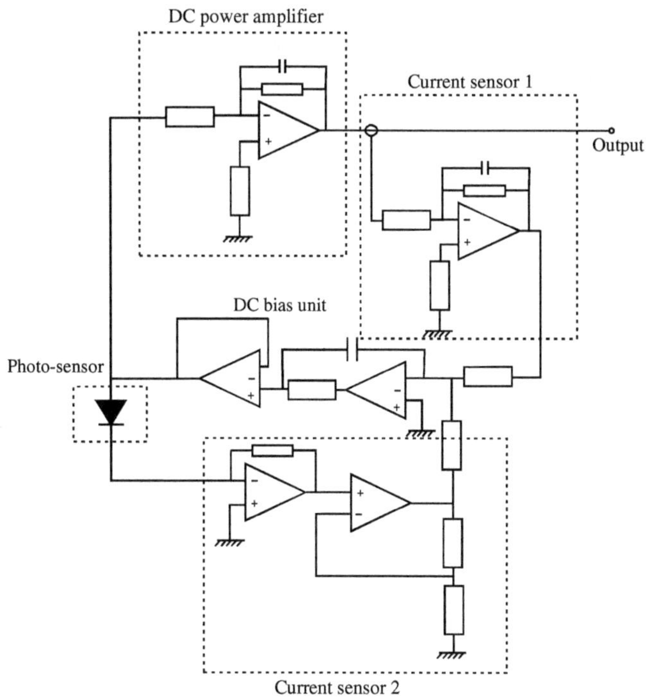

Fig. 6.3 Scheme of the PV source emulator proposed by Nagayoshi

The circuit uses a small PV cell as an I–V generator and provides a separate amplification of voltage and current to realize the operation as a PV emulator. The scheme of the I–V magnifier is illustrated in Fig. 6.3.

It is possible to observe that a DC power amplifier increases the output voltage of the small PV cell (obtained by a pn photo-sensor illuminated by a LED light), while the operational point is controlled by the feedback signal coming from a current sensor placed at the simulator output. In detail, the output current of the PV emulator I_{out} and the pn photo-sensor current I_{ph} is monitored by the current sensors 1 and 2, respectively. The output voltage of the current sensor 2 is kept equal to the voltage output of the current sensor 1 at any time by the feedback control. Therefore the current gain is given by:

$$A_I = I_{out}/I_{ph} = G_2/G_1 \tag{6.1}$$

where

$$G_1 = V_{s1}/I_{out} \tag{6.2}$$

$$G_2 = V_{s2}/I_{ph} \tag{6.3}$$

Being G_1 and G_2 the gains of the current sensors 1 and 2, respectively and V_{s1} and V_{s2} the output voltages from the current sensors 1 and 2, respectively. According to these equations, it is possible to observe that the current gain is independent of the voltage gain, defined by the power amplifier.

In this system, the temperature dependence of solar source is implemented by a temperature control of the photo sensor.

Several I–V curves were obtained by the described PV emulator using a power variable resistor as a load. These curves are proportionally magnified to x-axis by changing the voltage gain of the DC power amplifier and to y-axis by changing the current gain.

The output on series connection of emulation basic units has suggested the possibility for the system to emulate shading effects on a PV source.

The two above described realizations are both based on a physical sample of the source to be emulated. The use of the operational amplifier allows voltage and current delivered by the source to be amplified, their maximum values are limited only by the amplifier performance. Moreover, the operational amplifier has a good bandwidth to reproduce correctly transients. On the other hand, parasitic phenomena, tied to junction capacitance or wires inductance, can be reproduced correctly only with reference to the PV cell used as sample. As a matter of fact, the series (parallel) connection of PV cells lessens (increases) the junction capacitance value and increases (lessens) the parasitic inductance; hence, a mere multiplication of the voltage or of the current cannot take into account these effects. Finally, partial shading phenomena can be emulated only by using more emulators.

In order to overcome the limited flexibility of common PV emulation methods in readily simulating the influence of weather and PV source parameters, Easwarakhanthan et al. (1986) proposed a microcomputer controller emulator based on a programmable DC voltage source, a programmable multimeter and a power amplifier. In this system the operating points are approached for changes in load, solar irradiance, temperature, and in various PV module/array parameters according to a linear interpolation iterative technique.

In detail, a computer calculates the output current and voltage at the intersection between the I–V curve of the PV source and a straight line between two points subsequently processed on the load operating curve. Then it programs the DC voltage generator to apply the calculated voltage to the load, acquires the measured current by the multimeter, and compares current with that calculated.

A waiting loop is enabled when the relative difference between measured and computed currents reaches a desired limit.

A principle scheme of this emulator is shown in Fig. 6.4.

It should be observed that, the operation of the described system requires an a priori knowledge of the connected load operating curve.

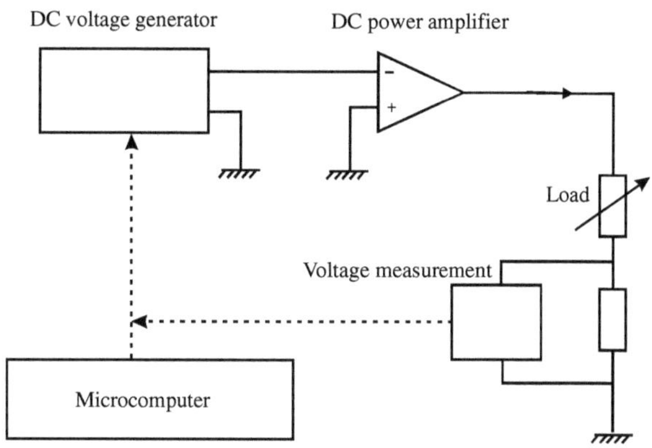

Fig. 6.4 Principle scheme of the PV source emulator proposed by Easwarakhanthan et al

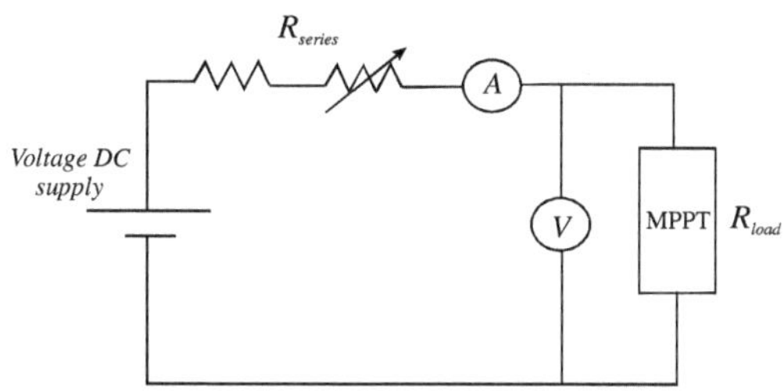

Fig. 6.5 Principle scheme of the PV source emulator proposed by Mukerjee and Dasgupta

6.2.1.2 Circuits Based on Linear Regulators

In this kind of circuits, the output voltage is obtained by a DC supply and a series component that causes a voltage drop. As a consequence, the output voltage is always lower than the voltage of the DC supply. This effect can be obtained by a power BJT that is polarized in active region and works as a class A amplifier. However, a similar effect, although quite rough, can be also obtained with a series resistor.

A low cost circuit used as a PV emulator for testing MPPT has been proposed by Mukerjee and Dasgupta (2007). It is based on a variable voltage DC power supply with a variable resistor at its output. This circuit gives a behavior with a peak on the power curve, similar to that of a PV generator, allowing the tracker to lock on to.

A principle scheme of the system is given in Fig. 6.5.

With reference to Fig. 6.5, if the resistance indicated as R_{series} is fixed, when R_{load} varies from its minimum to its maximum value, the voltage across it plotted against current gives a linear trend with a negative slope.

The power dissipated on R_{load} reaches its maximum when the voltage drop on such a resistance is about half the applied DC voltage, corresponding to the V_{oc} of the emulated PV generator. Therefore, the P–V curve gets a maximum similar to the P–V curve generated by a PV source. The MPPT will track this maximum.

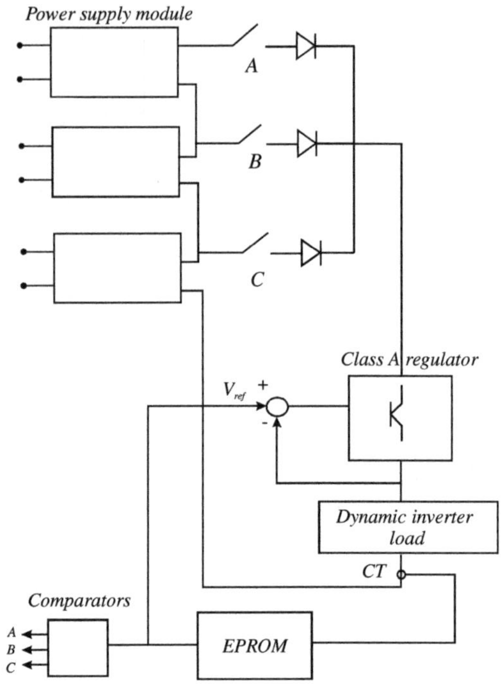

Fig. 6.6 Scheme of the PV source emulator proposed by Lloyd et al

In order to take into account variations in temperature, the value of the input DC supply voltage is modified, considering that, when temperature changes, the open circuit voltage changes significantly.

For increasing or decreasing the short-circuit current, the value of R_{series} is suitably modified, assuring that a finite resistance is always in series with the power supply, to avoid its possible damage.

An interesting solution using the combination of a class A regulator and several DC supplies that can be connected to the regulator by switches has been proposed by Lloyd et al. (2000). In particular, a modular design has been followed to setup a PV source emulator, whose single power unit is capable of delivering a maximum power of 400 W.

In order to take into account the real PV source's behavior, a set of precalculated I–V curves for specified values of solar irradiance and temperatures has been used. These curves are loaded into a EPROM used as a look-up-table (LUT).

The I–V curves have been calculated on the basis of a double diode model with reference to both a polycrystalline and amorphous silicon (a-Si) PV panels.

The hardware implementation of this system is depicted in Fig. 6.6.

The class A regulator produces an output voltage, according to the I–V characteristics contained within the EPROM, to test the inverter supplied by the PV source.

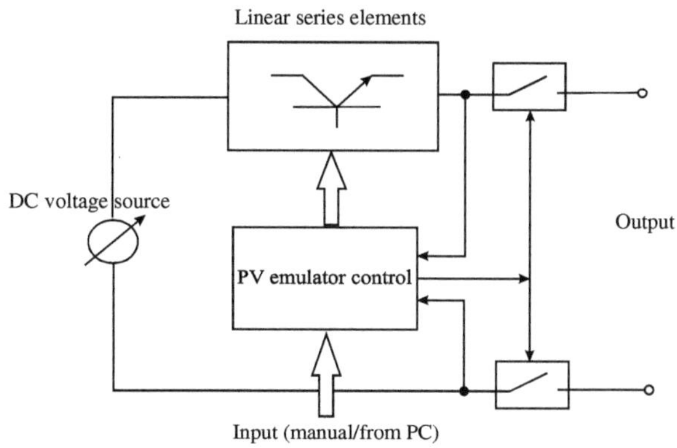

Fig. 6.7 Schematic description of the PV source emulator proposed by Haeberlin and Borgna

The regulator is provided with a current control allowing additional power units to be connected in parallel, to obtain higher output current and a close tolerance current sharing.

The static switches, indicated as A, B, and C in Fig. 6.7, are automatically selected to make the regulator operate within its range, as the working point sweeps through the I–V curve. Power supply modules can be connected in series or parallel to meet the I–V specifications of the PV inverter under test.

This emulation system permits two operation modes, as:

I. Current regulator: in this case the voltage is measured and input to the EPROM that gives a desired value of the load current.
II. Voltage regulator: in this case the EPROM gives the reference voltage, V_{ref}.

In the first case the measured voltage is used to switch the power supplies while, in the second case, the reference (or desired) voltage is used.

An interesting high power linear PV source emulator (for powers up to tens of kW) is proposed by Haeberlin and Borgna (2004) to be used for measurement on PV inverters, including MPPT efficiency.

It is based on a controlled linear current source characterized by high stability and no interaction with the power electronic circuits under test.

The principle scheme of such a device is depicted in Fig. 6.7.

It is possible to observe the presence of two series isolation switches used to provide galvanic isolation at the output of the PV emulator when it is not working.

The choice of a linear design, in this case, is considered appropriate for two main reasons: the better dynamic response and the absence of internally generated high frequency voltages, since no PWM switching occurs, that is an advantage as far as the electromagnetic compatibility is concerned.

On the other hand, the linear design implies a serious drawback; in particular, if the emulator is operated close to the short-circuit condition with high values of the voltage source a big amount of power will be dissipated in the linear series elements.

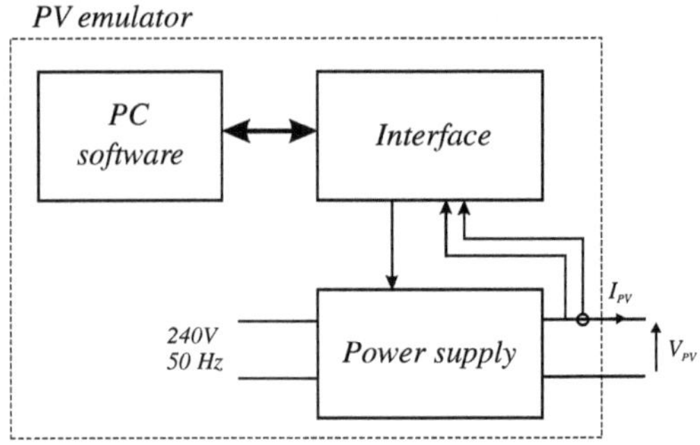

Fig. 6.8 Block diagram of the PV source emulator proposed by Khouzam and Hoffman

Anyway, this condition is uncommon and the power losses can be reduced by an appropriate choice of the feeding source voltage level and by an appropriate reduction of the current close to the short circuit point. These power losses, however, remain higher than the power lost in switching converters described in the next section.

In this emulator the power variation is imposed by varying the operating current. This operating mode reproduces solar irradiance variation with constant temperature.

6.2.1.3 Circuits Based on Switching Converters

The PV source emulators, whose hardware structure is based on power electronic converters, are the majority in the technical literature. The reason for their success relies in the following considerations.

A DC linear amplifier needs, in general, a large heatsink and its conversion power losses approach the 100 % when the PV emulator supplies the short-circuit current. In a DC/DC converter the power devices are operated considering only two fundamental states: the *on* state in which the voltage of the power switch is null and the *off* state in which, in turn, the current is null. In both cases, in principle, there is no dissipated power; only during transitions between *on* and *off* states and vice versa a low amount of power, dependent on the switching frequency is dissipated. As a consequence, the overall efficiency is expected to be very high and a smaller size emulation unit can be set up.

A PV emulator realization based on a switched mode step-down converter, operating under computer control, is due to Khouzam and Hoffman (1996). This system can emulate a PV generator in real time, accounting for the characteristics of any array, its size, as well as specific irradiance and temperature conditions, and PV aging.

All these data can be entered into the computer, so that the converter output is controlled to reproduce the corresponding operating point.

In detail, the system is composed of a preregulator, a switch-mode step-down DC/DC power converter producing up to 100 W (power supply), a computer interface and a modeling and control software.

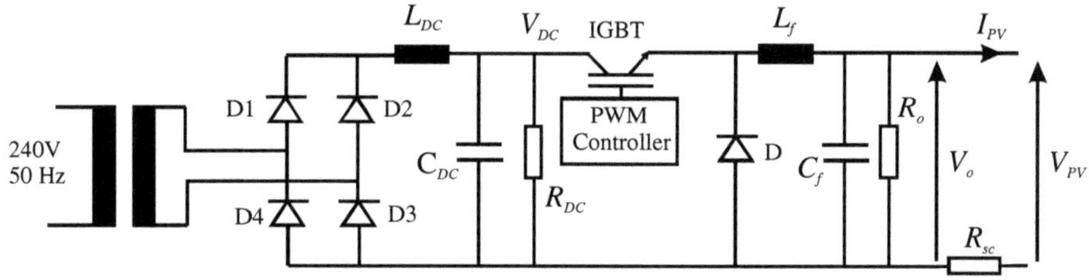

Fig. 6.9 Details of the power supply block (PV emulator proposed by Khouzam and Hoffman)

The novelty of the approach in the setup of this emulator was the use of a mathematical modeling to describe the output characteristic of the PV source.

Figure 6.8 shows the block diagram of the PV emulator.

The details of the block indicated as power supply in Fig. 6.8 are illustrated in Fig. 6.9. In particular, the mains voltage is stepped down by an isolation transformer, rectified, and smoothed to create an unregulated DC voltage, V_{DC}; this voltage is the input for the DC/DC step-down switching converter, operated at 50 kHz, that gives a regulated DC voltage ranging from 0 to 100 V.

The shunt resistor R_o is used to allow a continuous conduction mode operation of the DC/DC converter (i.e., an operation where the current in the converter inductor never goes to zero between switching cycles) even for I_{PV} equal to zero.

The series resistor R_{sc} is used to provide a degree of current limiting if the output is in short circuit.

By implementing the PV model equations in the control software, the output voltage of the DC/DC step-down converter is driven to follow the law $V_{PV} = f(I_{PV})$. In detail, the output current of the converter is firstly measured, then the corresponding theoretical voltage is calculated.

The process is repeated iteratively and within a few milliseconds until the correct output voltage and current are within a small tolerance for a given load and a specified set of PV parameters. As for these parameters, as well as solar irradiance, temperature, and aging conditions, they can all be selected and modified before and during operation. Furthermore, the PV array I–V curves may be defined in terms of either theoretical array parameters or interpolation between measured values of a real array.

Another interesting topology of PV source emulator has been proposed by Sanchis et al. (2003). Its main idea is to both measure the real and dynamic evolution of the characteristic curves of a PV generator at outdoor operation and then to reproduce them at the laboratory for inverters and MPPT testing.

A global description of the system is given in Fig. 6.10, where it is possible to observe that the whole system consists of an electronic converter, a microcontroller and a data storage unit. A measuring card is also included in the power converter with the aim to both measure, by suitable sensors, the I–V curves of a PV array and to implement the control loops.

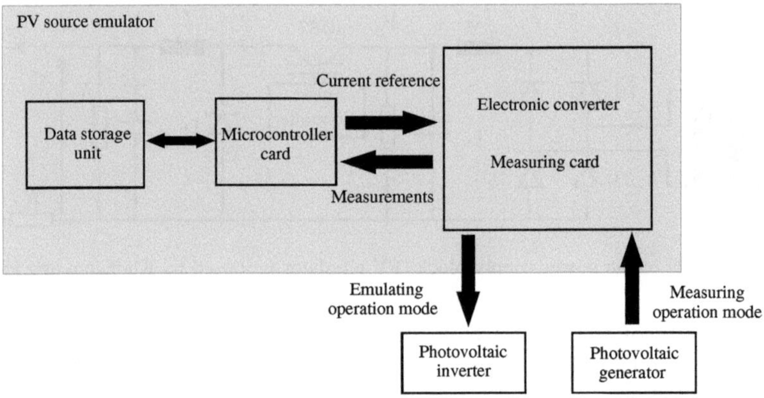

Fig. 6.10 Global description of the PV source emulator proposed by Sanchis et al

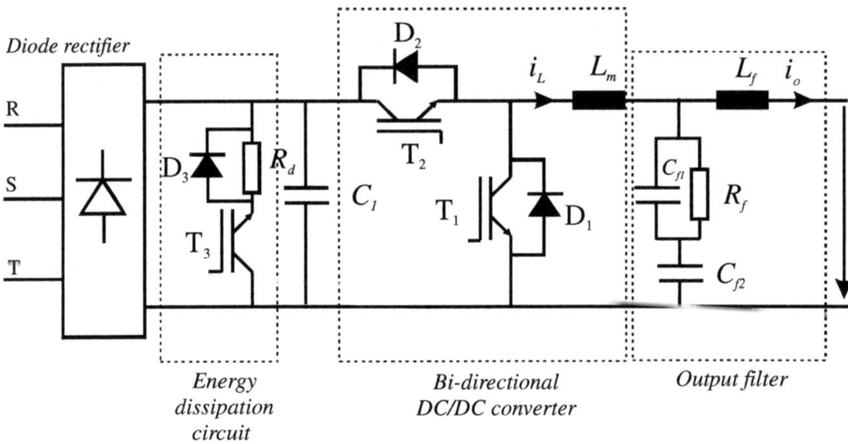

Fig. 6.11 Circuit scheme of the power stage of the PV source emulator proposed by Sanchis et al

The circuit scheme of the emulator's power stage is given in Fig. 6.11.

It can be noted that the emulator is realized by a bi-directional IGBT-based DC/DC converter that allows two different operation modes, i.e., measurement and emulation, with the evident advantage to combine both the planned working modes in a single equipment.

When the system runs in emulating mode, the power converters behave as a power source; in particular, it is operated as a buck converter supplied by the power grid via the diode rectifier.

On the contrary, when the system runs in measuring mode, the converter behaves as a variable loads and it is operated as a boost converter, supplied by a PV generator.

The system prototype can measure and emulate the behavior of a PV generator up to 15 kW and has the capability to measure continuously three time per second

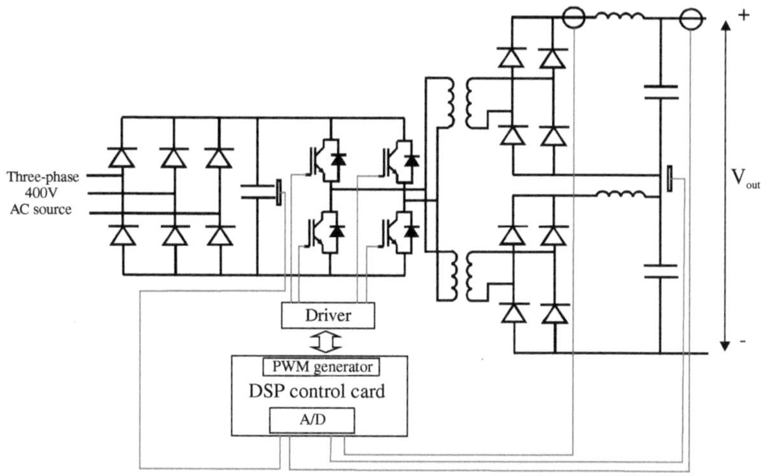

Fig. 6.12 Circuit scheme of the PVAE proposed by Martín-Segura et al

the full I–V curves up to seven PV arrays, that means to store approximately 10800 curves per hour in the database.

The usefulness of this equipment is the possibility to carry out a systematic characterization and testing of PV inverters and MPPT under any varying and nonideal operating condition.

A PV emulator based on a topology different from a single stage DC/DC converter is proposed by Martín-Segura et al. (2007). In particular, this photovoltaic array emulator (PVAE) is a multi-stage DC/DC converter composed of a DC/AC full bridge inverter, a high frequency transformer, and an uncontrolled rectifier for each secondary transformer winding. With this choice, a more secure operation, due to galvanic isolation, and high voltages are allowed.

The high output voltage availability allows PV arrays configurations, like those used in centralized technology inverter (see Sect. 3.5, Fig. 3.19), to be emulated.

The electrical scheme of the emulator is shown in Fig. 6.12. The full bridge converter is fed by a rectified three phase 400 V supply. In order to increase the output voltage two transformers, with the primary connected in parallel and the secondary in series, are used. The inverter is based on IGBTs commutated at 18 kHz. The converter output is obtained on the basis of a defined, simplified I–V curve suitably implemented on a DSP control card. The emulator prototype has a rated power of 4 kW and allows testing PV inverters with an input 0–650 V and 0–7 A.

Recently, PV emulators based on DC/DC converters controlled by Field Programmable Gate Arrays (FPGAs) have been proposed. The system developed by Koutroulis et al. (2009) is worth to be considered.

The hardware structure is based on a simple buck converter whose control is obtained by an FPGA-based unit, operating according to the PWM principle. In Fig. 6.13 the scheme of this PV simulator is shown.

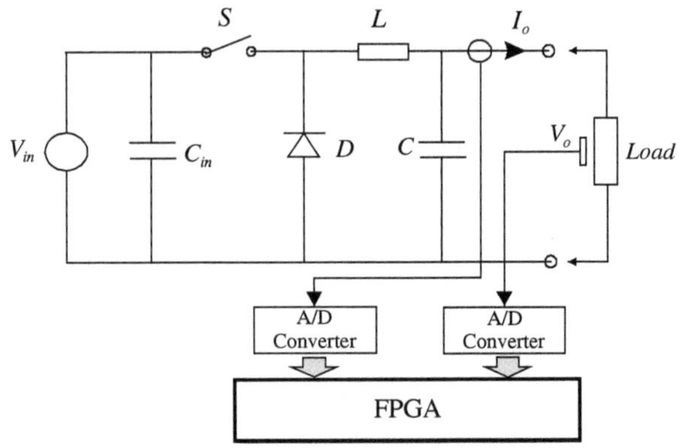

Fig. 6.13 Circuit scheme of the PVAE proposed by Koutroulis et al

It is possible to observe that it is composed of the DC/DC switching converter and the control system comprising the FPGA unit, the analog to digital (A/D) converters, and the voltage and current sensors.

The advantage of an FPGA in terms of flexibility, due to its reprograming capability and reconfigurable logic, allows the emulator to be adapted to any change. Therefore, the emulation of several types of PV generators is permitted.

In general, FPGA boards are characterized by fast computational features and make the implementation of parallel architecture feasible, thus obtaining a reduction of the control algorithm execution time.

If compared with processor-based implementations of PV emulators, the FPGA-based system contributes to a rapid prototyping since the VHDL code, used to configure the FPGA board, can be entered in a simulation routine to assess the control algorithm performance, before the hardware prototype construction.

Moreover, the high frequency operational features of the FPGA devices allow a PWM control unit architecture giving an increased power converter switching frequency to be implemented. Therefore, a reduction of the converter size (and cost) is achievable since it is primarily tied to the inductor magnetic material size which is, in turn, inversely proportional to the converter switching frequency.

The considered emulator reproduces the I–V curve of a given PV generator operating under any condition of irradiance and temperature. It has the flexibility to be operated either with real-time measured values of T and G, coming from temperature and solar irradiance sensors or with user-defined values of T and G, entered to the PV emulator in a digital format through a PC port. The developed prototype uses a power MOSFET rated 60 V and 50 A and a PWM switching frequency of about 50 kHz.

From this brief survey of PV emulators' topologies, it is possible to conclude that the most used solution is based on the use of DC/DC switching converters. As a matter of fact, this solution permits a simple realization, a good efficiency, and an accurate reproduction of the I–V curves both in steady state and dynamic conditions.

6.2.2 Control Stage: PV Behavior Implementation

The control law of a PV emulator is oriented to establish, for any environmental and load condition, an operating point lying on an I–V characteristic of the real photovoltaic array to be emulated.

As previously stated, theoretical models of a PV source based on equivalent circuits can be used to describe a complete set of I–V characteristics. A further option consists on the use of a database set up by measuring I–V curves in different environmental conditions, as a reference.

Another way to regulate the emulator output, implemented by some authors that have developed PV emulators based on operational amplifiers, is obtained by forcing the operating point of the sample cell by auxiliary circuits. This approach makes without the direct evaluation of either the I–V mathematical relationship or the characteristics' measurement

According to this method, the electrical behavior of the PV generator is considered by imposing a change in the operating point through a suitable and independent amplification of current and voltage at the output of a small PV cell, as explained in Sect. 6.2.1.1.

In the more common and promising PV emulators realized by switching converters, the output regulation is mostly based on I–V curves. With reference to these PV emulators, some general issues related to the control implementation are given in the next subsection.

6.2.2.1 I–V Curve-Based Regulation: Implementation Issues

The output of a PV emulator based on a switching converter is adjusted in real time by the control system so to reproduce the external I–V characteristics of a PV generator. As a result, from the PV inverter and load point of view, the PV emulator behaves as the PV source, as long as its characteristics are suitably programed (usually off-line) in the controller.

The general principle of the converter control can be schematized as shown in Fig. 6.14. In particular, for the most common case of a DC/DC converter, the duty cycle regulation is performed on the basis of the output current detection and the definition of reference voltage from the established I–V behavior of the PV source.

In the case shown in Fig. 6.14, the interface board has the task of transforming, isolating and filtering the current and voltage signals coming from the current, and voltage transducers, respectively.

Both current and voltage at the output of the converter are sent to the PWM generator as feedback signals, while the output current is entered in the control board. Here, the current is used to calculate the corresponding reference voltage which is sent to the PWM generator after A/D conversion and filtering.

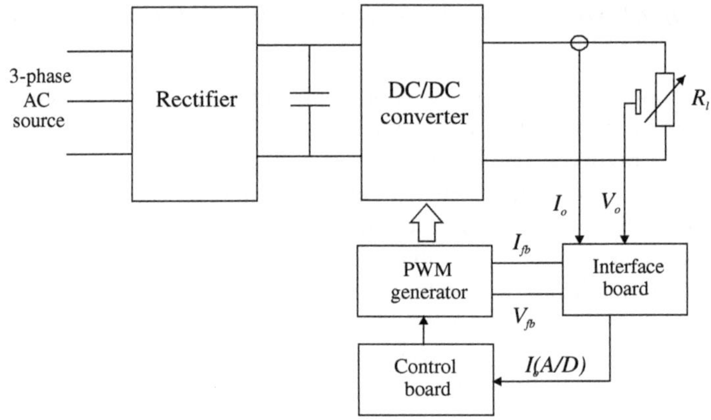

Fig. 6.14 General scheme of the converter control principle

Closed loop control is the preferred solution, even if the literature exhibits examples of open loop controls based on the real-time measurement of the load resistance, as in Zeng et al. (2002).

The PV source behavior definition within the control, i.e., the I–V curve representation, follows, as previously anticipated, two main approaches:

- the model-based approach;
- the measurement-based approach.

In the model-based representation of the PV characteristics, it is assumed that all the PV source theoretical parameters are available, therefore, the output voltage is calculated using these parameters and, in addition, solar irradiance and temperature.

In order to exploit the simplicity and flexibility of a linear model, some authors, simplify the PV source model used for the representation of the I–V curve. In particular, they consider the I–V characteristics as formed by the two lines obtained joining the remarkable points (V_{oc}, V_{MPP}, I_{MPP} and I_{sc}), according to Eqs. (6.4 and 6.5).

$$I = \frac{V_{MPP} - V_{oc}}{I_{MPP}} \cdot V + V_{oc} \tag{6.4}$$

$$I = \frac{V_{MPP}}{I_{MPP} - I_{sc}} \cdot V - \frac{V_{MPP} \cdot I_{sc}}{I_{MPP} - I_{sc}} \tag{6.5}$$

The shape of the linearized I–V curve is shown in Fig. 6.15.

In general, the use of simplifying approximations, when do not introduce significant errors, is advantageous since it implies a reduced calculation complexity. This simplification is especially applied when the resulting I–V curve is stored in a LUT, which is part of a digital control system with limited resources in terms of hardware capability.

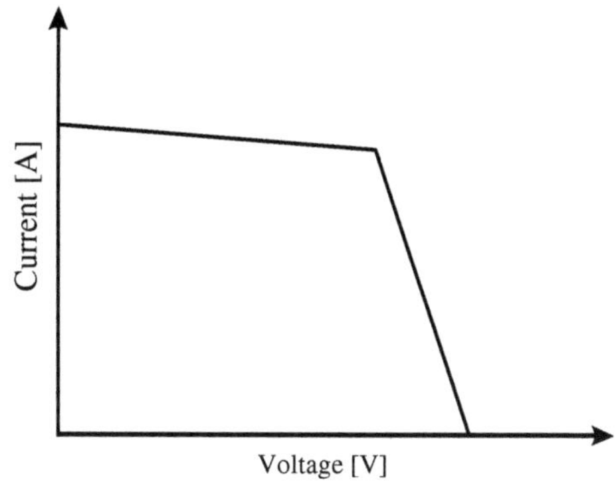

Fig. 6.15 Shape of the linearized I–V curve

In the measurement-based approach for I–V curve representation it is possible to distinguish:

- methods using interpolation between measured values of a real PV generator;
- methods using the full PV curve measurement.

In the first case (it is assumed that the theoretical parameters of the PV source are unavailable), three measured points are needed, i.e., the open circuit point, the short circuit point, and the maximum power point, being the measurements conducted in reference conditions for temperature and solar irradiance.

The resulting equations will be determined for any temperature and irradiance value.

In the second case, a controlled variable load (usually the PV inverter) is needed to make the PV generator completely cover the I–V characteristics. Moreover, with this approach, a suitable storage unit is required to store the measured data that will be organized and classified by a data processing unit to be used for emulation purpose. An appropriate quickness and capability to store a large amount of data are required to the used database.

Figure 6.16 illustrates the scheme of a systems used for the real-time measurement of a PV generator full I–V curves (it represents the measuring operation mode of the system proposed by Sanchis et al., described in Sect. 6.2.1.3).

Regardless of the chosen approach for the representation of the I–V curves, they have to be converted into a data table and suitably stored in a memory.

It should be observed that, the chosen word length of the calculated parameters comes from a trade-off between the controller hardware resources and the accuracy required to the PV emulator.

For example, in the PV emulator shown in Fig. 6.7, the I–V curves related to 64 different values of solar irradiance, between 0 and 1200 W/m^2, are calculated using a double diode PV model. Then they are stored into a 16-bit EPROM using 4096 words for each I–V curve.

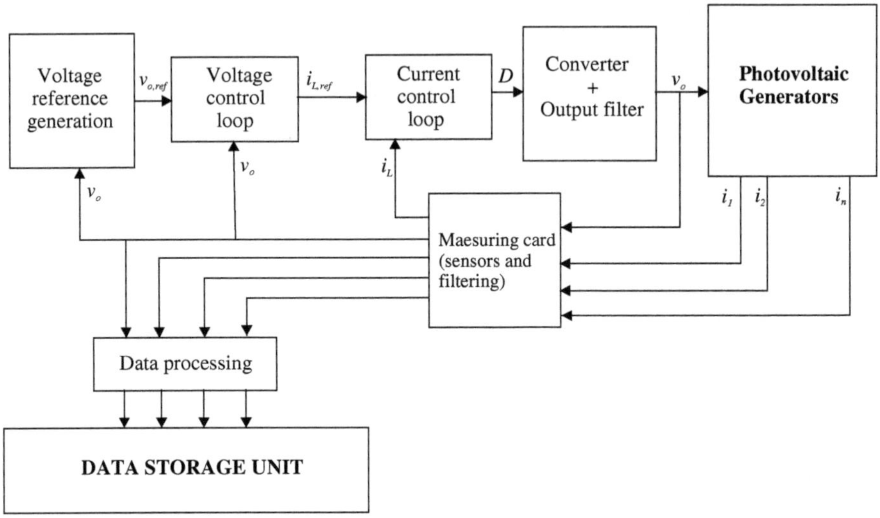

Fig. 6.16 Scheme of a systems used for the real-time measurement of a PV generator full I–V curves

On the other hand, for the PV emulator shown in Fig. 6.13, the I–V curve of an ideal PV module, represented according to a single diode model, is quantized in 8-bit integer format, according to the following relationships:

$$V_{l8m} = V\frac{(2^8 - 1)}{V_{\text{oc,stc}} - k_T\left(T_{C,\text{ref}} - 25\right)} \tag{6.6}$$

$$I_{l8m} = I\frac{(2^8 - 1)}{I_{\text{sc,stc}}\left(G_{\text{ref}}/1000\right)} \tag{6.7}$$

where V_{l8m} and I_{l8m} are the resulting integer values of the ideal PV module voltage and current, respectively, k_T is the open circuit voltage temperature coefficient expressed in V/ °C and $T_{C,\text{ref}}$, G_{ref} are the PV module reference temperature and solar irradiance expressed in [°C] and [W/m^2], respectively. Then, the resulting pairs of V_{l8m} and I_{l8m} are stored in a LUT implemented in the FPGA board.

Using this ideal I–V characteristic stored in a LUT, the characteristic of the PV module can be determined for any value of temperature and solar irradiance, as defined by the operator.

With specific regard to the implementation of the PV emulator control on a FPGA board, an interesting post-layout comparison has been done in Koutroulis et al. (2009) by considering several target FPGA devices, in terms of required resources, maximum PWM frequency, maximum sampling frequency, and cost.

The obtained results are given in Table 6.1.

It is possible to note that, in all cases, the implementation of the PV emulator control occupies a relatively small percentage of the available resources, thus

Table 6.1 Post layout comparison among different FPGA devices

FPGA device	Required resources (logic cells)	Maximum PWM signal frequency (kHz)	Maximum sampling frequency (kHz)	Cost ($)
Virtex XC2V1000	888 out of 27648 (3 %)	313.5	78.37	252
Spartan-3 XCES50	476 out of 1728 (27.5 %)	111.54	27.88	288
Spartan-3 XCES400	412 out of 8064 (5 %)	132.49	33.12	30
Virtex-44VLX80FF1148-11	1000 out of 80640 (1.2 %)	246.2	61.55	72
Spartan-3 XCES1000	LUTs: 1708 out of 15360 (11 %)	286.84	71.71	20

permitting the integration of additional power converter control operations in the same IC. Furthermore, due to the logic cells' availability, the PV emulator control can be implemented using more bits, thus increasing the accuracy of the emulator itself.

As for the maximum PWM frequency, it can be observed that any of the considered FPGA device allows the value of 100 kHz to be exceeded. Moreover, it should be observed that such values of PWM frequency, in general higher respect the PWM frequencies achievable by commercially available DSP or microcontrollers, can be obtained at a relatively low cost.

6.3 Dynamics and the Arbitrary Load Problem

In order to achieve an effective real-time operation, a PV emulator should exhibit an appropriate dynamic behavior. In particular, it should rapidly converge to the desired operating point when a modification of either the load or the environmental conditions occurs.

In the technical literature, response times of PV emulators up to tens of microseconds have been found, corresponding to step variations of load or solar irradiance. For example, in Matsukawa et al. (2003), proposing a 10 kW PV emulator based on a controlled active power load (APL), a response time of about 25 ms has been measured for a step variation in solar irradiance from 1000 to 500 W/m^2. The same emulator shows a response time of about 16 ms for current and about 25 ms for voltage, when a step variation of the applied load from 1 kΩ to 100 Ω occurs.

A good dynamic behaviour is particularly important in PV emulator systems including the I–V curve measurement in its operating modes, such as the system schematized in Fig. 6.10. Here, the need for making the PV generators cover continuously their characteristics, especially along the voltage axis, has suggested to include an additional voltage loop in the control board, as shown in Fig. 6.16. This loop acts as an outer control loop that provides the reference for the inner current loop. Both loops form a cascade control structure that allows a robust and accurate control of the voltage at the terminals of the PV source.

Thanks to a high switching frequency of the power converter, a cascade control structure characterized by a high dynamics is possible. Therefore, the I–V curves can be swept and measured in a few milliseconds to obtain a real-time measurement of their evolution.

In general, the design of the PV emulators' control principle is not a simple issue, since the control system should be insensitive to all involved parameters and the emulator circuit should operate correctly regardless, for example, of the value of the load capacitance.

In other words, since the main reason for using a PV emulator is to test PV inverters, a good dynamic behavior is required to avoid interference between the control system of the emulator and the control system of the supplied power converter. This issue is referred in the literature as the "arbitrary load problem".

The solution proposed by Ollila (1995), for a PV emulator system rated 1 kW and formed of a DC/DC step-down power converter, is based on the cascade connection of separate current and voltage controllers. Being the novelty of the approach, the special capability of the used controllers of tolerating high system gain variations.

In particular, it is observed that a low value of output emulator capacitance is required to obtain a fast dynamical response. Anyway, in presence of a loading inverter the cumulative capacitance at the emulator's output can significantly increase, leading to possible system gain variations up to 1000 times the original value.

In a DC/DC buck converter, a cascade control based on the use of conventional PIs, in particular, a fast PI-controller in the inner inductor current loop and a slower PI-controller in the outer capacitor voltage loop, allows to achieve a current limitation and a fast voltage response, as well.

Anyway, this control scheme presents a disadvantage. In detail, the integration time constant has to be set very slow to obtain a sufficient phase margin if the load capacitance is large.

On the other hand, the gain has to be small to guarantee a stable operation with low load capacitance, that implies the presence of long and large dynamical error.

Since the load is a priori undefined and no trimming is permitted in standard PI, these controllers are not considered as the best solution.

To overcome the above described issues, the use of adaptive, self tuning or fuzzy schemes is possible, which obviously implies computational demanding algorithms. On the other hand, for low cost applications, a cascade control using a voltage controller with a 45° phase lag independently from frequency is proposed whose transfer function is given in Eq. (6.8).

$$G_c(s) = \frac{1}{\sqrt{t_c s}} \tag{6.8}$$

This kind of controller is indicated as $1/\sqrt{s}$ controller.

In Eq. (6.8) t_c is the controller time constant.

The gain of the controller described by the transfer function Eq. (6.8) decreases 10 dB/decade regardless of frequency. If the system gain decreases of 20 dB/decade and has a 90° phase lag, the gain margin is infinite and the phase margin is 45° at any frequency. This leads to a slight overshoot and a fast damping in a closed loop step response, irrespective of the value of the load capacitance.

The overall closed loop bandwidth is determined by the controller and the system time constants.

In order to practically realize this kind of controller, zero-pole pairs can be used to approximate the transfer function within a given frequency interval. A truncated zero-pole form with one zero-pole pair per decade to approximate the controller transfer function is given in Eq. (6.9):

$$G_c(s) = \frac{t_c s + 1}{t_c s} \prod_{i=1}^{M} \frac{10^{-i} + t_c s}{10^{-1+1/2} + t_c s} \tag{6.9}$$

where M is the frequency range expressed in decades.

The approximation of the controller transfer function given in Eq. (6.9) is just suitable for several purposes but a better representation can be obtained by locating the poles and zeros, so that the average phase error is minimized in a given frequency interval. With this approach, a phase optimal solution is found assuming the presence of an additional pole (a measuring filter) in the system transfer function.

The normalized controller transfer function, in this case, will be in the form:

$$G_{c(p.u.)} = \frac{1}{\sqrt{s}} \approx k \frac{(s - z_1)(s - z_2)(s - z_3)}{s(s - p_1)(s - p_2)(s - p_3)} \tag{6.10}$$

The practical realization of the transfer functions in Eqs. (6.9) and (6.10) is quite simple and can be handled using operational amplifiers, as shown in Figs. 6.17 and 6.18.

The digital equivalents can be obtained using, for example, pole-zero matching techniques.

In order to control the PV emulator, two of the above described $1/\sqrt{s}$ controllers are used in a current–voltage cascade structure.

In the current loop the controller eliminates the need for adaptive control in discontinuous mode, while, in the voltage loop, it allows high values of load capacitance to be tolerated. Thus, the problem of the correct handling of an arbitrary loading converter is solved.

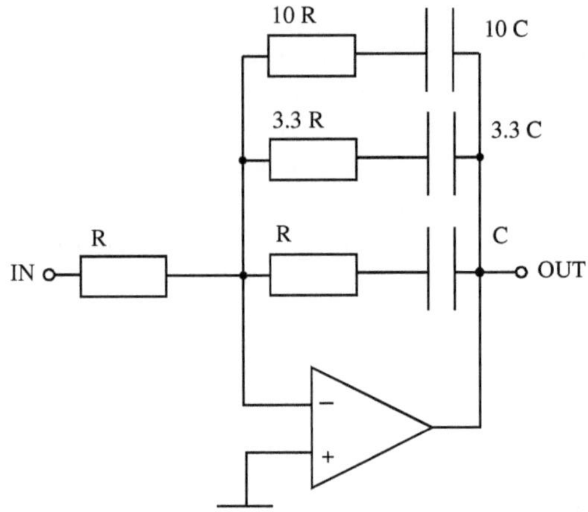

Fig. 6.17 Practical realization of the approximate $1/\sqrt{s}$ controller using a single operational amplifier

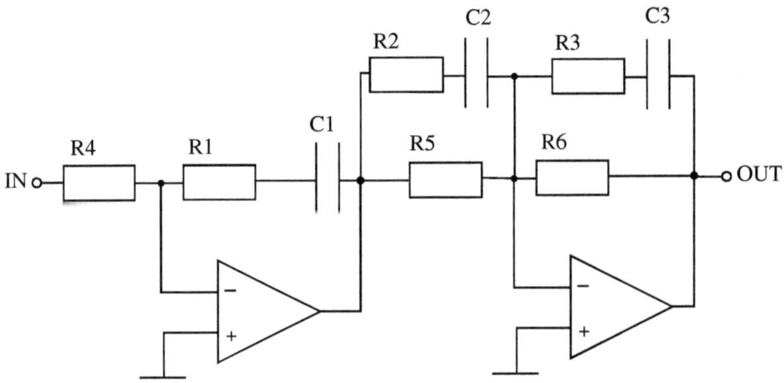

Fig. 6.18 Practical realization of the approximate $1/\sqrt{s}$ controller using two operational amplifiers

6.4 Non-Ideal Operating Conditions

The maximum flexibility and usefulness of a PV source emulator is achieved when it has the capability of reproducing correctly the PV generator electrical behavior also under nonideal operating conditions. As a matter of fact, they correspond to real and, sometimes, common situations. Therefore, testing PV inverters or MPPT techniques in such conditions gives a more realistic measure of the expected performance and can help in the improvement of current technologies and in the development of new ones.

Among the most common nonideal operating condition there is the partial shading of the PV generator.

Other possible situations are related to environmental conditions, such as rapidly changing atmospheric conditions, dawn, nightfall, etc., or to the status of the PV array such as aging, degradation due to high energy particle radiation damage (in space applications), and so on.

The natural way for taking into account these conditions is their suitable implementation in the PV emulator control stage dedicated to the representation of the I–V characteristics. This can be obtained by including the considered no-nidealities in the PV model (if the I–V curve representation is model based) or setting up a comprehensive database of measured I–V curves covering as much as possible the nonideal operating conditions to be reproduced (if the I–V curve representation is measurement based).

With reference to this last situation, it is worth observing that several nonideal conditions can be handled during emulation, if different test patterns, corresponding to different evolution of the PV generators under different weather conditions, are available from a database.

As far as the partial shading process is concerned, an example of PV source modeling approach including this phenomenon is described in Sect. 3.5, while Sect. 4.8.3 shows the Matlab® implementation of the PV source working under nonuniform illumination.

Khouzam and Hoffman (1996), whose PV emulator is schematized in Fig. 6.8, include in their PV model the effect of radiation damage, due to high energy particles trapped in the Van Allen Belt surrounding the earth. As a matter of fact, these particles penetrate the PV cells, disrupt the ordered lattice and introduce recombination centers, thus reducing the cells' performance.

A simplified modeling of the radiation damage is obtained by introducing the concept of an equivalent fluence, intended as the total number of monoenergetic particles per unit area producing the same level of damage as a spectrum of light particles. Using a the quantity of 1 MeV electrons, the effect of a particle on the short-circuit current of the PV cell is described by Eq. (6.11), due to Lush and Gray (1986).

$$I_{sc\psi} = I_{sc0} - C_i \log\left[1 + \frac{\psi}{\psi_x}\right] \tag{6.11}$$

where ψ is the irradiation fluence, ψ_x is the estimated fluence at the end of life (EOL) of the PV cell, I_{sc0} is the short-circuit current density in (mA/m^2) at the beginning of life (BOL) of the cell, i.e. at $\psi = 0$, $I_{sc\psi}$ is the short-circuit current density in (mA/m^2) at a fluence of ψ and C_i is a constant depending on the type of PV source.

A similar relationship for describing the voltage degradation has been proposed by Khouzam (1988) and used in voltage modeling by Lillington (1988), according to Eq. (6.12).

$$V_{oc} = V_{oc0} - C_v \log\left[1 + \left(\frac{\psi}{\psi_x}\right)^{0.8}\right] \tag{6.12}$$

where V_{oc0} is the open circuit voltage at BOL, V_{oc} is the open circuit voltage at a fluence ψ and C_v is a constant depending on the type of PV source.

Implementing Eqs. (6.11) and (6.12) in the equations that describe the PV model, an I–V curve representation for a PV generator including the radiation damage effect is obtained.

6.5 Rated Power

The output power capability of a PV emulator is related to the rated power of the power amplifier. In particular, the power capability can be increased by increasing its size.

On the other hand, in case of DC/DC switching converters, if power switching devices with power limitation at high switching frequencies are used, the output power of a PV emulator can be increased by using several, high switching frequency DC/DC converter units, in suitable connections, according to a modular concept.

In both cases, the use of a control unit only depending on the I–V relationship of the PV source to be emulated is particularly appreciable. As a matter of fact, if the control unit is not dependent on the power converter rated power, it is not required to be modified if the PV emulator capability is changed.

The technical literature exhibits switching DC/DC converter-based PV emulators whose rated power ranges between hundreds of W and tens of kW.

High power PV emulators based on controlled linear power sources have been realized as well, but this design implies the setup of bulky equipments in addition to all the necessary tricks to avoid high power dissipation.

6.6 Modularity

The desired output of a PV source emulator should be the same as a real PV generator provides to the subsequent conversion chain. On the other hand, the real PV source could be formed of a series/parallel connection of PV elementary cells or modules.

In order to obtain the maximum flexibility of the PV emulator in terms of current and voltage covered range, some authors introduce in the emulator's design the concept of modularity.

Modularity is a feature related to the possibility of increasing the voltage and current capabilities in an analogous way in which the number of PV emulation units is increased.

In other words, modularity means that additional PV modules or fields can be simulated by adding emulation units in series, if an increased voltage is required, or in parallel, if an increased current is needed. This feature allows, for example, a series of PV modules to be emulated by the series connection of the same number of emulators where each one implements the model of the module that can be subject to different solar irradiance, reproducing partial shading situations, as well.

The modular approach is promising and has been successfully adopted both within PV high power emulator based on linear amplifiers and on switching power converters.

6.7 Examples of Solutions Available on the Market

In addition to the PV emulators described in the technical literature, manufacturers have proposed different solutions for the PV source emulation, generally called Solar Array Simulators (SAS).

They usually adopt the concept of modularity, described in the previous section and consist of programmable DC power supplies which can be operated as individual equipment or in multi-channel configurations to meet higher power ratings requirements, according to a customized turn-key approach.

As an example, two PV emulator systems, proposed respectively by Elgar (Ametek) and Agilent, are described hereinafter.

The Elgar TerraSAS (TSAS) is a PV source emulator providing a turn-key approach to testing the MPPT characteristics for grid-connected inverters and DC charge controllers.

The TSAS emcompasses: programmable DC power supplies, a rack mounted controller, a keyboard, and a LCD display with control software and a local graphical user interface (GUI), output isolation and a PV simulation engine controlling the power supply. Thanks to this hardware structure, the TSAS can simulate most of test protocols or combination of events that a solar installation will be subjected to.

The key features of this system can be summarized as follows:

- Dynamic emulation of irradiance and temperature ranging from a clear day to cloud cover conditions;
- Possibility to ramp the voltage, temperature or irradiance level over a programed time interval;
- Read-back of voltage, current, irradiance level, and temperature setting;
- Tests for inverter Maximum Power Point Tracking (MPPT);
- Programmable I–V curves available for PV Inverter testing;
- Emulation of different types of solar cell material;
- Multi-Channel, up to 1 MW, with power supplies available in 1–15 kW increments.

The required performance for this PV emulator is obtained by high speed switching power supplies using Power MOSFETs, which typically switch very fastly. Higher switching frequency allows smaller output capacitors and inductors to be used, which is the key to an optimal high speed power supply design.

In Fig. 6.19 a photo of the TSAS is given.

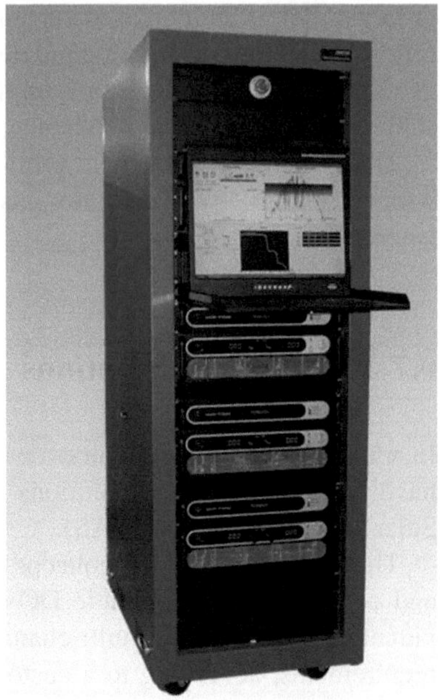

Fig. 6.19 View of the Elgar TerraSAS system (From Elgar TerraSAS. (Online) Available http://www.elgar.com/go/ts/TerraSAS_Datasheet_031910.pdf)

An advantage of this system is its modularity consisting in building blocks which can be programed by means of four parameters accounting for the I–V curve endpoints and for the current and voltage mode curve slopes.

More precisely, the only parameters required for a simulation are the open circuit voltage and short-circuit current, while the slope of the I–V curve can then be modified by the peak power parameters, V_{MPP} and I_{MPP}.

Changes to these parameters allow the shape of the I–V curve to be adapted to any fill factor between 0.5 and 1.

Once an I–V curve has been generated, changes to the irradiance or temperature level can be imposed by the operator so to test the behavior of a grid-connected inverter under realistic conditions. In general, PV inverters can be optimized for real MPP search modes, because shadowing and temperature changes can be simulated realistically.

The PV simulator has the ability to simulate both ideal I–V curves and irregular characteristics for peak power tracking when solar panels with different output characteristics are parallel connected.

With the PV emulator programed for different values of irradiance or temperature, the characteristic "multiple hump" I–V curve will result.

By programing the changes in irradiance and temperature in a table, dynamic simulation of compressed time profiles of a 24 h day can be run in a loop to simulate the alternation of day and night for long periods of time.

Long-term weather simulations can be run as well to determine the amount of energy delivered in a given situation.

The Agilent E4360 Modular SAS is a dual output programmable DC power source that performs the emulation of solar arrays under different conditions.

The E4360 SAS is basically a current source with very low output capacitance. It provides up to two outputs and up to 1.2 kW in a small mainframe.

This system is either available as an off-the-shelf instrument to be used as individual equipment or integrated into a full turn-key SAS system configured to meet the exact required specifications. For example, if greater output power and current are needed, the emulator units can be easily operated in parallel thanks to a firmware-based feature that allows the two unit channels to be treated as a single, synchronized channel with twice the output current, and power capability.

The key features of this SAS system are the following:

- Accurate emulation of any type of solar array;
- Small size;
- Modularity;
- Output power up to 600 W per output;
- Fast I–V curve change and fast recovery switching time;
- Easy to simulate environmental conditions;
- LAN, USB, and GPIB interfaces.

In Fig. 6.20 an example of a custom turn-key E4360 SAS-based system is illustrated.

The E4360 SAS provides three operating modes, i.e., simulator (SAS) mode, table mode, and fixed mode.

In particular, if the I–V curve of a PV generator has to be emulated, SAS or table modes are used. On the contrary, if a standard power supply is needed, fixed mode is used.

In simulator (SAS) mode, the E4360 SAS generates a I–V point table on the basis of an internal algorithm which approximates the I–V curve. This can be done via the I/O interfaces or from the front panel, where a PC is not needed. In order to establish a I–V curve in this mode, open circuit voltage, short-circuit current, and MPP are needed.

In table mode, the I–V curve is determined by a user-defined table of points with a minimum of three points, up to a maximum of 4000 points.

Up to 30 tables can be stored in each of the E4360 SAS built-in volatile and nonvolatile memory.

Current and voltage offsets can be applied to the selected Table (I–V curve) to simulate a change in the operating conditions of the PV generator.

Finally, fixed mode is the default mode when the unit is powered on. In this case, the unit has the rectangular I–V characteristics of a standard power supply.

Notwithstanding the certain value of such commercial solutions, their cost, especially for the emulation of high power PV generators, could represent a key point.

Fig. 6.20 Custom turn-key E4360 SAS-based system (From Agilent E4360 Modular solar array simulators, datasheet. (Online) Available http://cp.literature.agilent.com/litweb/pdf/5989-8485EN.pdf)

6.8 Conclusions

Photovoltaic industry is growing exponentially and the availability of specialized equipment for the laboratory emulation of PV generators is becoming essential for manufacturers and laboratories working in the field of PV plant-associated power electronics. In particular, since the trend moves toward more efficient PV inverters, the setup of effective and reliable PV emulators is becoming more and more a challenging issue.

The need of reproducing in laboratory the PV source behavior has been recognized since the end of 1970s. Several solutions have been devised. Firstly, emulators were based on a physical sample of the PV source and on operational amplifiers; then, more sophisticated models implemented on computers and power circuits based either on linear regulators or switching DC/DC converter have been proposed.

The emulators based on linear regulators are not affected by noise due to PWM operation and have a good dynamic behavior but their low efficiency requires good heatsink systems or variable DC supplies that could worsen the dynamics.

On the contrary, emulators based on switching DC/DC converters can deliver high powers with a good dynamic behavior by a correct choice of the switching frequency and of the feedback network.

This chapter presents a survey of PV emulators which have been proposed in the technical literature, including some examples of commercial solutions.

The most critical aspects related to PV emulators design and operation are put in evidence, such as power stage schemes, control features, dynamics, modularity, and so on. It has been definitely observed that the more common and promising modern PV emulators are realized by DC/DC switching converters, with output regulation based on suitable I–V curve representation.

On the basis of such elements, an indepth examination of the main topologies of DC/DC converters and of their related control techniques is considered particularly useful and will be the object of the next chapters.

Bibliography

Agilent E4360 Modular Solar Array Simulators, Models: E4360-62A, E4366-68A, Datasheet. (Online) Available http://cp.literature.agilent.com/litweb/pdf/5989-8485EN.pdf

Baert D (1979) Solar-cell panel simulation. Electron lett 15(2)

Bazzi AM, Klein Z, Sweeney M, Kroeger KP, Shenoy PS, Krein PT (2012) Solid-state solar simulator. IEEE Trans Ind Appl 48(4):1195–1202

Di Piazza MC, Vitale G (2010) Photovoltaic field emulation including dynamic and partial shadow conditions. Appl Energy 3(87):814–823

Elgar TerraSAS—Programmable Solar Array Simulator. (Online) Available http://www.elgar.com/go/ts/TerraSAS_Datasheet_031910.pdf

Haeberlin H, Borgna L (2004) A new approach for semi-automate measurement of PV inverters, especially MPP tracking efficiency, using linear PV array simulator with high stability. In: Proceedings 19th European photovoltaic solar energy conference, 2332–2336

Koutroulis E, Kalaitzakis K, Tzitzilonis V (2009) Development of an FPGA-based systemfor real-time simulation of photovoltaic modules. Microelectron J (40):1094–1102

Kouzam K, Hoffman K (1996) Real-time simulation of photovoltaic modules. Sol Energy 6(56):521–526

Lloyd SH, Smith GA, Infield DG (2000) Design and construction of a modular electronic photovoltaic simulator. In: Proceedings 8th international conference on power electronics and var speed drives, 120–123

Luque A, Hegedus S (eds) (2003) Handbook of photovoltaic science and engineering. Wiley, Hoboken

Martín-Segura G, Lopez-Mestre J, Teixido-Casas M, Sudria-Andreu A, (2007) Development of a photovoltaic array simulator systembased on a full-bridge structure. In: Proceedings 9th international conference electrical power quality and utilisation, 1–6

Matsukawa H, Koshiishi K, Koizumi H, Kurokawa K, Hamada M, Bo L (2003) Solar Energy mat solar cells (75):537–546

Mukerjee AK, Dasgupta N (2007) DC power supply used as photovoltaic simulator for testing MPPT algorihms. Renewable Energy (32):587–592

Nagayoshi H (2004) I-V curve simulation by multi-module simulator using I-V magnifier circuit. Sol Energy Mater Sol Cells (82):159–167

Ollila J (1995) A medium power PV-array simulator with a robust control strategy. In: Proceedings IEEE power electronic special conference 40–45

Sanchis P, Echeverria I, Ursua A, Alonso E, Gubia E, Marroyo L (2003) On the testing characterization, and evaluation of PV inverters and dynamic MPPT performance under real varying operating conditions. Prog Photovolt Res Appl (15):541–556

Solanki SC, Dubey S, Tiwari A (2009) Indoor simulation and testing of photovoltaic thermal (PV/T) air collectors. Appl Energy (86):2421–2428

Tiwari A, Barnwal P, Sandhu G, Sodha MS (2009) Energy metrics analysis of hybrid photovoltaic (PV) modules. Appl Energy (86):2615–2625

Zeng Q, Song P, Chang L (2002) A photovoltaic simulator based on a DC chopper. In: Proceedings of the 2002 IEEE canadian conference on electrical and computer engineering, 257–261

Chapter 7
DC/DC Power Converters

7.1 Introduction

Many industrial applications need a conversion of a voltage coming from a DC source into another DC voltage. A device that performs this kind of conversion is known as DC/DC converter.

Contrarily to the AC/AC conversion performed with high efficiency by transformers since the end of XIX century, DC/DC conversion has been made possible thanks to the use of power devices in switching operation; this kind of conversion is properly named *switching* DC/DC *conversion*, but it is often referred simply as DC/DC conversion.

First examples of switching converters trace back to 1950, they were mainly military and space applications even if some kind of vacuum-tube converters were used for example in car radios.[1]

At present, DC/DC switching converters are at the base of both the small electronic devices (mobile phones, portable computers etc.) and of industrial equipments. A switching DC/DC converter with a suitable control allows a reference voltage, smaller or greater than the input one, to be obtained and maintained even in presence of disturbances as load or input voltage variations. This reason makes the DC/DC converter as a good candidate to emulate a DC source in which the output voltage is tied to the supplied current like in a photovoltaic generator. To this aim, the control algorithm has to take into account the converter behavior; this is the subject of this chapter in which three main topologies of DC/DC converters are studied to define their static and dynamic performance.

---

[1] A vacuum tube requires a voltage of hundreds of voltage, in a car only the battery voltage of 12 V is available; as a consequence, in some luxury cars a switching circuit "ante litteram" was employed. This circuit, with a suitable vacuum tube, generates a square wave from the 12 V supply, this voltage is increased by a transformer and again rectified to obtain a high DC voltage for the radio receiver. This principle, adopting modern devices and transformers, is still utilized in some DC/DC converters, however, this is out of the scope of this book.

M. C. Di Piazza and G. Vitale, *Photovoltaic Sources*, Green Energy and Technology, DOI: 10.1007/978-1-4471-4378-9_7, © Springer-Verlag London 2013

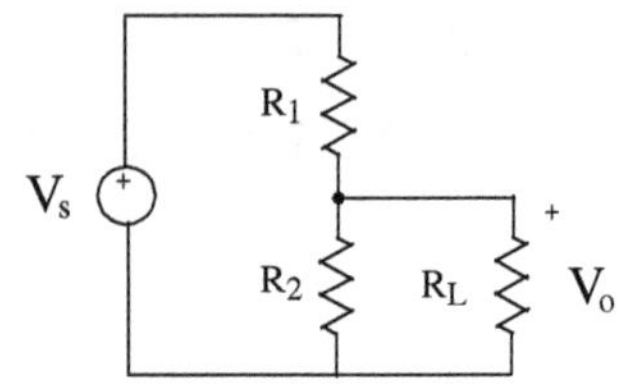

Fig. 7.1 Voltage divider supplied by a DC source

7.2 Linear Conversion

7.2.1 Linear Conversion by Voltage Divider

The simplest way to obtain a DC voltage by a DC source with a different voltage level consists on using a voltage divider, as shown in Fig. 7.1.

The output voltage is given by:

$$V_o = \frac{R_{2L}}{R_1 + R_{2L}} V_s \tag{7.1}$$

where $R_{2L} = R_2 R_L/(R_2 + R_L)$

The input and output powers are respectively given by:

$$P_s = \frac{V_s^2}{R_1 + R_{2L}} \tag{7.2}$$

$$P_o = \frac{V_o^2}{R_L} \tag{7.3}$$

By using Eq. 7.1 the efficiency conversion is obtained.

$$\eta = \frac{P_o}{P_s} = \frac{V_o^2}{R_L} \frac{R_1 + R_{2L}}{V_s^2} = \frac{R_{2L}^2}{R_L(R_1 + R_{2L})} \tag{7.4}$$

The output voltage does not depend on the load resistance value if $R_L \gg R_2$; in this case, the efficiency is equal to:

$$\eta_{R_L \gg R_2} = \frac{R_2^2}{R_L(R_1 + R_2)} \tag{7.5}$$

The obtained value is very low, it is due to the presence of the term $R_2/R_L \ll 1$.

By setting $R_1 = R$, $R_2 = R_L = 2R$, an output voltage equal to the input voltage divided by two is obtained, in this case $\eta = 0.25$.

From what explained above, it is clear that a DC conversion by a voltage divider presents some drawbacks:

- A DC voltage higher than the input voltage cannot be obtained;
- The output voltage depends on the load, in general;
- The efficiency is very poor.

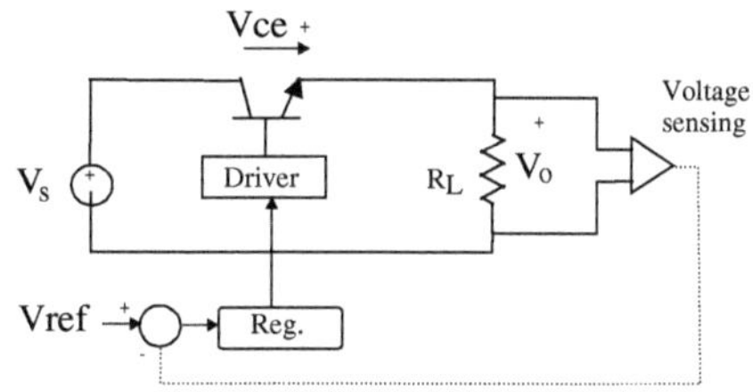

Fig. 7.2 Linear conversion circuit by a series regulator

Finally, it can be easily verified that the percent variation of the input voltage implies the same percent variation of the output voltage; it means that for a voltage divider:

$$\frac{\partial V_o/V_o}{\partial V_s/V_s} = 1 \tag{7.6}$$

Voltage dividers are however, utilized, for example, for voltage sensing applications where the power losses are negligible.

7.2.2 Linear Conversion by Series Regulator

In a linear conversion by series regulators the output voltage, lower than the input one, is obtained subtracting a voltage by the input generator.

The voltage to be subtracted is obtained by a power BJT, that is series connected between the source and the load, as collector to emitter voltage. To this aim, a suitable voltage is applied to the base of the BJT; in general this voltage is derived by a feedback action: an error signal is obtained by comparing the desired voltage with the real voltage and it is processed by a regulator. A principle circuit is sketched in Fig. 7.2.

Once the stability is assured by suitably designing the regulator, the load current will flow through the power BJT and the output voltage is given by:

$$V_o = V_s - V_{ce} \tag{7.7}$$

Being the current the same for the supply generator, the power BJT and the load, the same relationship holds for the power balance:

$$P_s = P_{BJT} + P_o \tag{7.8}$$

and the efficiency is given by:

$$\eta = \frac{P_o}{P_s} = \frac{V_o}{V_s} \tag{7.9}$$

In general, the described circuit has the advantage of a regulation of the output voltage; however, only a step down conversion is possible and the efficiency

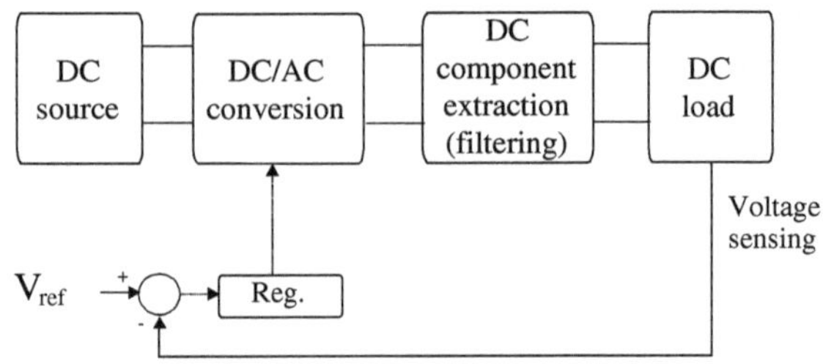

Fig. 7.3 Block diagram of a switching converter

remains low because all the power supplied by the source that it is not utilized by the load have to be dissipated by the power BJT. Then, this kind of circuit is utilizable only for low power.

7.3 Switching Conversion

In the above described circuits the main drawback, i.e., the low efficiency is due to the presence of a series connected device that subtracts a voltage by the source so that the remaining voltage represents the desired one. In this device (either the resistor R_1 in the voltage divider or the BJT in linear regulator), both voltage and current exist at the same time causing loss dissipation.

Switching conversion, on the contrary, is based on the use of a power electronic switch used in switching operation, it means the presence of only two fundamental states: the *on* state in which the voltage of the power switch is null and its current is imposed by external circuitry and the *off* state in which the current of the power switch is null and its voltage is imposed by external circuitry. In these cases, the dissipated power is null being null the product of current and voltage in the device. Only during transitions between *on* and *off* states a low amount of power is dissipated; this quantity depends on the nonnull time required to pass from a status to the other time and it is proportional to the switching frequency, nevertheless the overall efficiency is expected to be very high.

A block diagram of a switching converter is shown in Fig. 7.3. The DC source supplies the DC/AC conversion unit in which a periodical waveform is generated. The DC component is extracted by filtering and applied to the load. In general, the DC/AC conversion stage has a control input by which the output value can be regulated. For feedback systems, the obtained output value is compared with the reference one, the error is processed by a regulator whose output is applied to the control input of the DC/AC converter.

It should be noted that the DC/AC conversion implies the generation of harmonics that are not present in the supply, therefore it requires a nonlinear device and, in principle, the whole circuit has to be studied as a nonlinear circuit.

The power switch is operated periodically and the ratio of the time in which the switch is in *on* state divided by the switching period represents the control input of

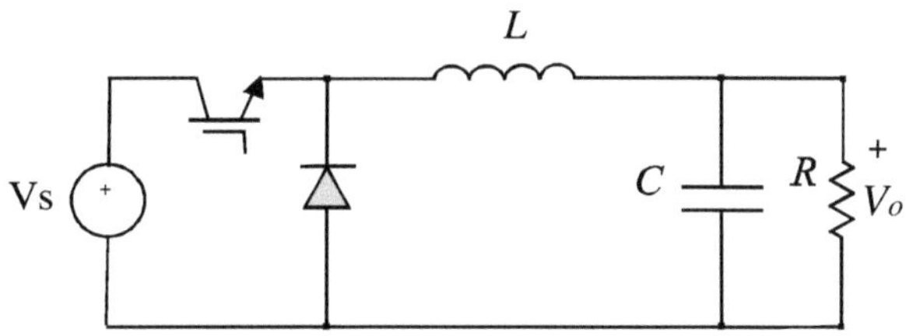

Fig. 7.4 Electrical circuit of a buck converter

the DC/AC conversion stage. By this input, the output value can be changed both in open loop and by feedback action.

By a suitable choice of the circuit topology, an output voltage with amplitude greater or smaller or both compared to the input voltage value can be obtained.

7.4 Buck Converter

The buck converter allows a DC voltage lower than the input voltage to be obtained. The electric circuit is shown in Fig. 7.4; the power switch is represented as an IGBT but other components as MOSFETs can be utilized.

With reference to Fig. 7.3, the DC/AC conversion stage is represented by the IGBT and the diode, the filter is a second-order one with an inductor and a capacitor, the load is assumed as purely resistive.

7.4.1 Continuous Operating Mode

The switch is operated in *on/off* state periodically with period T_s. When the switch is in *on* state, the diode is inversely biased and the circuit operates as the scheme of Fig. 7.5. When the switch is in *off* state, the supply voltage is not connected and the diode is directly biased: the circuit operates as the scheme of Fig. 7.6.

The following analysis is performed under the hypothesis that the current in the inductance flows continuously (this operating condition is defined as continuous operating mode) and by neglecting the parasitic resistive components of the inductor and of the capacitor.

The circuit can be studied as a succession of two linear circuits, one corresponding to the *on* state in a time interval T_{ON}, and the other corresponding to the *off* state in a time interval T_{OFF}. It follows that $T_{ON} + T_{OFF} = T_s$; the ratio $T_{ON}/T_s = D$ is indicated as the *duty cycle* of the circuit.

Finally, two sets of equations can be written for each state (the current and voltage in passive dipoles are chosen considering the dissipated power as positive).

During the *on* state the circuit is described by the equations:

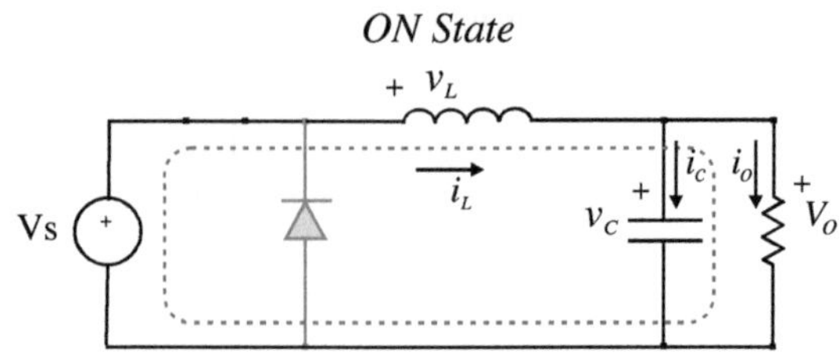

Fig. 7.5 Buck converter equivalent circuit in *on* state

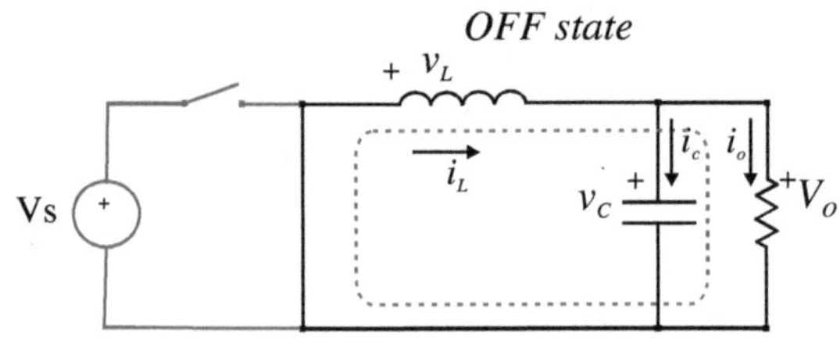

Fig. 7.6 Buck converter equivalent circuit in *off* state

$$\begin{cases} V_s = L\dfrac{\mathrm{d}i_L}{\mathrm{d}t} + v_c \\ V_o = v_c \\ i_L = i_c + i_o \end{cases} \tag{7.10}$$

During the *off* state the corresponding equations are achievable by Eq. 7.10 setting $V_s = 0$.

$$\begin{cases} 0 = L\dfrac{\mathrm{d}i_L}{\mathrm{d}t} + v_c \\ V_o = v_c \\ i_L = i_c + i_o \end{cases} \tag{7.11}$$

During T_{ON}, the inductor voltage is constant and equal to $V_s - V_o$, that is a positive value, its current rises linearly and flows toward the load. During T_{OFF}, the inductance voltage is $-V_o$, the current decreases linearly but flows always toward the load. The mean current of the inductance corresponds to the load current. All representative waveforms are sketched in Fig. 7.7, where the notation $\langle i_L \rangle$ stands for the current average value.

The static gain of the circuit can be obtained by considering that, in steady state condition, it holds that:

$$i_L(0) = i_L(T_s) \tag{7.12}$$

By integrating during T_s the inductance equation it follows that:

$$\int_0^{Ts} \mathrm{d}i_L = 0 = \int_0^{Ts} \frac{1}{L} v_L \mathrm{d}t \tag{7.13}$$

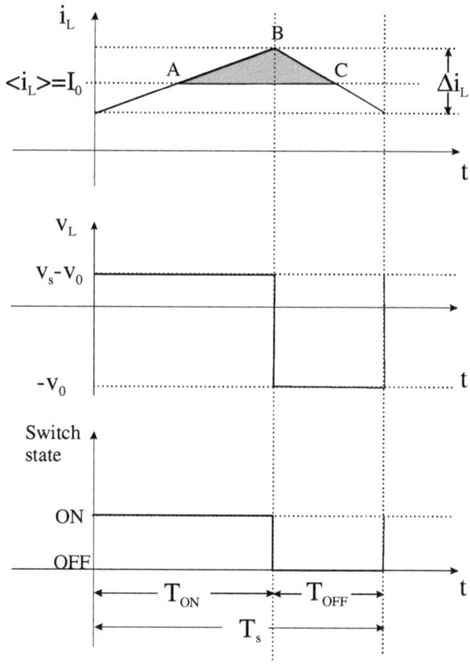

Fig. 7.7 Time domain waveforms in continuous operating mode during a switching period: inductor current (*top*), inductor voltage (*middle*), switch state (*bottom*)

The first integral is null through the Eq. 7.12 and the last member of Eq. 7.13 gives:

$$(V_s - V_o)T_{\mathrm{ON}} = V_o T_{\mathrm{OFF}} = V_o(T_s - T_{\mathrm{ON}}) \tag{7.14}$$

Finally, by considering the first and the last term:

$$\frac{V_o}{V_s} = D \tag{7.15}$$

The Eq. 7.15 says that the static gain of the buck converter is equal to the duty cycle. The output voltage is such that

$$V_s \leq V_o \leq 0 \tag{7.16}$$

The Eq. 7.16 contains equalities, they correspond to $D = 0$ and $D = 1$, however in this case and no switching operation occurs.

Assuming a loss-less circuit, the input supplied by the source is totally given to the load and the current gain is obtained:

$$\frac{I_o}{I_s} = \frac{1}{D} \tag{7.17}$$

In continuous operating mode, the current variation on the inductance results as a variation of the capacity charge while the DC component is given to the load.

With reference to Fig. 7.7, the increase of the charge in a period corresponds to the triangle ABC and the voltage variation on the load is given by:

$$\Delta V_o = \frac{\Delta Q}{C} = \frac{1}{C}\frac{1}{2}\frac{\Delta I_L}{2}\frac{T_s}{2} \tag{7.18}$$

The current variation in Eq. 7.18 can be evaluated during T_{OFF} as:

$$\Delta I_L = \frac{V_o}{L} T_{\mathrm{OFF}} = \frac{V_o}{L}(1-D)T_s \tag{7.19}$$

By substituting Eq. 7.19 into Eq. 7.18:

$$\Delta V_o = \frac{T_s}{8C}\frac{V_o}{L}(1-D)T_s \tag{7.19a}$$

The relative voltage variation is given by:

$$\frac{\Delta V_o}{V_o} = \frac{T_s^2}{8}\frac{1}{LC}(1-D) = \frac{\pi^2}{2}(1-D)\left(\frac{f_c}{f_s}\right)^2 \tag{7.20}$$

where:

$$f_c = \frac{1}{2\pi\sqrt{LC}} \tag{7.21}$$

The output voltage ripple could be minimized by choosing $f_c \ll f_s$. It depends on the duty cycle and it is maximum when $D \approx 0$

It should be noted that the above described analysis has been performed under the hypothesis that the output voltage is lower than the input one and that its ripple can be treated as a second-order approximation. These hypotheses are verified at the end of the analysis. It is a typical approach adopted for switching converters that, although being nonlinear circuits, they admit a unique solution.

7.4.2 Discontinuous Operating Mode

In the last section, it has been shown that the inductor current, during T_{OFF}, decreases linearly and that, its slope depends on the inductance value. A low inductance value corresponds to a low stored energy: when, during T_{OFF}, this energy ends the current reach the null value and from this time to the end of T_s the load is supplied only by the capacitor. In this case, the discontinuous operating mode occurs.

The inductor current, in the transition between continuous and discontinuous mode, is sketched in Fig. 7.8, where the notation $<i_L>$ stands for the current average value.

In particular, Fig. 7.8a shows the inductor current in continuous mode: the load current is high and the inductor current does not reach null value. Figure 7.8b

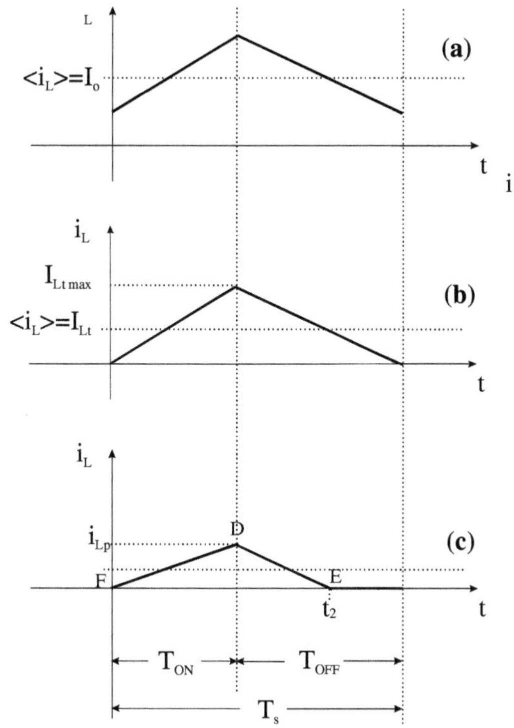

Fig. 7.8 Current waveforms in transition from continuous to discontinuous operating mode

shows the threshold in which the inductor current is null just at the end of T_s. For lower values of the load current the situation of Fig. 7.8c occurs, it corresponds to the discontinuous conduction mode.

This situation suggests that, during running conditions, the buck converter can change its operating mode depending on load variations. A high value of load resistance causes a low mean load current that can lead to the discontinuous conduction mode.

With reference to Fig. 7.8b the threshold current I_{Lt} can be calculated as:

$$I_{\text{Lt}} = \frac{1}{2} I_{\text{Lt max}} = \frac{1}{2L} T_{\text{ON}} (V_s - V_o) = \frac{DT_s}{2L} (V_s - V_o) \tag{7.22}$$

From Eq. 7.22 two expressions for I_{Lt}, can be obtained in case V_s or V_o is, respectively, constant:

$$I_{\text{Lt}} = \frac{V_s T_s}{2L} D(1 - D) \tag{7.23a}$$

$$I_{\text{Lt}} = \frac{V_o T_s}{2L} (1 - D) \tag{7.23b}$$

The inductance value which defines the threshold is defined as critical inductance L_C, it corresponds to a load current I_{Lc}. This value can be calculated neglecting the power losses in the converter. The supplied power is given by:

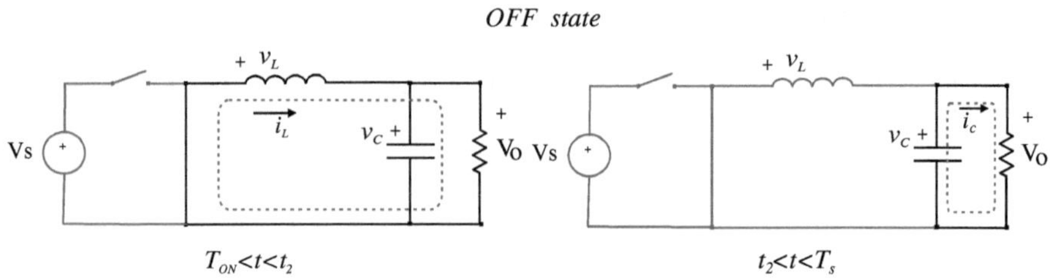

Fig. 7.9 Equivalent circuits in discontinuous operating mode during T_{OFF}

$$P_s = V_s I_s = V_s I_{\mathrm{Lc}} D \tag{7.24}$$

where the supplied current is considered equal to the load current multiplied by the duty cycle and the load current is equal to the mean inductance current I_{Lc}.

The load power is given by:

$$P_o = \frac{V_o^2}{R} \tag{7.25}$$

By equating Eq. 7.24 and Eq. 7.25, using for the mean inductance current Eq. 7.22 and considering that the converter operation is loss-less, the following equation is obtained:

$$V_s I_{\mathrm{Lc}} D = V_s \frac{(V_s - V_o)D^2}{2f_s L_c} = \frac{V_o^2}{R} \tag{7.26}$$

By taking into account Eq. 7.15, it follows that:

$$L_c = \frac{R(1-D)}{2f_s} \tag{7.27}$$

From Eq. 7.27 it is evident that the critical inductance is proportional to the load and inversely proportional to the switching frequency. If the inductance value is fixed, the critical load resistance can be defined as:

$$R_c = \frac{2f_s L}{(1-D)} \tag{7.28}$$

During T_{OFF} two equivalent circuits are recognizable. The first one holds for $T_{\mathrm{ON}} < t < t_2$, where t_2 is the time instant in which the inductor current reaches the null value, and the second holds for $t_2 < t < T_s$ in which the inductance current is null and the load is supplied only by the capacitor. The two circuits are shown in Fig. 7.9.

A new relationship for the static gain has to be calculated. The mean current on the load can be calculated by integrating the inductance current for $0 < t < t_2$ (it corresponds to the triangle area DEF in Fig. 7.8) and dividing for the switching period, this value is equal to the load voltage divided by its resistance.

$$\frac{1}{T_s}\int_0^{t_2} i_L(t)\mathrm{d}t = \frac{1}{2}I_{\mathrm{Lp}}t_2 f_s = \frac{V_o}{R} \tag{7.29}$$

where I_{Lp} is the maximum value of the inductor current. Moreover, considering that:

$$(V_s - V_o) = L\frac{I_{\mathrm{Lp}}}{T_{\mathrm{ON}}} \tag{7.30}$$

Rewriting Eq. 7.14 as:

$$(V_s - V_o)T_{\mathrm{ON}} = V_o(t_2 - T_{\mathrm{ON}}) \tag{7.31}$$

It follows that:

$$V_s T_{\mathrm{ON}} f_s - V_o t_2 f_s = 0 \tag{7.32}$$

Now, obtaining the value $t_2\, f_s$ from Eq. 7.29, the value I_{Lp} from Eq. 7.30, the value of R from Eq. 7.27, substituting them into Eq. 7.32 and solving for V_o, the following second-order equation is achieved:

$$V_o^2 + V_o\left[V_s\frac{D^2 L_c}{(1-D)L}\right] - \left[V_s^2\frac{D^2 L_c}{(1-D)L}\right] = 0 \tag{7.33}$$

The solution of Eq. 7.33 represents the buck converter gain in discontinuous operating mode ($L < L_c$).

$$\frac{V_o}{V_s} = \frac{D\left(\sqrt{D^2 + \frac{4L(1-D)}{L_c}} - D\right)}{\frac{2L(1-D)}{L_c}} \tag{7.34}$$

The Eq. 7.34 can be also expressed as:

$$\frac{V_o}{V_s} = \frac{2}{1 + \sqrt{1 + \frac{4L(1-D)}{D^2 L_c}}} \tag{7.35}$$

The derivative of the ratio between the output and input voltage calculated respect to the duty cycle is given by:

$$\frac{\partial}{\partial D}\left(\frac{V_o}{V_s}\right) = \frac{4\frac{L}{L_c}\frac{(2-D)}{D^3}}{\left[1 + \sqrt{1 + \frac{4L(1-D)}{D^2 L_c}}\right]^2\sqrt{1 + \frac{4L(1-D)}{D^2 L_c}}} \tag{7.36}$$

The time t_2 in which the inductance current becomes null is obtained by substituting Eq. 7.34 in Eq. 7.32.

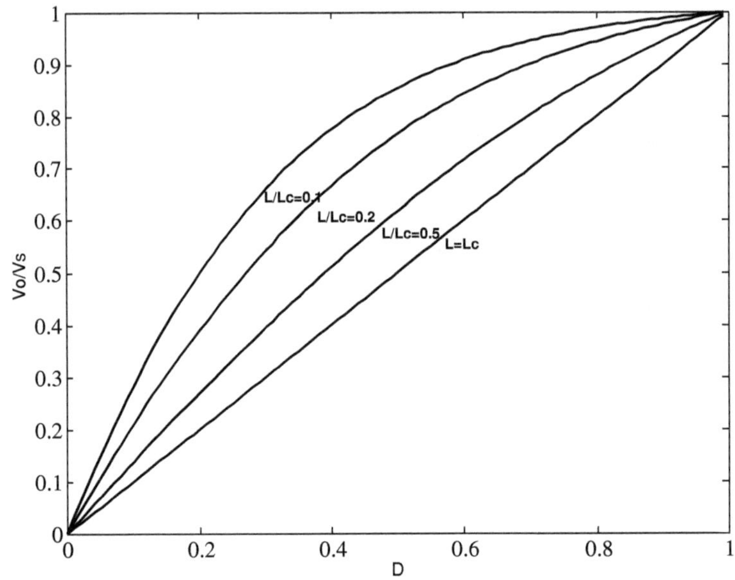

Fig. 7.10 V_o/V_s versus D in discontinuous conduction mode

$$t_2 = \frac{\frac{2L(1-D)}{DL_c f_s}}{\sqrt{1 + \frac{4L(1-D)}{L_c D^2}} - 1} \tag{7.37}$$

The ratio between the output and input voltage versus D is sketched in Fig. 7.10. A straight line is obtained for $L/L_c = 1$, in discontinuous mode, in general, for given D a higher value of V_o/V_s is obtained. In particular, the smaller is the ratio L/L_c the greater is the increase of V_o/V_s.

Figure 7.11 shows the derivative of the ratio between the output and input voltage. It should be noted that for low values of D a great variation of the derivative of the gain is observed, it implies that a small variation of D around a low value of D will produce a greater variation of the output voltage compared to the effect of the same small variation around a point with high D. The variation is as much higher as the ratio L/L_c is low.

Finally, Fig. 7.12 shows the quantity $\Delta = 1 - t_2/T_s$ versus D in discontinuous mode that corresponds to $(T_s - t_2)/T_s$; it is, in practice, the time interval in which there is no current in the inductor weighted by the switching period. It should be noted that, in continuous conduction mode, a null value is obtained as expected. In discontinuous mode, this interval is longer when the ratio L/L_c is lower.

As explained before, the buck converter exhibits different behavior in continuous or in discontinuous mode. In practical applications, the crossing from continuous and discontinuous (or vice versa) is avoided to preserve stability.

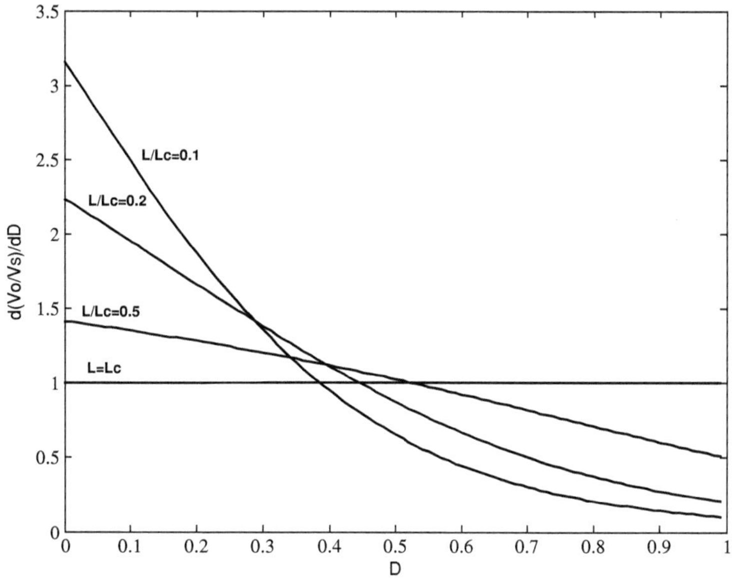

Fig. 7.11 Derivative of V_o/V_s versus D in discontinuous conduction mode

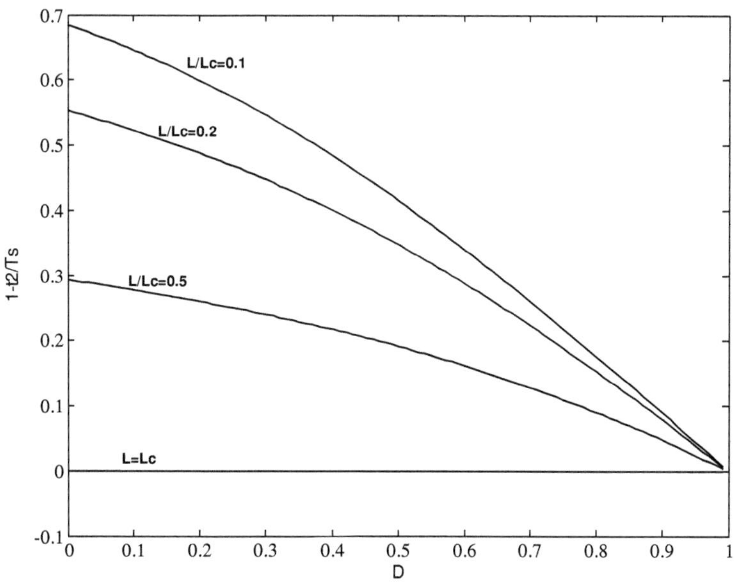

Fig. 7.12 Plot of $1 - t_2/T_s$ versus D in discontinuous conduction mode

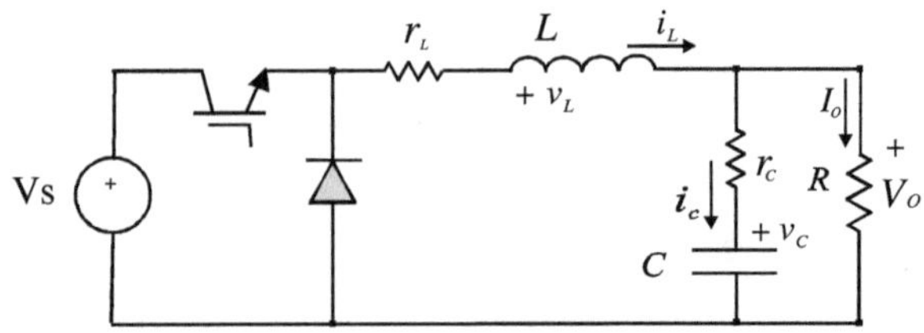

Fig. 7.13 Buck converter including parasitic parameters

The input current is always highly discontinuous, it rises from a null value to its maximum value in a very short time interval. This high value of di/dt, coupled with the parasitic inductance of the input circuit (including wires, printer circuit board etc.), causes voltage spikes on the switch. For this reason, an input capacitor is often placed near the power switch to avoid spikes propagation.

Finally, the current in the output capacitor is smoothed for the presence of the inductor and does not exhibit large variations.

7.4.3 Continuous Operating Mode Including Parasitic Parameters

In a real circuit, two additional parameters have to be taken into account: the parasitic resistance of the inductor r_L and the parasitic resistance of the capacitor r_C. The corresponding circuit is shown in Fig. 7.13.

A new set of equations can be written for the *on* state and for the *off* state respectively.

During T_{ON} the circuit equations are:

$$\begin{cases} V_s = L\dfrac{\mathrm{d}i_L}{\mathrm{d}t} + r_L i_L + r_c i_c + v_c \\ V_o = v_c + r_c i_c \\ i_L = i_c + I_o \end{cases} \tag{7.38}$$

It is clear that by neglecting both r_L and r_C the Eq. 7.10 is obtained.

By taking into account the constitutive equation of the capacitor and including the LKV equation, written for the loop formed by the load resistance and the capacitor including its parasitic resistance:

$$V_o = R\left(i_L - C\frac{\mathrm{d}v_c}{\mathrm{d}t}\right) \tag{7.39}$$

The Eq. 7.38 can be rewritten by using the state variables i_L, v_C, and the load resistance R.

$$\begin{cases} V_s = L\dfrac{di_L}{dt} + r_L i_L + R\left(i_L - C\dfrac{dv_c}{dt}\right) \\ 0 = -v_c - Cr_c\dfrac{dv_c}{dt} + R\left(i_L - C\dfrac{dv_c}{dt}\right) \end{cases} \tag{7.40}$$

Finally, by solving for the time derivative of the state variables:

$$\begin{cases} \dfrac{di_L}{dt} = -\dfrac{Rr_c + Rr_L + r_c r_L}{L(R + r_c)} i_L - \dfrac{R}{L(R + r_c)} v_c + \dfrac{1}{L} V_s \\ \dfrac{dv_c}{dt} = \dfrac{R}{C(R + r_c)} i_L - \dfrac{1}{C(R + r_c)} v_c \end{cases} \tag{7.41}$$

The output voltage can be expressed versus the state variables as well:

$$V_0 = \frac{Rr_C}{R + r_C} i_L + \frac{R}{R + r_C} v_C \tag{7.42}$$

The corresponding equations during T_{OFF} are obtained by Eq. 7.41 by setting V_s equal to zero.

$$\begin{cases} \dfrac{di_L}{dt} = -\dfrac{Rr_c + Rr_L + r_c r_L}{L(R + r_c)} i_L - \dfrac{R}{L(R + r_c)} v_c \\ \dfrac{dv_c}{dt} = \dfrac{R}{C(R + r_c)} i_L - \dfrac{1}{C(R + r_c)} v_c \end{cases} \tag{7.43}$$

These last three equations can be well approximated neglecting the product $r_L r_C$ and taking into account that $R \gg r_L$ and $R \gg r_C$:

$$\begin{cases} \dfrac{di_L}{dt} = -\dfrac{r_c + r_L}{L} i_L - \dfrac{1}{L} v_c + \dfrac{1}{L} V_s \\ \dfrac{dv_c}{dt} = \dfrac{1}{C} i_L - \dfrac{1}{CR} v_c \end{cases} \tag{7.41a}$$

$$V_0 = r_C i_L + v_C \tag{7.42a}$$

$$\begin{cases} \dfrac{di_L}{dt} = -\dfrac{r_c + r_L}{L} i_L - \dfrac{1}{L} v_c \\ \dfrac{dv_c}{dt} = \dfrac{1}{C} i_L - \dfrac{1}{CR} v_c \end{cases} \tag{7.43a}$$

7.5 Boost Converter

The DC/DC boost converter allows an output voltage higher than the input one to be achieved. In such a converter, there are the same components of the buck

Fig. 7.14 Boost converter

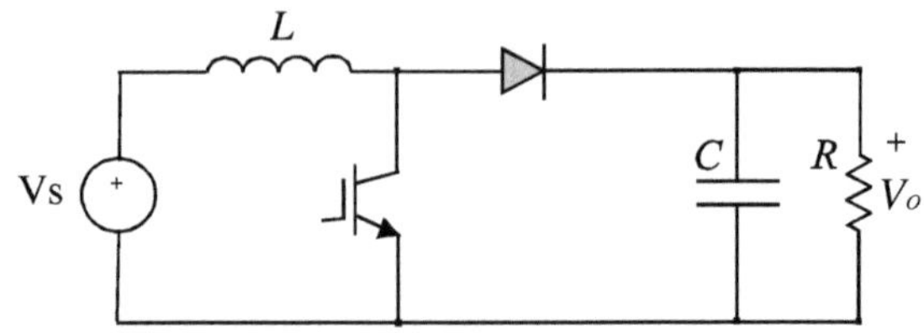

Fig. 7.15 Boost converter equivalent circuit in *on* state

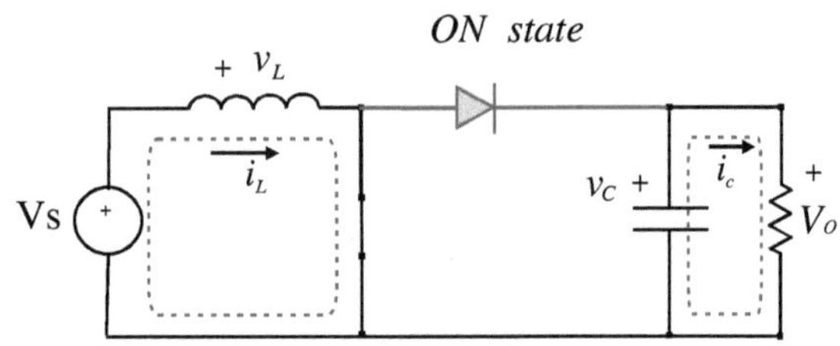

Fig. 7.16 Boost converter equivalent circuit in *off* state

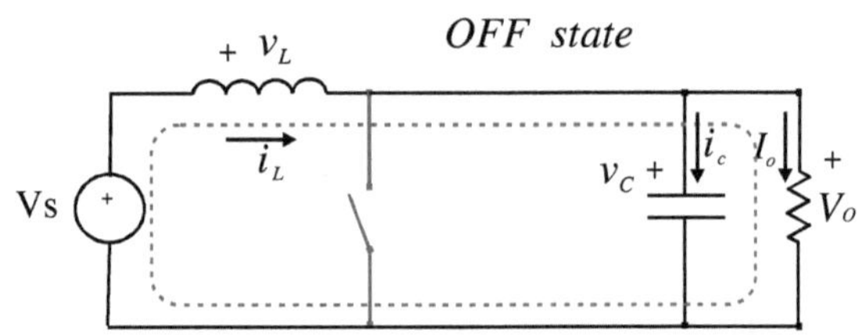

converter but with different topology. The voltage generator is connected to the inductor that, in turn, is connected by a power switch to the zero voltage reference and by a diode to a parallel formed by the load resistor and a capacitor. Differently from the buck converter, the voltage source generator supplies continuously to the converter, and the current presents a smaller ripple.

The circuit of the boost converter is shown in Fig. 7.14.

7.5.1 Continuous Operating Mode

When the power switch is in *on* state, the diode is reverse biased and the load is supplied only by the capacitor. When the power switch is in *off* state, the voltage generator supplies the inductor and the capacitor parallel connected to the load through the diode.

As in the buck converter, two different circuits, belonging to *on* and *off* state are respectively recognizable.

During the *on* state, the corresponding circuit, shown in Fig. 7.15, is described by the following equations:

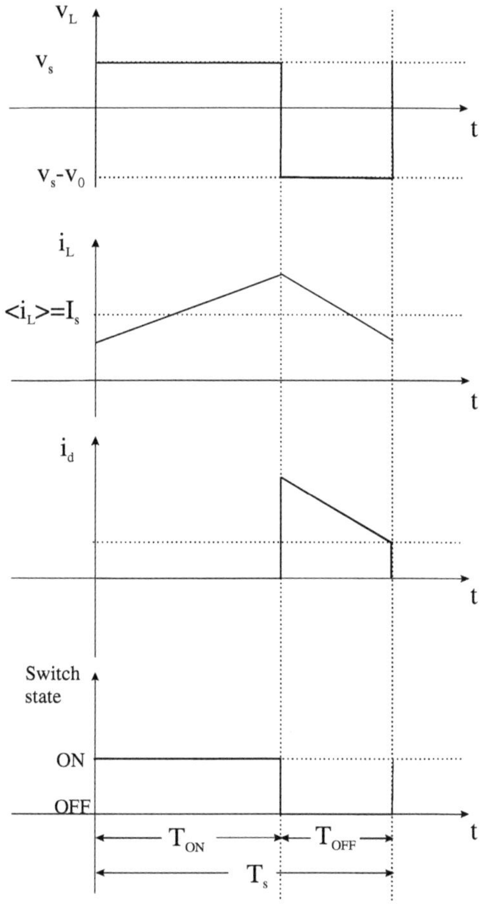

Fig. 7.17 Time domain waveforms in continuous conduction mode from *top* to *bottom*: inductor voltage, inductor current, diode current(i_d), switch state

$$\begin{cases} V_s = L\dfrac{di_L}{dt} \\ V_o = v_c \end{cases} \tag{7.44}$$

During the *off* state, the corresponding circuit, shown in Fig. 7.16, is described by the following equations:

$$\begin{cases} V_s = L\dfrac{di_L}{dt} + v_c \\ V_o = v_c \\ i_L = i_c + I_o \end{cases} \tag{7.45}$$

During T_{ON}, the inductor voltage is constant and equal to V_s, its current rises linearly. During T_{OFF}, the inductor voltage is $V_s - V_o$, its current decreases linearly and flows toward the load.

The mean current of the inductance corresponds to the input current. It should be noted that this situation is different from the buck converter in which the mean current of the inductance corresponds to the output current.

All representative waveforms are sketched in Fig. 7.17, where the notation $< I_L >$ stands for the current average value.

The static gain of the circuit can be obtained by considering that, in steady state condition, it is $i_L(0) = i_L(T_s)$. As a consequence, by integrating during T_s the inductor Eq. 7.13, it follows that:

$$V_s T_{ON} = (V_o - V_s) T_{OFF} \tag{7.46}$$

From which:

$$\frac{V_o}{V_s} = \frac{1}{1 - D} \tag{7.47}$$

Assuming a loss-less circuit, the ratio between input and output current is given by:

$$\frac{I_o}{I_s} = 1 - D \tag{7.48}$$

The output voltage ripple can be calculated assuming that the current ripple of the diode flows through the capacitor and the average value through the load. By calculating the charge variation in the capacitor during T_{ON}, it is possible to obtain:

$$\Delta V_o = \frac{\Delta Q}{C} = \frac{I_o T_{ON}}{C} = \frac{V_o}{R} \frac{T_s D}{C} \tag{7.49}$$

from which:

$$\frac{\Delta V_o}{V_o} = \frac{D}{f_s RC} \tag{7.50}$$

It should be noted that, differently from the buck converter, the output voltage ripple depends on the load value. In general, this value is greater compared with the buck because in boost converter the inductor is employed as energy storage and it does not take part in output filtering.

The peak to peak inductor current can be calculated by integrating the inductor equation during T_{ON}:

$$\Delta I_L = \frac{V_s}{L} T_{ON} = \frac{V_s}{L} D T_s \tag{7.51}$$

The same variation is required to the voltage generator, this variation is greater compared to the required one in the buck converter. As a matter of fact, in this last circuit maintaining the same value of $V_s T_s / L$, in buck converter it depends on a factor $D(1 - D)$ as it can be deduced from Eq. 7.19.

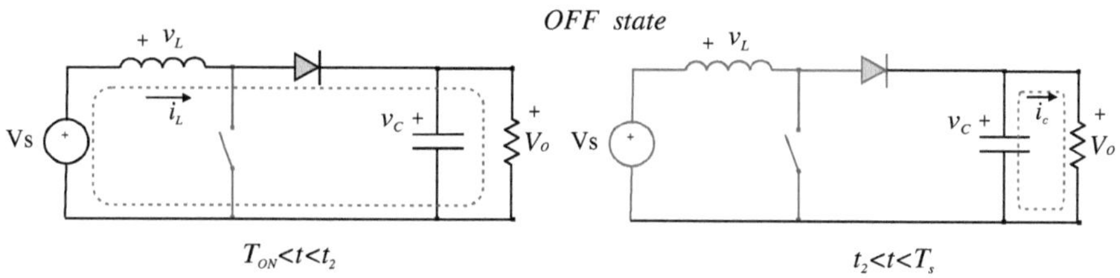

Fig. 7.18 Equivalent circuits in discontinuous operating mode during T_{OFF}

Fig. 7.19 Time domain waveforms in discontinuous conduction mode: inductor voltage (*top*), inductor current (*middle*), switch state (*bottom*)

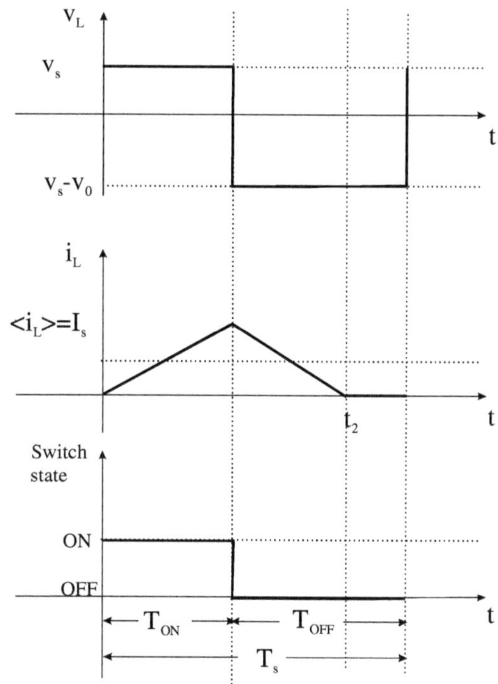

7.5.2 Discontinuous Operating Mode

In discontinuous mode, the inductor current becomes null at time t_2 with $T_{\mathrm{ON}} < t_2 \leq T_s$. In this case, during T_{OFF}, two equivalent circuits are recognizable as shown in Fig. 7.18. For $T_{\mathrm{ON}} < t \leq t_2$ the circuit behaves as in continuous mode. For $t_2 < t \leq T_s$ the inductance stored energy is null and the current is null as well. The load is supplied only by the capacitor. The corresponding waveforms are sketched in Fig. 7.19.

At the boundary between continuous and discontinuous mode $t_2 = T_s$ and the corresponding threshold current is given by:

$$I_{\mathrm{Lt}} = \frac{\Delta I_L}{2} = \frac{V_s D}{2L_c f_s} \tag{7.52}$$

where L_c is the critical inductance.

Assuming that the input power is equal to the output power:

$$V_s I_s = V_s I_{\mathrm{Lt}} = \frac{V_o^2}{R} \tag{7.53}$$

The critical inductance can be calculated by substituting Eq. 7.52 in Eq. 7.53 considering that the mean input current is equal to the average inductance current, I_{Lt} and taking into account Eq. 7.47.

$$L_c = \frac{RD(1-D)^2}{2f_s} \tag{7.54}$$

For given load resistance and switching frequency, the value of L_c must be chosen greater than the maximum value that can be reached by Eq. 7.54 for varying D to assure the continuous operation mode.

The critical load resistance is defined as:

$$R_c = \frac{2f_s L_c}{D(1-D)^2} \tag{7.55}$$

So that, when the load resistance is greater than the critical resistance, the operation becomes discontinuous.

To calculate the static gain during discontinuous mode, Eq 7.46 can be recalculated as:

$$V_s T_{\mathrm{ON}} + (V_s - V_o)(t_2 - T_{\mathrm{ON}}) = 0 \tag{7.56}$$

Considering that the average input current is equal to the average inductor current:

$$I_L = \frac{I_{L\max}}{2} = \frac{V_s D}{2Lf_s} \tag{7.57}$$

where $I_{L\max}$ is the maximum value of the inductor current.

The output current is obtained by weighting the inductor current during $(t_2 - T_{\mathrm{ON}})$

$$I_o = \frac{V_o}{R} = I_L \frac{t_2 - T_{\mathrm{ON}}}{T_s} \tag{7.58}$$

By Eqs. 7.57, 7.58, and 7.54 it follows that:

$$\frac{t_2 - T_{\mathrm{ON}}}{T_s} = \frac{V_o}{R}\frac{2Lf_s}{V_s D} = \frac{V_o}{V_s}\frac{L}{L_c}(1-D)^2 \tag{7.59}$$

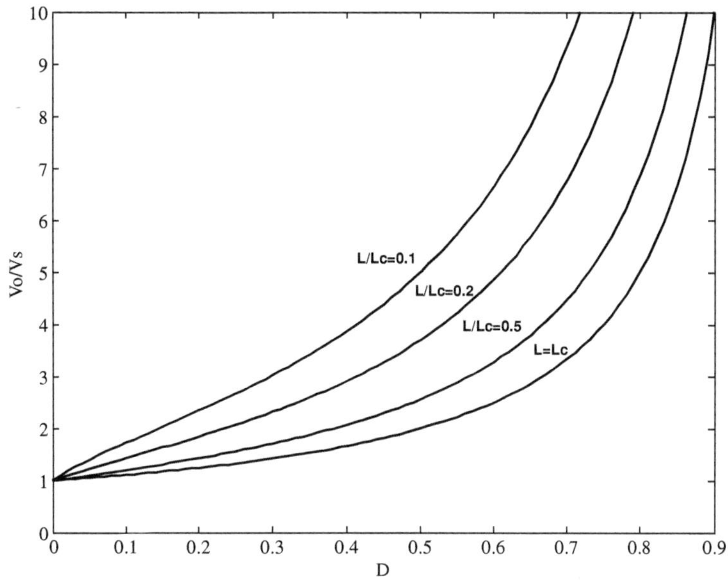

Fig. 7.20 Gain versus duty cycle in discontinuous mode

The Eq. 7.59 now can be substituted into Eq. 7.56 obtaining:

$$V_s^2 D - (V_o - V_s)V_o \frac{L}{L_c}(1-D)^2 = 0 \tag{7.60}$$

The Eq. 7.60 can be rearranged as a second-order equation:

$$\left(\frac{V_o}{V_s}\right)^2 - \left(\frac{V_o}{V_s}\right) - \frac{L_c}{L}\frac{D}{(1-D)^2} = 0 \tag{7.61}$$

whose solution gives the static gain in discontinuous mode:

$$\frac{V_o}{V_s} = \frac{1 + \sqrt{1 + \frac{4DL_c}{L(1-D)^2}}}{2} \tag{7.62}$$

From Eq. 7.59 the time in which the current is null can be obtained by using Eq. 7.62.

$$\frac{t_2}{T_s} = D + \frac{1 + \sqrt{1 + \frac{4DL_c}{L(1-D)^2}}}{2}\frac{L}{L_c}(1-D)^2 \tag{7.63}$$

Figure 7.20 shows the gain versus the duty cycle in discontinuous mode. It should be noted that for low values of the ratio L/L_c a great boosting is obtained with the same value of D.

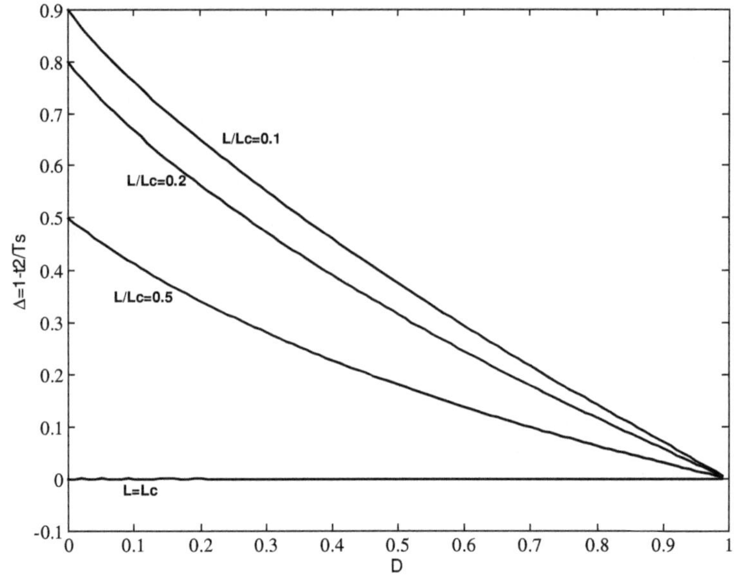

Fig. 7.21 Δ versus duty cycle in discontinuous mode

Fig. 7.22 Boost converter including parasitic parameters

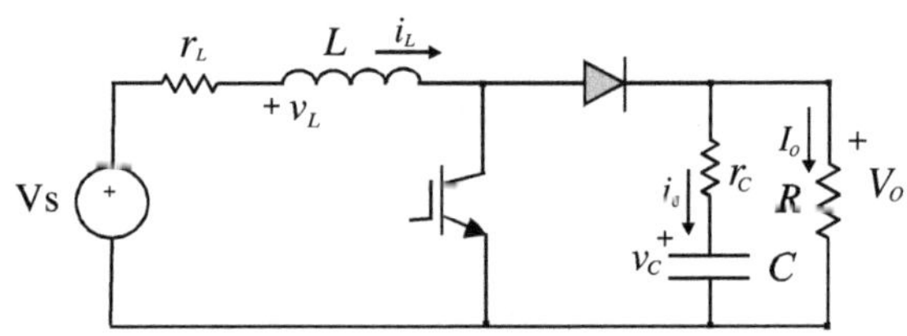

Figure 7.21 shows the quantity $\Delta = 1 - \frac{t_2}{T_s}$ versus D. This represents the time interval in which the inductor current is null weighted by T_s. In continuous mode Δ is null, in discontinuous mode for the same value of D, it is as much higher as lower is the ratio L/L_c because it corresponds to a low stored energy in the inductor.

7.5.3 Continuous Operating Mode Including Parasitic Parameters

The circuit of the boost converter including parasitic resistances of the capacitor and of the inductor is shown in Fig. 7.22.

During T_{ON} this circuit is described by the following equations:

$$\begin{cases} V_s = L\dfrac{\mathrm{d}i_L}{\mathrm{d}t} + r_L i_L \\ v_c + r_c C\dfrac{\mathrm{d}v_c}{\mathrm{d}t} = -RC\dfrac{\mathrm{d}v_c}{\mathrm{d}t} \end{cases} \tag{7.64}$$

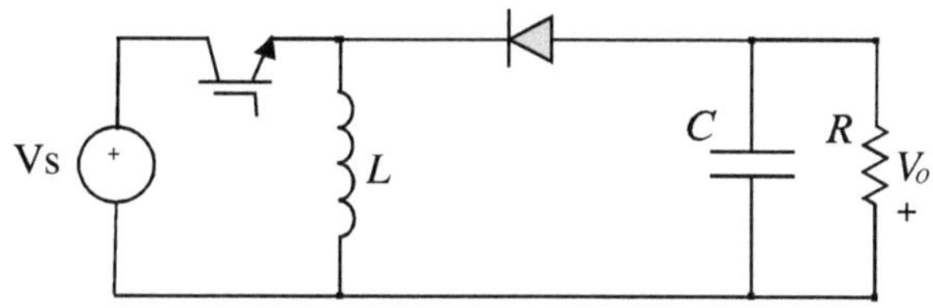

Fig. 7.23 Buck-boost converter

Solving respect to the state variables derivative, considering the capacitor equation and that $R \gg r_c$ the following equations are obtained:

$$\begin{cases} \frac{\mathrm{d}i_L}{\mathrm{d}t} = -\frac{r_L}{L} i_L + \frac{1}{L} V_s \\ \frac{\mathrm{d}v_c}{\mathrm{d}t} = -\frac{v_c}{RC} \end{cases} \tag{7.65}$$

The output voltage can be obtained versus the capacitor voltage as:

$$V_o = v_c + r_c \frac{V_o}{R} \tag{7.66}$$

This last expression can be further approximated:

$$V_o = v_c \left(\frac{R}{R - r_c} \right) \cong v_c \tag{7.67}$$

During T_{OFF} the equivalent circuit is the same of the buck converter in *on* state; as a consequence, the corresponding equations are equal to Eq. 7.41a; it follows that the boost converter, during T_{OFF}, considering its parasitic parameters, is described by the equations:

$$\begin{cases} \frac{\mathrm{d}i_L}{\mathrm{d}t} = -\frac{r_c + r_L}{L} i_L - \frac{1}{L} v_c + \frac{1}{L} V_s \\ \frac{\mathrm{d}v_c}{\mathrm{d}t} = \frac{1}{C} i_L - \frac{1}{CR} v_c \end{cases} \tag{7.68}$$

7.6 Buck-Boost Converter

The buck-boost converter is used to obtain an output voltage either higher or lower than the input one. The topology is similar to that of the buck but the diode and the inductor are exchanged. The diode has the cathode in common with the inductor and with the power switch. The electrical scheme is illustrated in Fig. 7.23.

Differently from buck and boost converter, the output voltage is negative respect to the zero reference voltage.

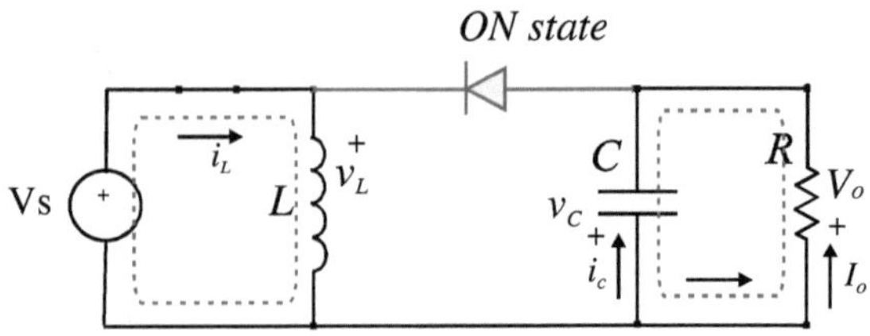

Fig. 7.24 Buck-boost converter equivalent circuit during T_{ON}

7.6.1 Continuous Operating Mode

When the power switch is in *on* state, the voltage source supplies the inductor, the diode is reverse biased and the load is supplied only by the capacitor.

The equivalent circuit is sketched in Fig. 7.24, it is described by the following equations:

$$\begin{cases} V_s = L\dfrac{\mathrm{d}i_L}{\mathrm{d}t} \\ V_o = v_c \end{cases} \tag{7.69}$$

It should be noted that, in this condition, the buck-boost presents the same circuit of the boost in *on* state.

During the *off* state, the inductor current flows through the parallel circuit formed by the capacitor and the load. The equivalent circuit is sketched in Fig. 7.25, it is described by the following equations:

$$\begin{cases} 0 = L\dfrac{\mathrm{d}i_L}{\mathrm{d}t} + v_c \\ V_o = v_c \\ i_L = i_c + I_o \end{cases} \tag{7.70}$$

The gain of the buck-boost converter can be calculated by integrating the inductor Eq. 7.13 during T_s and taking into account that it is $i_L(0) = i_L(T_s)$.

$$V_s T_{ON} - V_o T_{OFF} = 0 \tag{7.71}$$

From which:

$$\frac{V_o}{V_s} = \frac{D}{1-D} \tag{7.72}$$

The Eq. 7.72 says that the output voltage is lower that the input one when the duty cycle is $D < 0.5$. As above explained, V_o has the sign opposite to V_s.

The output versus input current ratio, neglecting the power losses, is given by:

$$\frac{I_o}{I_s} = \frac{1-D}{D} \tag{7.73}$$

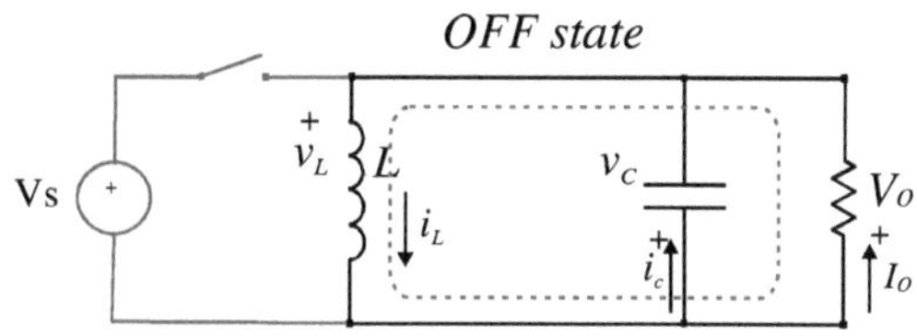

Fig. 7.25 Buck-boost converter equivalent circuit during T_{OFF}

The output voltage ripple is given by the charge variation in the capacitor; this value can be calculated as the product of the load current and T_{ON}.

$$\Delta V_o = \frac{\Delta Q}{C} = \frac{I_o T_{\mathrm{ON}}}{C} = \frac{V_o}{R}\frac{T_s D}{C} \tag{7.74}$$

Finally, the relative voltage output variation is given by:

$$\frac{\Delta V_o}{V_o} = \frac{D}{f_s RC} \tag{7.75}$$

The inductor current ripple can be calculated by integrating the inductor equation during T_{ON}.

$$\Delta I_L = \frac{V_s}{L} T_{\mathrm{ON}} = \frac{V_s}{L} D T_s \tag{7.76}$$

Both the relative voltage output variation and the current ripple expressions are equal to those obtained for the boost converter.

7.6.2 Discontinuous Operating Mode

In discontinuous mode, the inductor current becomes null before the turning on of the power switch. At the threshold, this current is null only at the instant $t = T_s$. The corresponding inductor current I_{Lt} is given by:

$$I_{\mathrm{Lt}} = \frac{\Delta I_L}{2} = \frac{1}{2}\frac{1}{L} V_s T_{\mathrm{ON}} = \frac{V_s D}{2 L_c f_s} \tag{7.77}$$

where L_c is the critical inductance.

The mean current supplied by the voltage generator is equal to:

$$I_s = \frac{T_{\mathrm{ON}}}{T_s} I_L = D I_L \tag{7.78}$$

Now, by assuming that the input power is equal to the output power [see (7.53)]

$$V_s \frac{V_s D^2}{2 f_s L_c} = \frac{V_o^2}{R} \tag{7.79}$$

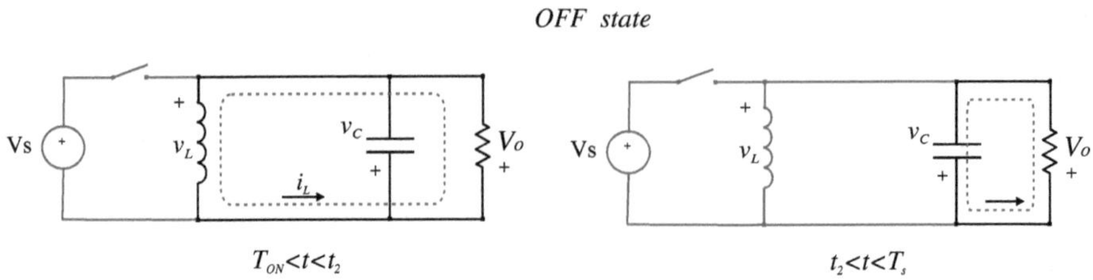

Fig. 7.26 Equivalent circuits in discontinuous operating mode during T_{OFF}

from which:

$$L_c = \frac{RD^2V_s^2}{2f_sV_o^2} = \frac{R(1-D)^2}{2f_s}. \tag{7.80}$$

The static gain in discontinuous mode is obtained by (7.79) with L in place of L_c.

$$\frac{V_o}{V_s} = \sqrt{\frac{RD^2}{2f_sL}} = D\sqrt{\frac{L_c}{L(1-D)^2}} \tag{7.81}$$

During T_{OFF}, in discontinuous mode, two circuits are recognizable: they are sketched in Fig. 7.26.

7.6.3 Continuous Operating Mode Including Parasitic Parameters

Considering the parasitic resistances of the capacitor and of the inductor in the buck-boost converter, the circuit of Fig. 7.27 is obtained.

During T_{ON}, the equivalent circuit is the same of the corresponding one of the boost during T_{ON} and the equations are the same of Eq. 7.65 rewritten here.

$$\begin{cases} \dfrac{\mathrm{d}i_L}{\mathrm{d}t} = -\dfrac{r_L}{L}i_L + \dfrac{1}{L}V_s \\ \dfrac{\mathrm{d}v_C}{\mathrm{d}t} = -\dfrac{v_C}{RC} \end{cases} \tag{7.82}$$

The output voltage versus the capacitor voltage can be approximated as:

$$V_o = v_C\left(\frac{R}{R-r_c}\right) \cong v_C \tag{7.83}$$

During T_{OFF}, the circuit is equivalent to the buck converter in *off* state, setting $V_s = 0$. By Eq. 7.38, with the position $V_s = 0$, the following equations are obtained:

Fig. 7.27 Buck-boost converter including parasitic parameters

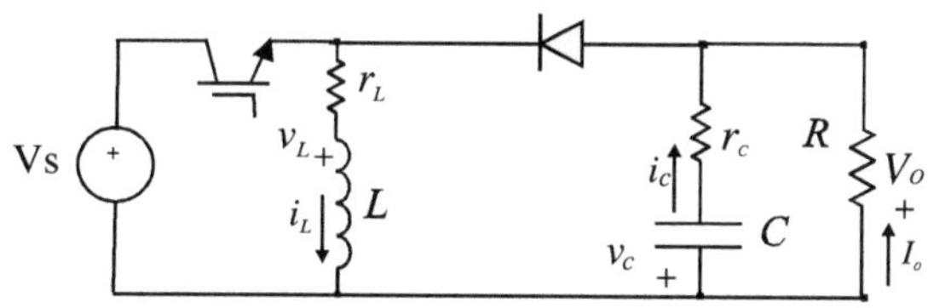

$$\begin{cases} 0 = L\dfrac{\mathrm{d}i_L}{\mathrm{d}t} + r_L i_L + r_c i_c + v_c \\ V_o = v_c + r_c i_c \\ i_L = i_c + I_o \end{cases} \tag{7.84}$$

Considering that

$$V_o = R\left(i_L - C\frac{\mathrm{d}v_c}{\mathrm{d}t}\right) \tag{7.85}$$

The Eq. 7.84 can be rewritten by using the state variables i_L, v_C and the load resistance R.

$$\begin{cases} \dfrac{di_L}{dt} = -\dfrac{r_c + r_L}{L} i_L - \dfrac{1}{L} v_c \\ \dfrac{dv_c}{dt} = \dfrac{1}{C} i_L - \dfrac{1}{CR} v_c \end{cases} \tag{7.86}$$

7.7 Comparison Among Buck, Boost, and Buck-Boost Topologies

In Fig. 7.28, all the studied circuits are shown. There are respectively the buck, the boost, and the buck-boost converter in *on* and in *off* state. It should be noted that the buck converter in *on* state is equivalent to the buck converter circuit in *off* state. Boost and buck-boost in *on* state present the same topology and finally the buck converter in *off* state has the same circuit as the buck-boost in *off* state; in particular, they can be achieved by boost in *off* state by setting $V_s = 0$.

Table 7.1 shows the static gain in continuous and discontinuous mode, the inductor current ripple and the relative voltage ripple for the buck, boost, and buck-boost converter, respectively.

As shown in the dedicated section, in general, the gain increases in discontinuous mode for lower values of L/L_C. The boost and buck-boost converters exhibit a higher value of inductor current ripple compared to that of the buck; moreover their relative voltage ripple depend on load value; on the contrary, in buck converter, this value can be lessened by choosing $f_c \ll f_s$.

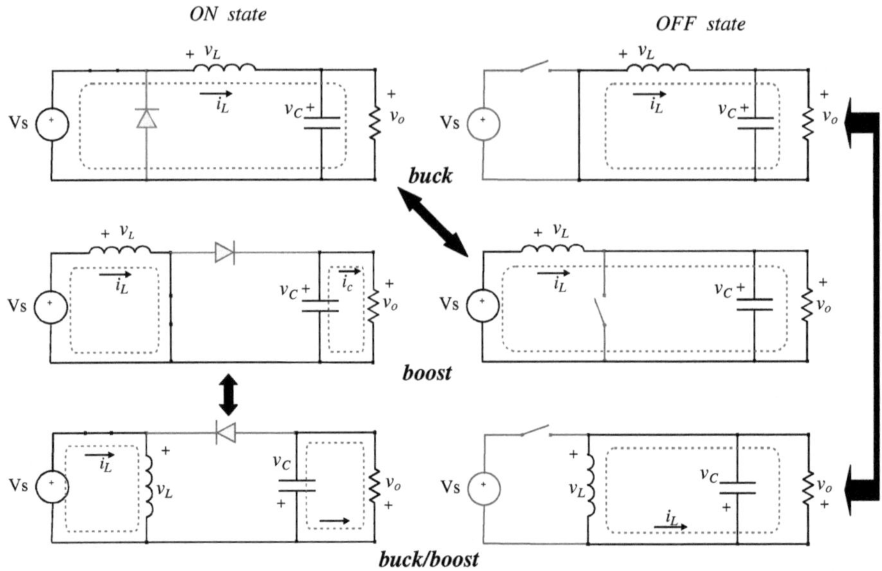

Fig. 7.28 Comparison of buck, boost, and buck-boost topologies in *on* and in *off* state

Table 7.1 Outline of the static gain in continuous and discontinuous mode, the inductor current ripple and the relative voltage ripple for the buck, boost, and buck-boost converter

	$\frac{V_o}{V_s}$ continuous mode	$\frac{V_o}{V_s}$ discontinuous mode	ΔI_L	$\frac{\Delta V_o}{V_o}$
Buck	D	$\frac{2}{1+\sqrt{1+\frac{4L(1-D)}{D^2 L_c}}}$	$\frac{V_s}{L} D(1-D)T_s$	$\frac{\pi^2}{2}(1-D)\left(\frac{f_c}{f_s}\right)^2$
Boost	$\frac{1}{1-D}$	$\frac{1+\sqrt{1+\frac{4DL_c}{L(1-D)^2}}}{2}$	$\frac{V_s}{L} DT_s$	$\frac{D}{f_s RC}$
Buck/boost	$\frac{D}{1-D}$	$D\sqrt{\frac{L_c}{L(1-D)^2}}$	$\frac{V_s}{L} DT_s$	$\frac{D}{f_s RC}$

7.8 Non-Ideal Behavior of Devices and Their Influence on the DC/DC Converter Operation

7.8.1 Inductor

In the above described DC/DC converters, the inductor has been considered as an ideal device with a parasitic series connected resistor. This last component models the winding resistance, it implies power losses and voltage drop with consequent

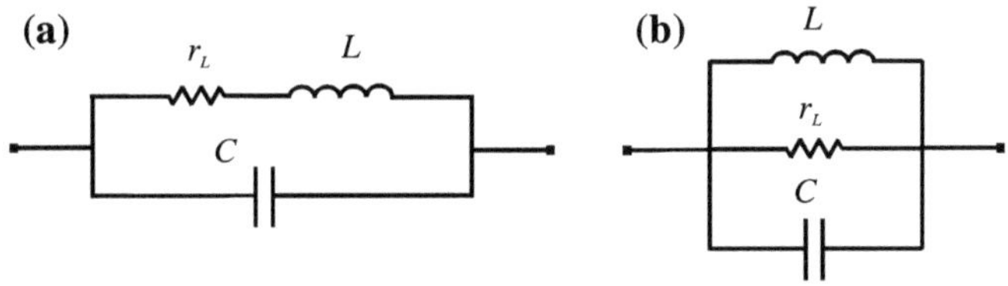

Fig. 7.29 Inductor equivalent model for air core material (**a**) and magnetic core material (**b**)

lessening of the output voltage of the converter. Moreover, the pole/zero values are modified as shown in Sects. 7.10–7.12.

In general, this model is sufficiently accurate. However if the operating frequency is increased, capacitive coupling among windings become important; in this case, two equivalent circuits can be adopted according to whether air core material (Fig. 7.29a) or magnetic core material (Fig. 7.29b) is used.

In both cases, resonant circuits are obtained. A self resonance frequency (SRF) can be defined as the value in which the inductive reactance equals the capacitive reactance. Above the SRF the inductor behaves like a capacitor, for this reason inductors are commonly used below the SRF. Figure 7.30 shows the bode diagram of a real inductor, with air core material, having $L = 2$ mH, $r_L = 0.8\ \Omega$, $C = 25$ pF. The curve is superimposed with the ideal inductor impedance (dotted line). It should be noted that at high frequency the capacitive effects cause a lessening of the overall impedance and, the phase is different from +90° of the ideal inductor as well.

7.8.2 Capacitor

The parasitic behavior of the capacitor includes the effective series resistance (ESR) and the effective series inductance (ESL). The equivalent model is shown in Fig. 7.31.

The parasitic elements are not always constant, in general they depend on the capacitor type (tantalum, ceramic, electrolytic, etc.) and on the operating frequency. In a DC/DC switching converter, the ESL produces spikes due to the capacitor current derivative, while the ESR causes voltage drop due to the converter inductor ripple that flows through the capacitor. This results in an additional ripple on the output voltage equal to $\Delta v_o = (\mathrm{ESR})\Delta I_L$, where ΔI_L is the ripple current in the converter inductor.

In Fig. 7.32, the bode diagram of a real capacitor with $C = 33\ \mu\mathrm{F}$, $r_C = 0.61\ \Omega$, $L = 0.3\ \mu\mathrm{H}$, superimposed with the ideal capacitor impedance (dotted line) is shown. It is important to remark that the model with only the parasitic series resistance is quite accurate for DC/DC converter normal operating frequency.

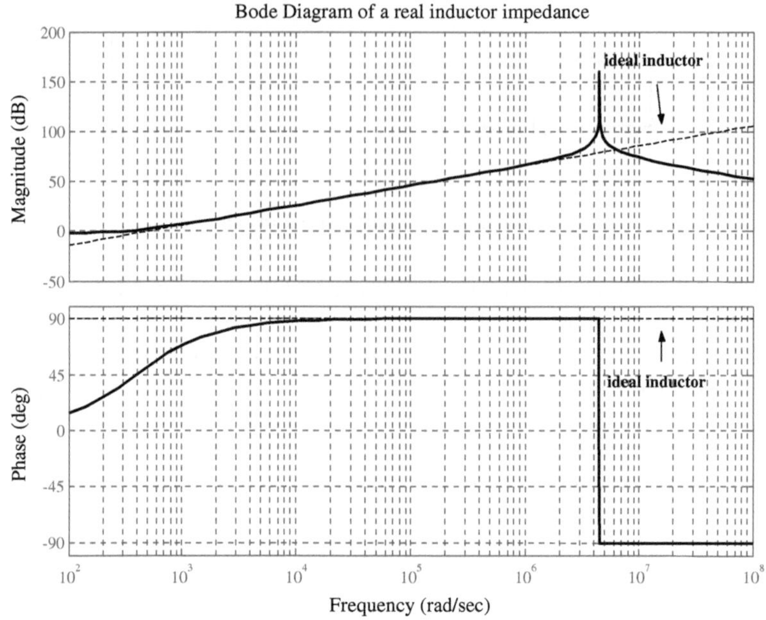

Fig. 7.30 Bode diagram of a real inductor superimposed with the ideal one

Fig. 7.31 Capacitor equivalent model

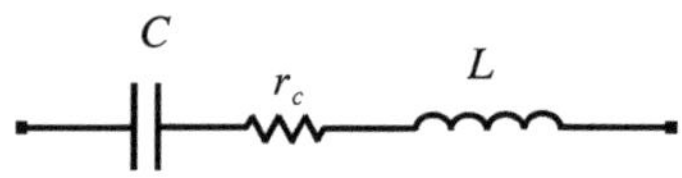

7.8.3 Diode

In the above described DC/DC converters, the diode has been treated as an uncontrolled switch that behaves like a short-circuit in *on* state and an open circuit in *off* state. The corresponding I–V characteristic is shown in Fig. 7.33a.

On the basis of the p–n junction behavior, described in Sect. 2.11, it can be observed that the real diode can be approximated as shown by the dotted line in Fig. 7.33b. During the *off* state, the diode behaves always like an open circuit, during *on* state it exhibits a constant voltage drop given by V_γ (coming from V_{bi} introduced in Sect. 2.11) and a series equivalent resistance. The equivalent model is given in Fig. 7.34.

The diode non ideality implies power losses during conduction (corresponding to T_{OFF} of the DC/DC converter) and a voltage drop with consequent lessening of the converter output voltage. The I–V relationship of a real diode is given by:

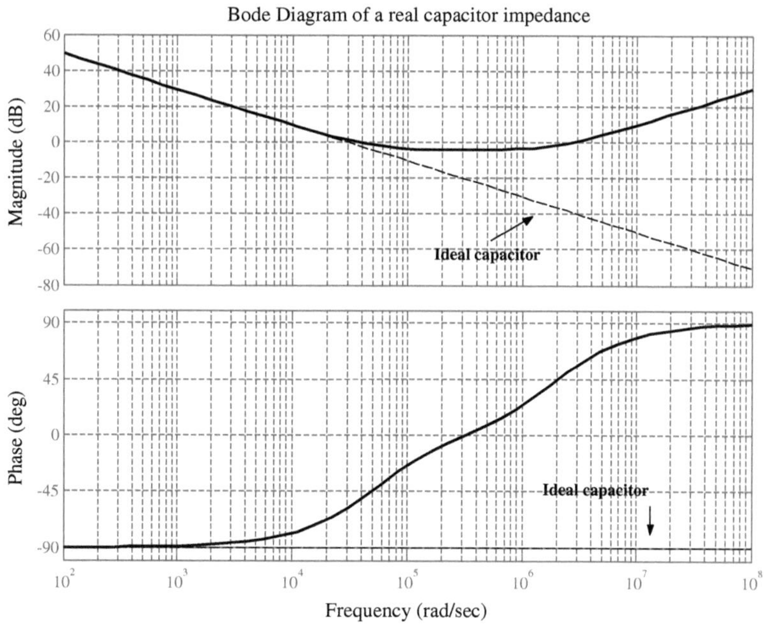

Fig. 7.32 Bode diagram of a real capacitor superimposed with the ideal one

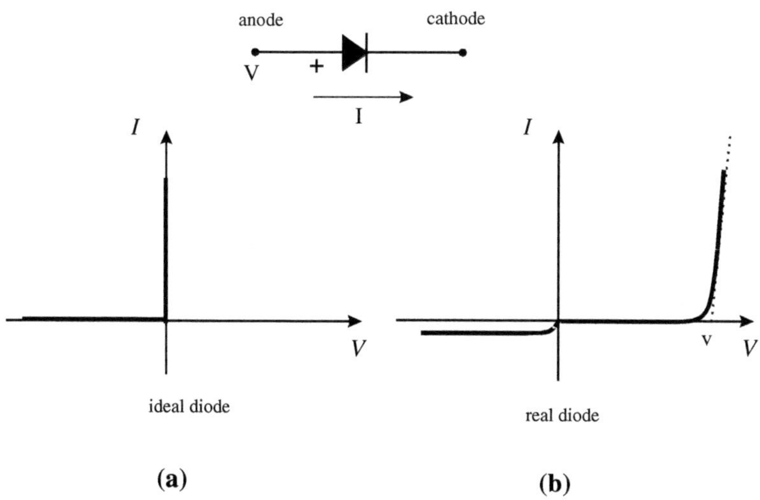

Fig. 7.33 Ideal (**a**) and real (**b**) diode characteristic

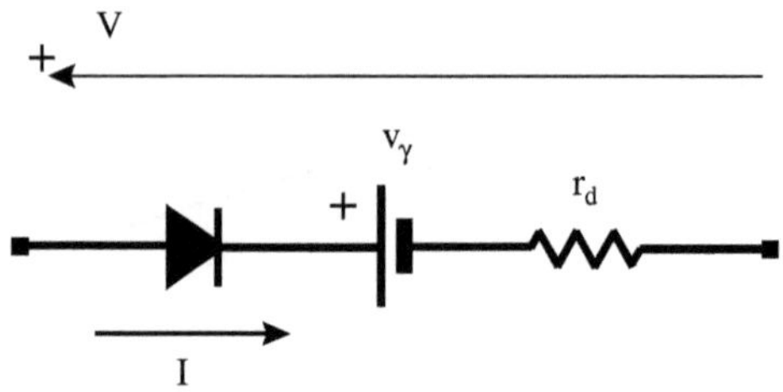

Fig. 7.34 Equivalent real diode circuit

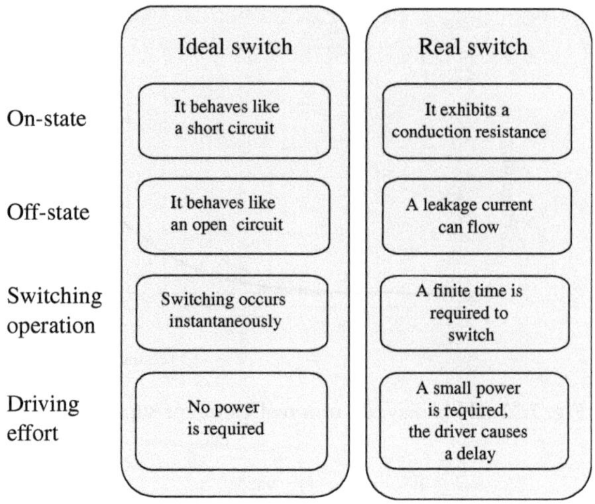

Fig. 7.35 Difference between the ideal switch and the real one

$$V = v_\gamma + Ir_d \tag{7.87}$$

For power diodes $V_\gamma \approx 1$ V. Finally, the diode current is smoothed during transients for the presence of parasitic junction capacitance.

7.8.4 Power Switch

In the previous analysis, the power switch has been considered as an ideal device that behaves like a controlled switch. The difference between the ideal switch and the real device is shown in Fig. 7.35. The ideal switch behaves like a short-circuit in *on* state, it commutates from *on* to *off* and vice versa in a null time and no effort is required to drive it from a state to another.

In a real device, during conduction, a voltage drop due to a non-null conduction resistance occurs; during blocking state a leakage current can flow, commutation requires a finite time and finally an appropriate circuit is required to drive correctly the device.

The first aspect influences the DC/DC converter output voltage, as a matter of fact, considering, for example, the buck converter, the inductor voltage differs from what considered in Eq. 7.14; in particular, this equation becomes:

$$[(V_s - V_o) - I_o(r_{\text{on}} + r_L)]T_{\text{ON}} = [V_o + v_d + I_o(r_L + r_d)]T_{\text{OFF}} \tag{7.88}$$

where r_{on} is the conduction resistance of the power switch, r_d and v_d are respectively the conduction resistance and the voltage drop; assuming that $r_{\text{on}} \approx r_d$ and that $I_o = V_o/R$, Eq. 7.88 gives:

$$V_o = \frac{DV_s - (1 - D)v_d}{\left(1 + \frac{r_{\text{on}} + r_L}{R}\right)} \tag{7.89}$$

From Eq. 7.89, it is possible to notice that the output voltage is lowered compared to the case with ideal components. To obtain the same value of the ideal case a higher D is needed.

The presence of the conduction resistance causes conduction losses; the mean dissipated power $\langle P_{\text{ON}} \rangle$during a switching period is given by:

$$\langle P_{\text{ON}} \rangle = \frac{I_{\text{ON}} V_{\text{ON}} T_{\text{ON}}}{T_S} = I_{\text{ON}}\, V_{\text{ON}} T_{\text{ON}} f_s \tag{7.90}$$

where I_{ON} is the conduction current, V_{ON} the voltage drop of the device, T_{ON} the conduction time and $f_s = 1/T_s$ the switching frequency.

During *off* state the current leakage current is negligible and practically there is no power dissipation.

During the commutation from conduction to interdiction, the voltage across the device rises from V_{ON} to the maximum value and the current falls to zero. In the transient both the voltage and the current are non-null at the same time resulting in power losses.

The lost energy during a transition is given with a good approximation by the area of the triangle defined by the voltage rising, the current falling and having as the base the sum of the rising time of the voltage (t_{rv}) and the falling time of the current (t_{fi}); it is depicted in Fig. 7.36. Commutation is usually delayed respect to the command (t_d) because of the electronic circuitry needed for the command of the power devices.

The expression of the lost energy is given by:

$$E_{\text{ON}\to\text{OFF}} = \frac{1}{2} V_{\text{OFF}} I_{\text{ON}} \left(t_{\text{rv}} + t_{\text{fi}}\right) \tag{7.91}$$

where V_{OFF} is the voltage acting on the power switch during the *off* state and I_{ON} represents its current during the *on* state. The analysis of Fig. 7.36 shows that the faster the commutation, the lower the area of the aforementioned triangle, the lower the switching losses in the device.

Similar considerations are valid for the transition from interdiction to conduction, considering now rising time of the current (t_{ri}) and falling time of the voltage (t_{fv}). So, the mean power loss $<P_{\text{SW}}>$ can be expressed as:

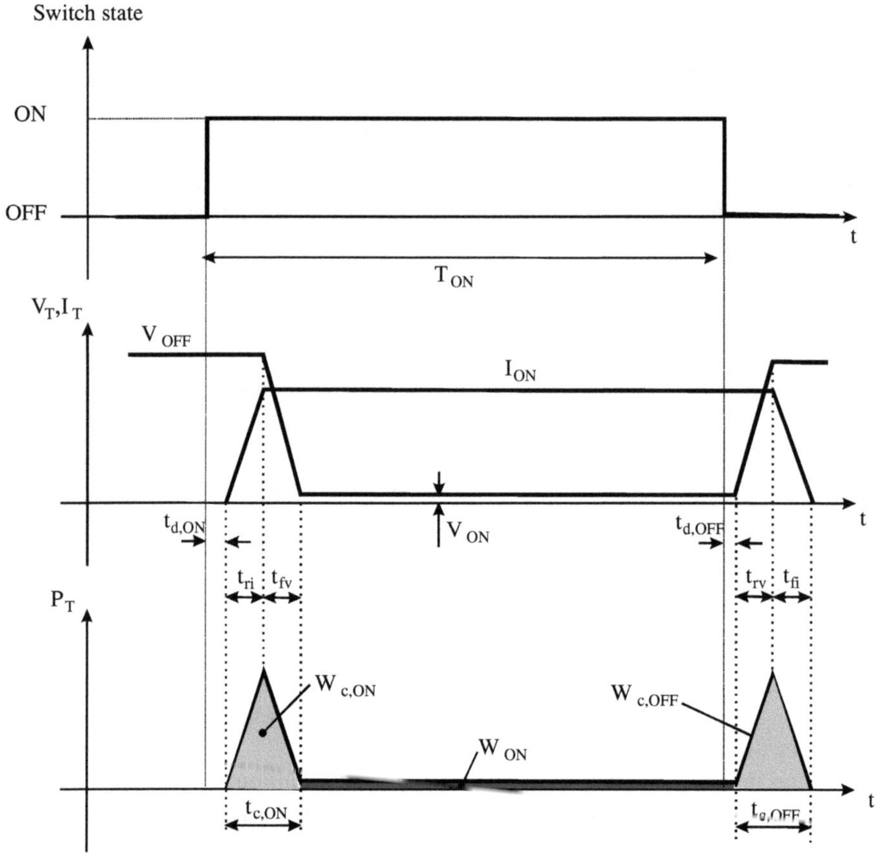

Fig. 7.36 Time-domain waveforms for a real power switch in operating conditions: switch state (*top*), voltage and current of the device (*middle*), dissipated power (*bottom*)

$$\langle P_{SW}\rangle = \frac{V_{OFF}I_{ON}}{2}\frac{[(t_{rv}+t_{fi})+(t_{ri}+t_{fv})]}{T_s} = \frac{V_{OFF}I_{ON}}{2}\cdot[(t_{rv}+t_{fi})+(t_{ri}+t_{fv})]f_s \tag{7.92}$$

It should be noted that both the computed powers depend on switching frequency. The higher f_s, the higher the power losses. A higher f_s could imply a degradation of the converter electromagnetic compatibility; on the other hand, the dynamic performances are improved and a smaller size of the filtering reactive devices is required.

It should be noted that the recent availability on the market of fast devices with low values of switching times has contributed to the switching losses reduction.

7.9 State Space Representation

A continuous-time linear system can be described by using the state space representation in the following form:

$$\begin{cases} \dot{x}(t) = \boldsymbol{A}x(t) + \boldsymbol{B}u(t) \\ y(t) = \boldsymbol{C}x(t) + \boldsymbol{D}u(t) \end{cases} \tag{7.93}$$

where $\boldsymbol{x}(t)$ is the state variables vector, $\boldsymbol{u}(t)$ is the sources vector, $\boldsymbol{y}(t)$ is the outputs vector and $\boldsymbol{A}$, $\boldsymbol{B}$, $\boldsymbol{C}$, $\boldsymbol{D}$ are the coefficient matrices.

By equating to zero the first equation of Eq. 7.93, if $\boldsymbol{A}$ admits inverse form, a DC solution giving the state variables versus the sources can be obtained.

$$\boldsymbol{x} = -\boldsymbol{A}^{-1}\boldsymbol{B}\boldsymbol{u} \tag{7.94}$$

The first equation of Eq. 7.93 can be transformed into Laplace domain:

$$s\boldsymbol{X}(s) = \boldsymbol{A}\boldsymbol{X}(s) + \boldsymbol{B}\boldsymbol{U}(s) \tag{7.95}$$

From which the state variables versus sources are obtained:

$$\boldsymbol{X}(s) = (s\boldsymbol{I} - \boldsymbol{A})^{-1}\boldsymbol{B}\boldsymbol{U}(s) \tag{7.96}$$

Where $\boldsymbol{I}$ is an identity matrix with the same dimension of $\boldsymbol{A}$.

In the same way, starting from the second equation of Eq. 7.93:

$$\boldsymbol{Y}(s) = [\boldsymbol{C}(s\boldsymbol{I} - \boldsymbol{A})^{-1}\boldsymbol{B} + \boldsymbol{D}]\boldsymbol{U}(s) \tag{7.97}$$

7.10 State Space Averaging

A DC/DC converter can be analyzed by a state space representation considering it as a succession of two linear circuits, one corresponding to *on* state and the other to *off* state.

By identifying with subscript "ON" the coefficients matrices during T_{ON} and with subscript "OFF" the coefficients matrices during T_{OFF} respectively, the two following representations are valid:

$$\begin{cases} \dot{\boldsymbol{x}}(t) = \boldsymbol{A}_{\mathrm{ON}}\boldsymbol{x}(t) + \boldsymbol{B}_{\mathrm{ON}}\boldsymbol{u}(t) & 0 \le t < T_{\mathrm{ON}} \\ \dot{\boldsymbol{x}}(t) = \boldsymbol{A}_{\mathrm{OFF}}\boldsymbol{x}(t) + \boldsymbol{B}_{\mathrm{OFF}}\boldsymbol{u}(t) & T_{\mathrm{ON}} \le t < T_s \end{cases} \tag{7.98a}$$

$$\begin{cases} \boldsymbol{y}(t) = \boldsymbol{C}_{\mathrm{ON}}\boldsymbol{x}(t) + \boldsymbol{C}_{\mathrm{ON}}\boldsymbol{u}(t) & 0 \le t < T_{\mathrm{ON}} \\ \boldsymbol{y}(t) = \boldsymbol{C}_{\mathrm{OFF}}\boldsymbol{x}(t) + \boldsymbol{C}_{\mathrm{OFF}}\boldsymbol{u}(t) & T_{\mathrm{ON}} \le t < T_s \end{cases} \tag{7.98b}$$

It should be noted that the two equations in Eq. 7.98a and in Eq. 7.98b have the state variables, the input and the output vectors in common.

The state space averaging is obtained by multiplying the first of Eq. 7.98a for D and the second for $(1 - D)$ and summing them.

The aim of this technique is to obtain small signal transfer functions allowing the converter to be studied as a linear device with reference to small perturbations around the operating point.

$$\dot{\boldsymbol{x}}(t) = [\boldsymbol{A}_{\mathrm{ON}}D + \boldsymbol{A}_{\mathrm{OFF}}(1 - D)]\boldsymbol{x}(t) + [\boldsymbol{B}_{\mathrm{ON}}D + \boldsymbol{B}_{\mathrm{OFF}}(1 - D)]\boldsymbol{u}(t) \tag{7.99}$$

The Eq. 7.99 can be now written as Eq. 7.93, where:

$$\begin{cases} \boldsymbol{A} = \boldsymbol{A}_{\mathrm{ON}}D + \boldsymbol{A}_{\mathrm{OFF}}(1 - D) \\ \boldsymbol{B} = \boldsymbol{B}_{\mathrm{ON}}D + \boldsymbol{B}_{\mathrm{OFF}}(1 - D) \end{cases} \tag{7.100}$$

In a similar way[2]:

$$\mathbf{y}(t) = [\mathbf{C}_{\mathrm{ON}}D + \mathbf{C}_{\mathrm{OFF}}(1 - D)]\mathbf{x}(t) + [\mathbf{D}_{\mathrm{ON}}D + \mathbf{D}_{\mathrm{OFF}}(1 - D)]\mathbf{u}(t) \tag{7.101}$$

$$\begin{cases} \boldsymbol{C} = \boldsymbol{C}_{\mathrm{ON}}D + \boldsymbol{C}_{\mathrm{OFF}}(1 - D) \\ \boldsymbol{D} = \boldsymbol{D}_{\mathrm{ON}}D + \boldsymbol{D}_{\mathrm{OFF}}(1 - D) \end{cases} \tag{7.102}$$

The Eq. 7.99 and Eq. 7.101 are nonlinear equation that can be linearized around an operating point by considering the variables as the sum of a DC and an AC component. Assuming that the state vector is i-dimensional, the input vector is j-dimensional and the output vector is k-dimensional; and indicating with the subscript "0" the DC components and with "$\sim$" the AD components, it means to consider:

$$\begin{cases} x_i = x_{i0} + \tilde{x}_i \\ u_j = u_{j0} + \tilde{u}_j \\ y_k = y_{k0} + \tilde{y}_k \\ D = d + \tilde{d} \end{cases} \tag{7.103}$$

By substituting Eq. 7.103 in Eq. 7.99:

$$\dot{\tilde{\boldsymbol{x}}} = \left[\boldsymbol{A} + \boldsymbol{A}_{ON}\tilde{d} + \boldsymbol{A}_{OFF}\tilde{d}\right](\boldsymbol{x}_0 + \tilde{\boldsymbol{x}}) + \left[\boldsymbol{B} + \boldsymbol{B}_{ON}\tilde{d} + \boldsymbol{B}_{OFF}\tilde{d}\right](\boldsymbol{u}_0 + \tilde{\boldsymbol{u}}) \tag{7.104}$$

where $\boldsymbol{A}$ and $\boldsymbol{B}$ are defined by Eq. 7.100, where $D = d$.

By rearranging Eq. 7.104 and neglecting the products of AC terms, it follows:

[2] The reader should not confuse the matrix $\boldsymbol{D}$ (in bold case) in Eq. 7.93 with the duty cycle D. In the following, the matrix $\boldsymbol{D}$ will not be considered.

$$\tilde{\dot{x}} = \boldsymbol{A}\boldsymbol{x}_0 + \boldsymbol{B}\boldsymbol{u}_0 + \boldsymbol{A}\tilde{x} + [(\boldsymbol{A}_{\mathrm{ON}} - \boldsymbol{A}_{\mathrm{OFF}})\boldsymbol{x}_0 + (\boldsymbol{B}_{\mathrm{ON}} - \boldsymbol{B}_{\mathrm{OFF}})\boldsymbol{u}_0]\tilde{d} + \boldsymbol{B}\tilde{\boldsymbol{u}} \quad (7.105)$$

The first two terms in the second member correspond to the DC solution obtained by the first equation of Eq. 7.93

$$\boldsymbol{A}\boldsymbol{x}_0 + \boldsymbol{B}\boldsymbol{u}_0 = 0 \quad (7.106)$$

It follows that:

$$\begin{aligned}\tilde{\dot{x}} &= \boldsymbol{A}\boldsymbol{x} + [(\boldsymbol{A}_{\mathrm{ON}} - \boldsymbol{A}_{\mathrm{OFF}})\boldsymbol{x}_0 + (\boldsymbol{B}_{\mathrm{ON}} - \boldsymbol{B}_{\mathrm{OFF}})\boldsymbol{u}_0]\tilde{d} + \boldsymbol{B}\tilde{\boldsymbol{u}} \\ &= \boldsymbol{A}\tilde{\boldsymbol{x}} + \boldsymbol{B}\tilde{\boldsymbol{u}} + \boldsymbol{E}\tilde{d}\end{aligned} \quad (7.107)$$

In a similar way, from the second equation of Eq. 7.93, considering that the $\boldsymbol{D}$ matrix is null:

$$\boldsymbol{y}(t) = [\boldsymbol{C}_{\mathrm{ON}}D + \boldsymbol{C}_{\mathrm{OFF}}(1 - D)]\boldsymbol{x}(t) \quad (7.108)$$

$$(\boldsymbol{y}_0 + \tilde{\boldsymbol{y}}) = \left[\boldsymbol{C}_{\mathrm{ON}}\left(d + \tilde{d}\right) + \boldsymbol{C}_{\mathrm{OFF}}\left(1 - d - \tilde{d}\right)\right](\boldsymbol{x}_0 + \tilde{\boldsymbol{x}}) \quad (7.109)$$

By neglecting the products of AC terms and considering the DC solution of the second Eq. 7.93, it is obtained:

$$\boldsymbol{y}_0 = \boldsymbol{C}\boldsymbol{x}_0 \quad (7.110)$$

It follows that:

$$\tilde{\boldsymbol{y}} = \boldsymbol{C}\tilde{\boldsymbol{x}} + \left(\boldsymbol{C}_{\mathrm{ON}}\tilde{d} - \boldsymbol{C}_{\mathrm{OFF}}\tilde{d}\right)\boldsymbol{x}_0 \quad (7.111)$$

where C is defined by Eq. 7.102, with $D = d$.

Considering Eq. 7.110 and Eq. 7.106:

$$\mathbf{y}_0 = \mathbf{C}\mathbf{x}_0 = -\mathbf{C}\mathbf{A}^{-1}\mathbf{B}\mathbf{u}_0 \quad (7.112)$$

For a system with a unique input (V_s) and output (V_o):

$$\frac{V_o}{V_s} = -\boldsymbol{C}\boldsymbol{A}^{-1}\boldsymbol{B} \quad (7.113)$$

This last equations gives the DC relationship between input and output.

The Eq. 7.107 and Eq. 7.111 can be transformed into Laplace domain and considering, for simplicity, a system with one input ($u = u_0 + \tilde{u} = V_s + \tilde{v}_s$) and one output ($y = y_0 + \tilde{y} = V_o + \tilde{v}_o$).

The Eq. 7.107 in the Laplace domain becomes:

$$s\tilde{\boldsymbol{X}}(s) = \boldsymbol{A}\tilde{\boldsymbol{X}}(s) + [(\boldsymbol{A}_{\mathrm{ON}} - \boldsymbol{A}_{\mathrm{OFF}})\boldsymbol{x}_0 + (\boldsymbol{B}_{\mathrm{ON}} - \boldsymbol{B}_{\mathrm{OFF}})V_s]\tilde{d}(s) + \boldsymbol{B}\tilde{v}_s(s) \quad (7.114)$$

Eq. 7.114 can be rewritten as:

$$\tilde{X}(s) = [s\boldsymbol{I} - \boldsymbol{A}]^{-1}[(\boldsymbol{A}_{ON} - \boldsymbol{A}_{OFF})\boldsymbol{x}_0 + (\boldsymbol{B}_{ON} - \boldsymbol{B}_{OFF})V_s]\tilde{d}(s) + [s\boldsymbol{I} - \boldsymbol{A}]^{-1}\boldsymbol{B}\tilde{v}_s(s) \tag{7.115}$$

The Eq. 7.111 in the Laplace domain has the following form:

$$\tilde{v}_o(s) = \boldsymbol{C}\tilde{\boldsymbol{X}}(s) + (\boldsymbol{C}_{ON} - \boldsymbol{C}_{OFF})\boldsymbol{x}_0\tilde{d}(s) \tag{7.116}$$

By substituting Eq. 7.115 in Eq. 7.116, the final expression of the output small variations versus the duty cycle and the input small variations are obtained.

$$\begin{aligned}\tilde{v}_0(s) = \boldsymbol{C}[s\boldsymbol{I} - \boldsymbol{A}]^{-1}[(\boldsymbol{A}_{ON} - \boldsymbol{A}_{OFF})\boldsymbol{x}_0 + (\boldsymbol{B}_{ON} - \boldsymbol{B}_{OFF})V_s]\tilde{d}(s) \\ + (\boldsymbol{C}_{ON} - \boldsymbol{C}_{OFF})\boldsymbol{x}_0\tilde{d}(s) + \boldsymbol{C}[s\boldsymbol{I} - \boldsymbol{A}]^{-1}\boldsymbol{B}\tilde{v}_s(s)\end{aligned} \tag{7.117}$$

The transfer function of the output versus the duty cycle can be obtained by setting $\tilde{v}_s = 0$.

$$\begin{aligned}\frac{\tilde{v}_0(s)}{\tilde{d}(s)} = \boldsymbol{C}[s\boldsymbol{I} - \boldsymbol{A}]^{-1}[(\boldsymbol{A}_{ON} - \boldsymbol{A}_{OFF})\boldsymbol{x}_0 + (\boldsymbol{B}_{ON} - \boldsymbol{B}_{OFF})V_s] \\ + (\boldsymbol{C}_{ON} - \boldsymbol{C}_{OFF})\boldsymbol{x}_0\end{aligned} \tag{7.118}$$

The transfer function of the output versus the input voltage can be obtained by setting $\tilde{d} = 0$.

$$\frac{\tilde{v}_0(s)}{\tilde{v}_s(s)} = \boldsymbol{C}[s\boldsymbol{I} - \boldsymbol{A}]^{-1}\boldsymbol{B} \tag{7.119}$$

7.11 State Space Averaging of Buck Converter

The analysis of the buck converter is performed on the basis of the implicit time domain equations of the circuits during T_{ON} and T_{OFF} with the aim of identifying the coefficient matrices.

Considering Eq. 7.41a and Eq. 7.42a, here rewritten for the sake of clarity,

$$\begin{cases}\dfrac{di_L}{dt} = -\dfrac{r_c + r_L}{L}i_L - \dfrac{1}{L}v_c + \dfrac{1}{L}V_s \\ \dfrac{dv_c}{dt} = \dfrac{1}{C}i_L - \dfrac{1}{CR}v_c\end{cases}$$

$$V_o = r_C i_L + v_C$$

during T_{ON}, the time domain equations can be obtained from Eq. 7.41a; the corresponding matrices are:

$$\begin{cases} \boldsymbol{A}_{\text{ON}} = \begin{bmatrix} -\frac{r_c+r_L}{L} & -\frac{1}{L} \\ \frac{1}{C} & -\frac{1}{CR} \end{bmatrix} \\ \boldsymbol{B}_{\text{ON}} = \begin{bmatrix} 1/L \\ 0 \end{bmatrix} \\ \boldsymbol{C}_{\text{ON}} = [r_C \quad 1] \\ \boldsymbol{D}_{\text{ON}} = 0 \end{cases} \tag{7.120}$$

During T_{OFF}, the time domain equations can be obtained from Eq. 7.41a, by setting $V_s = 0$. The corresponding matrices are:

$$\begin{cases} \boldsymbol{A}_{\text{OFF}} = \boldsymbol{A}_{\text{ON}} \\ \boldsymbol{B}_{\text{OFF}} = \begin{bmatrix} 0 \\ 0 \end{bmatrix} \\ \boldsymbol{C}_{\text{OFF}} = \boldsymbol{C}_{\text{ON}} \\ \boldsymbol{D}_{\text{OFF}} = 0 \end{cases} \tag{7.121}$$

From which:

$$\begin{cases} \boldsymbol{A} = \boldsymbol{A}_{\text{ON}} \\ \boldsymbol{B} = \boldsymbol{B}_{\text{ON}} D \\ \boldsymbol{C} = \boldsymbol{C}_{\text{ON}} \end{cases} \tag{7.122}$$

The Eq. 7.113 becomes:

$$V_0 = -\boldsymbol{C}_{\text{ON}} \boldsymbol{A}_{\text{ON}}^{-1} \boldsymbol{B}_{\text{ON}} D V_s \tag{7.123}$$

where

$$\boldsymbol{A}_{\text{ON}}^{-1} = \frac{LC}{1 + \frac{r_C+r_L}{R}} \begin{bmatrix} -\frac{1}{RC} & \frac{1}{L} \\ -\frac{1}{C} & -\frac{r_C+r_L}{L} \end{bmatrix} \tag{7.124}$$

$$\begin{aligned} \frac{V_s}{V_0} &= -\frac{RLC}{R + r_C + r_L} [r_C \quad 1] \begin{bmatrix} -\frac{1}{RC} & \frac{1}{L} \\ -\frac{1}{C} & -\frac{r_C+r_L}{L} \end{bmatrix} \begin{bmatrix} \frac{1}{L} \\ 0 \end{bmatrix} D \\ &= D \frac{R + r_C}{R + r_C + r_L} \end{aligned} \tag{7.125}$$

The Eq. 7.118 becomes:

$$\frac{\tilde{v}_0(s)}{\tilde{d}(s)} = \boldsymbol{C}_{\text{ON}} [s\boldsymbol{I} - \boldsymbol{A}_{\text{ON}}]^{-1} \boldsymbol{B}_{\text{ON}} V_s \tag{7.126}$$

where

$$[s\boldsymbol{I} - \boldsymbol{A}_{\mathrm{ON}}] = \begin{bmatrix} s + \frac{r_C + r_L}{L} & \frac{1}{L} \\ -\frac{1}{C} & s + \frac{1}{RC} \end{bmatrix} \tag{7.127}$$

It follows that:

$$\det[s\boldsymbol{I} - \boldsymbol{A}_{\mathrm{ON}}] = \frac{s^2 RLC + s[L + RC(r_C + r_L)] + R}{RLC} \tag{7.128}$$

where it has been assumed $R \gg (r_c + r_L)$.

Considering that

$$[s\boldsymbol{I} - \boldsymbol{A}_{\mathrm{ON}}]^{-1} = \frac{1}{\det[s\boldsymbol{I} - \boldsymbol{A}_{\mathrm{ON}}]} \begin{bmatrix} s + \frac{1}{RC} & -\frac{1}{L} \\ \frac{1}{C} & s + \frac{r_C + r_L}{L} \end{bmatrix} \tag{7.129}$$

it is obtained:

$$\begin{aligned} \frac{\tilde{v}_O(s)}{\tilde{d}(s)} &= \frac{V_s}{\det[s\boldsymbol{I} - \boldsymbol{A}_{\mathrm{ON}}]} [r_C \quad 1] \begin{bmatrix} s + \frac{1}{RC} & -\frac{1}{L} \\ \frac{1}{C} & s + \frac{r_C + r_L}{L} \end{bmatrix} \begin{bmatrix} \frac{1}{L} \\ 0 \end{bmatrix} = \\ & V_s \frac{sCr_C + 1}{LC\left[s^2 + s\left(\frac{1}{RC} + \frac{r_C + r_L}{L}\right) + \frac{1}{LC}\right]} \end{aligned} \tag{7.130}$$

The Eq. 7.119 becomes:

$$\frac{\tilde{v}_O(s)}{\tilde{v}_s(s)} = \boldsymbol{C}_{\mathrm{ON}} [s\boldsymbol{I} - \boldsymbol{A}_{\mathrm{ON}}]^{-1} \boldsymbol{B}_{\mathrm{ON}} D \tag{7.131}$$

Equation 7.131 is the same of the Eq. 7.126 by substituting V_s with D; then, from Eq. 7.130, it follows that:

$$\frac{\tilde{v}_O(s)}{\tilde{v}_s(s)} = D \frac{sCr_C + 1}{LC\left[s^2 + s\left(\frac{1}{RC} + \frac{r_C + r_L}{L}\right) + \frac{1}{LC}\right]} \tag{7.132}$$

It should be noted that Eq. 7.130 does not depend on the duty cycle D.

7.12 State Space Averaging of Boost Converter

In *on* state the boost converter circuit is described by Eqs. 7.65 and 7.67, rewritten here for the sake of clarity.

$$\begin{cases} \dfrac{\mathrm{d}i_L}{\mathrm{d}t} = -\dfrac{r_L}{L} i_L + \dfrac{1}{L} V_s \\ \dfrac{\mathrm{d}v_c}{\mathrm{d}t} = -\dfrac{v_c}{RC} \end{cases}$$

$$V_o = v_c \left(\frac{R}{R - r_c} \right) \cong v_c$$

In *off* state the corresponding equations are those obtained for the buck converter in *on* state (see Sect. 7.4.3), i.e., Eq. 7.41a and Eq. 7.42a, rewritten here.

$$\begin{cases} \dfrac{\mathrm{d}i_L}{\mathrm{d}t} = -\dfrac{r_c + r_L}{L} i_L - \dfrac{1}{L} v_c + \dfrac{1}{L} V_s \\ \dfrac{\mathrm{d}v_c}{\mathrm{d}t} = \dfrac{1}{C} i_L - \dfrac{1}{CR} v_c \end{cases}$$

$$V_o = r_c i_L + v_c$$

On the basis of such equations, the following coefficients matrices are obtained:

$$\begin{cases} \boldsymbol{A}_{\mathrm{ON}} = \begin{bmatrix} -\frac{r_L}{L} & 0 \\ 0 & -\frac{1}{CR} \end{bmatrix} \\ \boldsymbol{B}_{\mathrm{ON}} = \begin{bmatrix} 1/L \\ 0 \end{bmatrix} \\ \boldsymbol{C}_{\mathrm{ON}} = \begin{bmatrix} 0 & 1 \end{bmatrix} \\ \boldsymbol{D}_{\mathrm{ON}} = 0 \end{cases} \tag{7.133}$$

$$\begin{cases} \boldsymbol{A}_{\mathrm{OFF}} = \begin{bmatrix} -\frac{r_c + r_L}{L} & -\frac{1}{L} \\ \frac{1}{C} & -\frac{1}{CR} \end{bmatrix} \\ \boldsymbol{B}_{\mathrm{OFF}} = \begin{bmatrix} 1/L \\ 0 \end{bmatrix} \\ \boldsymbol{C}_{\mathrm{OFF}} = \begin{bmatrix} r_c & 1 \end{bmatrix} \\ \boldsymbol{D}_{\mathrm{OFF}} = 0 \end{cases} \tag{7.134}$$

$$\begin{cases} \boldsymbol{A} = \boldsymbol{A}_{\mathrm{ON}} D + \boldsymbol{A}_{\mathrm{OFF}} (1 - D) = \begin{bmatrix} \frac{-r_c(1-D) - r_L}{L} & -\frac{1}{L}(1 - D) \\ \frac{1}{C}(1 - D) & -\frac{1}{CR} \end{bmatrix} \\ \boldsymbol{B} = \boldsymbol{B}_{\mathrm{ON}} D + \boldsymbol{B}_{\mathrm{OFF}} (1 - D) = \boldsymbol{B}_{\mathrm{OFF}} = \begin{bmatrix} 1/L \\ 0 \end{bmatrix} \\ \boldsymbol{C} = \boldsymbol{C}_{\mathrm{ON}} D + \boldsymbol{C}_{\mathrm{OFF}} (1 - D) = \begin{bmatrix} r_c(1 - D) & 1 \end{bmatrix} \end{cases} \tag{7.135}$$

By substituting the obtained coefficient matrices in Eq. 7.113, the DC relationship between the output voltage and the input voltage is obtained.

$$\frac{V_o}{V_s} = -\boldsymbol{C}\boldsymbol{A}^{-1}\boldsymbol{B} = -\frac{\boldsymbol{1}}{\det \boldsymbol{A}}\begin{bmatrix} r_c(1-D) & 1 \end{bmatrix}\begin{bmatrix} -\frac{1}{CR} & \frac{1}{L}(1-D) \\ -\frac{1}{C}(1-D) & \frac{-r_c(1-D)-r_L}{L} \end{bmatrix}\begin{bmatrix} 1/L \\ 0 \end{bmatrix} \tag{7.136}$$

where

$$\det \boldsymbol{A} = \frac{r_c(1-D) + r_L + R(1-D)^2}{RLC} \tag{7.137}$$

It follows that

$$\frac{V_o}{V_s} = \frac{R(1-D)}{(1-D)^2 R + r_L + r_c(1-D)} \tag{7.138}$$

where the approximation $R \approx R + r_c$ is done.

In order to obtain the output voltage to duty cycle transfer function, Eq. 7.118 is used.

$$[s\boldsymbol{I} - \boldsymbol{A}] = \begin{bmatrix} s - \frac{-r_c(1-D)-r_L}{L} & \frac{1}{L}(1-D) \\ -\frac{1}{C}(1-D) & s + \frac{1}{CR} \end{bmatrix} \tag{7.139}$$

$$\det[s\boldsymbol{I} - \boldsymbol{A}] = \frac{s^2 RLC + s\{L + RC[r_c(1-D) + r_L]\} + [r_c(1-D) + r_L] + R(1-D)^2}{RLC} \tag{7.140}$$

$$[s\boldsymbol{I} - \boldsymbol{A}]^{-1} = \frac{1}{\det[s\boldsymbol{I} - \boldsymbol{A}]}\begin{bmatrix} \frac{sRC+1}{RC} & -\frac{1}{L}(1-D) \\ \frac{1}{C}(1-D) & \frac{sL+r_c(1-D)+r_L}{L} \end{bmatrix} \tag{7.141}$$

Considering that $\boldsymbol{B}_{\mathrm{ON}} = \boldsymbol{B}_{\mathrm{OFF}}$, Eq. 7.118 becomes:

$$\frac{\tilde{v}_o(s)}{\tilde{d}(s)} = \boldsymbol{C}[s\boldsymbol{I} - \boldsymbol{A}]^{-1}(\boldsymbol{A}_{\mathrm{ON}} - \boldsymbol{A}_{\mathrm{OFF}})\boldsymbol{x}_0 + (\boldsymbol{C}_{\mathrm{ON}} - \boldsymbol{C}_{\mathrm{OFF}})\boldsymbol{x}_0 \tag{7.142}$$

where

$$(\boldsymbol{A}_{\mathrm{ON}} - \boldsymbol{A}_{\mathrm{OFF}})\boldsymbol{x}_0 = \begin{bmatrix} \frac{r_c}{L} & \frac{1}{L} \\ -\frac{1}{C} & 0 \end{bmatrix}\begin{bmatrix} i_{L0} \\ v_{c0} \end{bmatrix} \tag{7.143}$$

$$(\boldsymbol{C}_{\mathrm{ON}} - \boldsymbol{C}_{\mathrm{OFF}})\boldsymbol{x}_0 = \begin{bmatrix} -r_c & 0 \end{bmatrix}\begin{bmatrix} i_{L0} \\ v_{c0} \end{bmatrix} \tag{7.144}$$

It should be noted that, differently from the buck, the steady state inductor current and voltage capacitor have to be utilized. The (7.143) becomes:

$$(\boldsymbol{A}_{\mathrm{ON}} - \boldsymbol{A}_{\mathrm{OFF}})\boldsymbol{x}_0 = \begin{bmatrix} \frac{r_c}{L} & \frac{1}{L} \\ -\frac{1}{C} & 0 \end{bmatrix} \begin{bmatrix} \frac{1}{R(1-D)} \\ 1 \end{bmatrix} V_o = \begin{bmatrix} \frac{r_C+R(1-D)}{RL(1-D)} \\ -\frac{1}{RC(1-D)} \end{bmatrix} V_o \tag{7.145}$$

$$\begin{aligned}
&\boldsymbol{C}[s\boldsymbol{I}-\boldsymbol{A}]^{-1}(\boldsymbol{A}_{\mathrm{ON}} - \boldsymbol{A}_{\mathrm{OFF}})\boldsymbol{x}_0 \\
&= \frac{1}{\det[s\boldsymbol{I}-\boldsymbol{A}]}\begin{bmatrix} r_c(1-D) & 1 \end{bmatrix} \begin{bmatrix} \frac{sRC+1}{RC} & -\frac{1}{L}(1-D) \\ \frac{1}{C}(1-D) & \frac{sL+r_c(1-D)+r_L}{L} \end{bmatrix} \begin{bmatrix} \frac{r_C+R(1-D)}{RL(1-D)} \\ -\frac{1}{RC(1-D)} \end{bmatrix} V_o \\
&= \frac{1}{\det[s\boldsymbol{I}-\boldsymbol{A}]}\begin{bmatrix} r_c(1-D) & 1 \end{bmatrix} \begin{bmatrix} \frac{sRC[r_C+R(1-D)]+r_C+2R(1-D)}{R} \\ R(1-D)^2 - sL - r_L \end{bmatrix} \frac{V_o}{RLC(1-D)} \\
&= \frac{1}{\det[s\boldsymbol{I}-\boldsymbol{A}]}\left\{ \begin{array}{c} s\{r_C C(1-D)[r_C + R(1-D)] - L\} + \\ + \left[(2r_C + R)(1-D)^2 - r_L + \frac{r_C^2}{R}(1-D)\right] \end{array} \right\} \frac{V_o}{RLC(1-D)} \\
&\cong \frac{1}{\det[s\boldsymbol{I}-\boldsymbol{A}]}\left\{ \begin{array}{c} s\{r_C C(1-D)[r_C + R(1-D)] - L\} + \\ + \left[(2r_C + R)(1-D)^2 - r_L\right] \end{array} \right\} \frac{V_o}{RLC(1-D)}
\end{aligned} \tag{7.146}$$

Being

$$(\boldsymbol{C}_{\mathrm{ON}} - \boldsymbol{C}_{\mathrm{OFF}})\boldsymbol{x}_0 = \begin{bmatrix} -r_c & 0 \end{bmatrix} \begin{bmatrix} i_{L0} \\ v_{c0} \end{bmatrix} = -\frac{r_c}{R(1-D)} V_o \tag{7.147}$$

Eq. 7.142 with Eqs. 7.146 and 7.147 give:

$$\begin{aligned}
\frac{\tilde{v}_o(s)}{\tilde{d}(s)} &= \frac{1}{\det[s\boldsymbol{I}-\boldsymbol{A}]}\left\{ \begin{array}{c} s\{r_c C(1-D)[r_c + R(1-D)] - L\} + \\ + \left[(2r_c + R)(1-D)^2 - r_L\right] \end{array} \right\} \frac{V_o}{RLC(1-D)} - \frac{r_c}{R(1-D)} V_o \\
&= \left(\frac{s\{r_c RC(1-D)[r_c + R(1-D)] - RL\} + R\left[(2r_c + R)(1-D)^2 - r_L\right]}{s^2 RLC + s\{L + RC[r_c(1-D) + r_L]\} + [r_c(1-D) + r_L] + R(1-D)^2} - r_c \right) \frac{V_o}{R(1-D)} \\
&= \left(\frac{-s^2 RLCr_c + s\left[r_c R^2 C(1-D)^2 - RL - r_c r_L RC\right] + \left[R^2(1-D)^2 - r_L R - r_c^2(1-D)\right]}{s^2 RLC + s\{L + RC[r_c(1-D) + r_L]\} + [r_c(1-D) + r_L] + R(1-D)^2} \right) \frac{V_o}{R(1-D)}
\end{aligned} \tag{7.148}$$

In calculations to obtain Eq. 7.148, the approximation $R + r_c \approx R$ is done.

On the basis of Eq. 7.119, the transfer function of the output voltage versus the input voltage can be determined.

$$\frac{\tilde{v}_o(s)}{\tilde{v}_s(s)} = \boldsymbol{C}[s\boldsymbol{I} - \boldsymbol{A}]^{-1}\boldsymbol{B}$$

$$= \frac{1}{\det[s\boldsymbol{I} - \boldsymbol{A}]}\begin{bmatrix} r_c(1-D) & 1 \end{bmatrix}\begin{bmatrix} \frac{sRC+1}{RC} & -\frac{1}{L}(1-D) \\ \frac{1}{C}(1-D) & \frac{sL+r_c(1-D)+r_L}{L} \end{bmatrix}\begin{bmatrix} 1/L \\ 0 \end{bmatrix}$$

$$= \frac{1}{\det[s\boldsymbol{I} - \boldsymbol{A}]}\begin{bmatrix} r_c(1-D) & 1 \end{bmatrix}\begin{bmatrix} \frac{sRC+1}{LRC} \\ \frac{1}{LC}(1-D) \end{bmatrix} = \frac{R(sr_cC+1)(1-D)}{LRC\det[s\mathrm{I} - \mathrm{A}]}$$

$$= \frac{R(sr_cC+1)(1-D)}{s^2RLC + s\{L + RC[r_c(1-D) + r_L]\} + [r_c(1-D) + r_L] + R(1-D)^2} \tag{7.149}$$

where the approximation $R + r_c \approx R$ is done.

7.13 State Space Averaging of Buck-Boost Converter

In on state, the circuit is the same of the boost in *on* state, then its equations are the Eqs. 7.65 and 7.67, recalled in Sect. 7.12 and rewritten here for clarity.

$$\begin{cases} \dfrac{\mathrm{d}i_L}{\mathrm{d}t} = -\dfrac{r_L}{L}i_L + \dfrac{1}{L}V_s \\ \dfrac{\mathrm{d}v_c}{\mathrm{d}t} = -\dfrac{v_c}{RC} \end{cases}$$

$$V_o = v_c\left(\frac{R}{R - r_c}\right) \cong v_c$$

As a consequence, the coefficient matrices in *on* state are the same as those obtained for the boost in *on* state, i.e., the matrices in Eq. 7.133, rewritten here for clarity

$$\begin{cases} \boldsymbol{A}_{\mathrm{ON}} = \begin{bmatrix} -\frac{r_L}{L} & 0 \\ 0 & -\frac{1}{CR} \end{bmatrix} \\ \boldsymbol{B}_{\mathrm{ON}} = \begin{bmatrix} 1/L \\ 0 \end{bmatrix} \\ \boldsymbol{C}_{\mathrm{ON}} = [0 \quad 1] \\ \boldsymbol{D}_{\mathrm{ON}} = 0 \end{cases}$$

In *off* state, the circuit can be deduced by the buck circuit in *on* state, setting V_s equal to zero. The corresponding equations are Eq. 7.150, obtained from Eq. 7.41a with the position $V_s = 0$, and Eq. 7.42a, recalled in Sect. 7.12 and rewritten here.

$$\begin{cases} \dfrac{di_L}{dt} = -\dfrac{r_c + r_L}{L} i_L - \dfrac{1}{L} v_c \\ \dfrac{dv_c}{dt} = \dfrac{1}{C} i_L - \dfrac{1}{CR} v_c \end{cases} \tag{7.150}$$

$$V_o = r_c i_L + v_c$$

The coefficient matrices in *off* state are deduced by the buck in the same state.

$$\begin{cases} \boldsymbol{A}_{\text{OFF}} = \begin{bmatrix} -\frac{r_c+r_L}{L} & -\frac{1}{L} \\ \frac{1}{C} & -\frac{1}{CR} \end{bmatrix} \\ \boldsymbol{B}_{\text{OFF}} = \begin{bmatrix} 0 \\ 0 \end{bmatrix} \\ \boldsymbol{C}_{\text{OFF}} = \begin{bmatrix} r_C & 1 \end{bmatrix} \\ \boldsymbol{D}_{\text{OFF}} = 0 \end{cases} \tag{7.151}$$

Comparing Eq. 7.151 with Eq. 7.134, it can be observed that only the $\boldsymbol{B}_{\text{OFF}}$ of the buck-boost is different from the corresponding boost matrix, having all its elements null.

As for matrices $\boldsymbol{A}$, $\boldsymbol{B}$ and $\boldsymbol{C}$, they can be calculated according to Eq. 7.100 and Eq. 7.102.

$$\begin{cases} \boldsymbol{A} = \boldsymbol{A}_{\text{ON}} D + \boldsymbol{A}_{\text{OFF}}(1-D) = \begin{bmatrix} \frac{-r_c(1-D)-r_L}{L} & -\frac{1}{L}(1-D) \\ \frac{1}{C}(1-D) & -\frac{1}{CR} \end{bmatrix} \\ \boldsymbol{B} = \boldsymbol{B}_{\text{ON}} D + \boldsymbol{B}_{\text{OFF}}(1-D) = \boldsymbol{B}_{\text{ON}} D = \begin{bmatrix} 1/L \\ 0 \end{bmatrix} D \\ \boldsymbol{C} = \boldsymbol{C}_{\text{ON}} D + \boldsymbol{C}_{\text{OFF}}(1-D) = \begin{bmatrix} r_c(1-D) & 1 \end{bmatrix} \end{cases} \tag{7.152}$$

It should be noted that the matrices $\boldsymbol{A}$ and $\boldsymbol{C}$ coincide with the corresponding matrices obtained for the boost.

On the other hand, the matrix $\boldsymbol{B}$ is equal to the corresponding one of the boost multiplied by the duty cycle D.

The static gain is obtained from Eq. 7.113

$$\frac{V_O}{V_s} = -\boldsymbol{C}\boldsymbol{A}^{-1}\boldsymbol{B} = -\frac{\boldsymbol{1}}{\det \boldsymbol{A}} \begin{bmatrix} r_c(1-D) & 1 \end{bmatrix} \begin{bmatrix} -\frac{1}{CR} & \frac{1}{L}(1-D) \\ -\frac{1}{C}(1-D) & \frac{r_c(D-1)-r_L}{L} \end{bmatrix} \begin{bmatrix} 1/L \\ 0 \end{bmatrix} D \tag{7.153}$$

the obtained expression is the same of that obtained for the boost converter multiplied by D; comparing Eq. 7.136 with Eq. 7.153 it follows that:

$$\frac{V_0}{V_s} = \frac{RD(1-D)}{(1-D)^2 R + r_L + r_c(1-D)} \tag{7.154}$$

where the approximation $R + r_c \approx R$ is done.

The output voltage to duty cycle transfer function is obtained by Eq. 7.118 that becomes:

$$\frac{\tilde{v}_0(s)}{\tilde{d}(s)} = \boldsymbol{C}[s\boldsymbol{I} - \boldsymbol{A}]^{-1}[(\boldsymbol{A}_{\rm ON} - \boldsymbol{A}_{\rm OFF})\boldsymbol{x}_0 + \boldsymbol{B}_{\rm ON} V_s] + (\boldsymbol{C}_{\rm ON} - \boldsymbol{C}_{\rm OFF})\boldsymbol{x}_0$$

The term $(\boldsymbol{A}_{\rm ON} - \boldsymbol{A}_{\rm OFF})\boldsymbol{x}_{\rm o}$ is given by:

$$(\boldsymbol{A}_{\rm ON} - \boldsymbol{A}_{\rm OFF})\boldsymbol{x}_0 = \begin{bmatrix} \frac{r_c}{L} & \frac{1}{L} \\ -\frac{1}{C} & 0 \end{bmatrix} \begin{bmatrix} \frac{D}{R(1-D)} \\ 1 \end{bmatrix} V_o = \begin{bmatrix} \frac{r_c D + R(1-D)}{RL(1-D)} \\ -\frac{D}{RC(1-D)} \end{bmatrix} V_o \tag{7.155}$$

Therefore,

$$\begin{aligned} (\boldsymbol{A}_{\rm ON} - \boldsymbol{A}_{\rm OFF})\boldsymbol{x}_0 + \boldsymbol{B}_{ON} V_s &= \begin{bmatrix} \frac{r_c + R(1-D)}{RL(1-D)} + \frac{(1-D)}{DL} \\ -\frac{D}{RC(1-D)} \end{bmatrix} V_o \\ &= \begin{bmatrix} \frac{r_c D^2 + RD(1-D) + R(1-D)^2}{RLD(1-D)} \\ -\frac{D}{RC(1-D)} \end{bmatrix} V_o \end{aligned} \tag{7.156}$$

Being

$$\begin{aligned} &\boldsymbol{C}[s\boldsymbol{I} - \boldsymbol{A}]^{-1} \\ &= \frac{1}{\det[s\boldsymbol{I} - \boldsymbol{A}]} \begin{bmatrix} r_c(1-D) & 1 \end{bmatrix} \begin{bmatrix} \frac{sRC+1}{RC} & -\frac{1}{L}(1-D) \\ \frac{1}{C}(1-D) & \frac{sL + r_c(1-D) + r_L}{L} \end{bmatrix} \\ &= \frac{1}{\det[s\boldsymbol{I} - \boldsymbol{A}]} \begin{bmatrix} \frac{r_c(1-D)(sRC+1) + R(1-D)}{RC} & \frac{-r_c(1-D)^2 + sL + r_c(1-D) + r_L}{L} \end{bmatrix} \end{aligned} \tag{7.157}$$

it follows that:

$$\begin{aligned} &\boldsymbol{C}[s\boldsymbol{I} - \boldsymbol{A}]^{-1}[(\boldsymbol{A}_{\rm ON} - \boldsymbol{A}_{\rm OFF})\boldsymbol{x}_0 + \boldsymbol{B}_{ON} V_s] \\ &= \frac{1}{\det[s\boldsymbol{I} - \boldsymbol{A}]} \begin{bmatrix} \frac{r_c(1-D)(sRC+1) + R(1-D)}{RC} & \frac{-r_c(1-D)^2 + sL + r_c(1-D) + r_L}{L} \end{bmatrix} \cdot \\ &\begin{bmatrix} \frac{r_C D^2 + RD(1-D) + R(1-D)^2}{RLD(1-D)} \\ -\frac{D}{RC(1-D)} \end{bmatrix} V_o \\ &= \frac{1}{\det[s\boldsymbol{I} - \boldsymbol{A}]} \frac{V_o}{RLC(1-D)D} \left\{ s\left[Cr_c^2(1-D)D^2 + Cr_c R(1-D)^2 - LD^2 \right] + R(1-D)^2 - r_L D^2 \right\} \\ &= \frac{s\left[Cr_c^2(1-D)D^2 + Cr_c R(1-D)^2 - LD^2 \right] + R(1-D)^2 - r_L D^2}{(1-D)D(s^2 RLC + s\{L + RC[r_c(1-D) + r_L]\} + [r_c(1-D) + r_L] + R(1-D))} V_o \end{aligned} \tag{7.158}$$

where the term $r_c D$ has been neglected respect to the term R.

Table 7.2 Outline of the static gain including parasitic parameters and of the output voltage versus duty cycle and input voltage for buck, boost, and buck-boost converters

	Buck
$\frac{V_o}{V_s}$	$D\frac{R+r_c}{R+r_c+r_L}$
$\frac{\tilde{v}_o(s)}{\tilde{d}(s)}$	$V_o\frac{sr_c+1}{LC\left[s^2+s\left(\frac{1}{RC}+\frac{r_c+r_L}{L}\right)+\frac{1}{LC}\right]}$
$\frac{\tilde{v}_o(s)}{\tilde{v}_s(s)}$	$D\frac{sr_c+1}{LC\left[s^2+s\left(\frac{1}{RC}+\frac{r_c+r_L}{L}\right)+\frac{1}{LC}\right]}$
	Boost
$\frac{V_o}{V_s}$	$\frac{R(1-D)}{(1-D)^2R+r_L+r_c(1-D)}$
$\frac{\tilde{v}_o(s)}{\tilde{d}(s)}$	$\left(\frac{-s^2RLCr_c+s[r_cR^2C(1-D)-RL-r_cr_LRC]+\left[R^2(1-D)^2-r_LR-r_c^2(1-D)\right]}{s^2RLC+s\{L+RC[r_c(1-D)+r_L]\}+[r_c(1-D)+r_L]+R(1-D)^2}\right)\frac{V_o}{R(1-D)}$
$\frac{\tilde{v}_o(s)}{\tilde{v}_s(s)}$	$\frac{R(sr_cC+1)(1-D)}{s^2RLC+s\{L+RC[r_c(1-D)+r_L]\}+[r_c(1-D)+r_L]+R(1-D)^2}$
	Buck-boost
$\frac{V_o}{V_s}$	$\frac{RD(1-D)}{(1-D)^2R+r_L+r_c(1-D)}$
$\frac{\tilde{v}_o(s)}{\tilde{d}(s)}$	$\left[\frac{s[Cr_c^2(1-D)D^2+Cr_cR(1-D)^2-LD^2]+R(1-D)^2-r_LD^2}{(1-D)D(S^2RLC+s\{L+RC[r_c(1-D)+r_L]\}+[r_c(1-D)+r_L]+R(1-D))}+\frac{r_cD}{R(1-D)}\right]V_o$
$\frac{\tilde{v}_o(s)}{\tilde{v}_s(s)}$	$\frac{R(sr_cC+1)(1-D)D}{s^2RLC+s\{L+RC[r_c(1-D)+r_L]\}+[r_c(1-D)+r_L]+R(1-D)^2}$

The term $(\boldsymbol{C}_{\text{ON}} - \boldsymbol{C}_{\text{OFF}})\boldsymbol{x}_o$ is given by:

$$(\boldsymbol{C}_{\text{ON}} - \boldsymbol{C}_{\text{OFF}})\boldsymbol{x}_0 = \begin{bmatrix} -r_c & 0 \end{bmatrix}\begin{bmatrix} i_{L0} \\ v_{c0} \end{bmatrix} = -\frac{r_cD}{R(1-D)}V_o \tag{7.159}$$

Therefore, the final expression of the output voltage to duty cycle transfer function is given by:

$$\frac{\tilde{v}_0(s)}{\tilde{d}(s)} = \left[\frac{s\left[Cr_c^2(1-D)D^2 + Cr_cR(1-D)^2 - LD^2\right] + R(1-D)^2 - r_LD^2}{(1-D)D(s^2RLC + s\{L + RC[r_c(1-D) + r_L]\} + [r_c(1-D) + r_L] + R(1-D))} + \right.$$
$$\left. - \frac{r_cD}{R(1-D)} \right] V_o =$$
$$= \left[\frac{-s^2RLCr_cD^2 + s\left[R^2Cr_c(1-D)^2 - RCr_cr_LD^2 - LRD^2\right] + R^2(1-D)^2 - RD^2[r_L + r_c(1-D)]}{(s^2RLC + s\{L + RC[r_c(1-D) + r_L]\} + [r_c(1-D) + r_L] + R(1-D))} \right] \cdot$$
$$\cdot \frac{1}{RD(1-D)} V_o \tag{7.160}$$

The transfer function of the output voltage versus the input voltage can be determined considering Eq. 7.119 written for the buck-boost. It can be easily noted that the same expression of the boost multiplied for D is obtained; from Eq. 7.149 it follows that:

$$\frac{\tilde{v}_0(s)}{\tilde{v}_s(s)} = \boldsymbol{C}[s\boldsymbol{I} - \boldsymbol{A}]^{-1}\boldsymbol{B}$$
$$= \frac{1}{\det[s\boldsymbol{I} - \boldsymbol{A}]} \begin{bmatrix} r_c(1-D) & 1 \end{bmatrix} \begin{bmatrix} \frac{sRC+1}{RC} & \frac{1}{L}(1-D) \\ \frac{1}{C}(1-D) & \frac{sL - r_c(D-1) + r_L}{L} \end{bmatrix} \begin{bmatrix} 1/L \\ 0 \end{bmatrix} D$$
$$= \frac{R(sr_CC + 1)(1-D)D}{s^2RLC + s\{L + RC[r_c(1-D) + r_L]\} + [r_c(1-D) + r_L] + R(1-D)^2} \tag{7.161}$$

7.14 Synopsis

Table 7.2 outlines, for buck, boost, and buck-boost converters: the static gain including parasitic parameters, the output voltage to duty cycle transfer function and the output voltage to input voltage transfer function.

The effect of parasitic parameters implies a lessening of the output voltage in open loop operating conditions. In the transfer functions all the converters behave as second order systems, where their pole/zero position is influenced by the load value and by the duty cycle as well.

7.15 Conclusions

DC/DC switching converters allow a DC voltage to be transformed into a regulated DC value. A lower voltage than the input one or a higher or both can be obtained by using buck, boost or buck-boost converter, respectively.

Best performance is obtained by a feedback control action because the effect of parasitic parameters implies a lessening of the output voltage in open loop operating conditions.

In order to design the appropriate control network, a small signal analysis is necessary to obtain the converter transfer function in the *s*-domain. Comparing the transfer functions of the different converters, it can be noted that they depend both on the converter duty cycle and load value. However, this dependency is simpler in the case of the buck converter, making it more easily controllable and, as a result, a good option for the PV source emulation purpose.

Bibliography

Ang S, Oliva A (2011) Power-switching converters. CRC press, Boca Raton FL

Gottleb MI (1984) Power supplies switching regulators inverters & converters. Tab Books, Blue Ridge summit, PA

Mohan M, Undeland TM, Robbins WP (1995) Power Electronics. Wiley, NY

Ramshaw RS (1994) Power electronics semiconductor switch. Chapman & Hall, London

Chapter 8
Feedback Control of the DC/DC Converters for PV Source Emulation

8.1 Negative Feedback Classical Control

In general, the aim of a voltage source like a DC/DC converter consists in obtaining a voltage equal to a desired value assumed as the reference value.

As explained in Chap. 7, in practical cases, the output of a DC/DC converter is different from the theoretical value. Several factors contribute to this phenomenon, some of these are tied to the parasitic parameters of the converter components, they are quite constant during operation but can vary from one device to another and should be measured. Other parameters as the duty cycle and the load value normally vary and modify the output value as well.

For these reasons, the open-loop operation is not satisfactory and a feedback action is required. In this case, the output is continuously measured and its value is compared to the reference one. Depending on the difference between them, a correction action is imposed to the converter.

Moreover, for a PV source emulation, in which the output voltage has to follow a reference that is related to the source to be emulated, a feedback action allows the desired dynamical behavior of the real source to be obtained too.

8.1.1 Closed-Loop Gain

Figure 8.1 shows a block diagram of a closed-loop system. The original system has a transfer function A; a feedback network with transfer function β is added.

The input of the closed-loop system is the desired output value V_{ref}; the obtained output V_o is processed by the feedback network, then the difference with the reference value, that represents the error signal, is the input of the system to be controlled.

M. C. Di Piazza and G. Vitale, *Photovoltaic Sources*, Green Energy and Technology, DOI: 10.1007/978-1-4471-4378-9_8, © Springer-Verlag London 2013

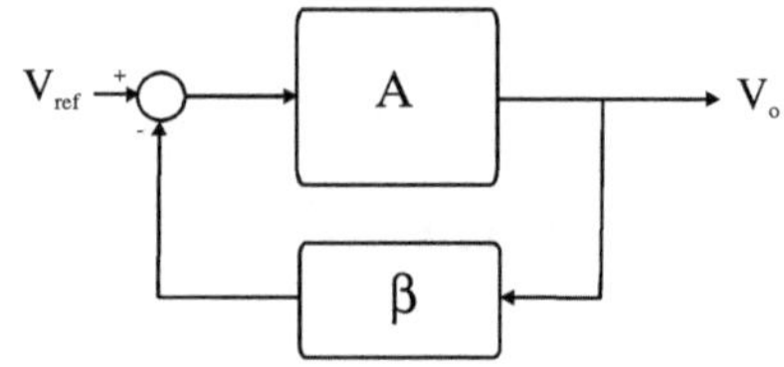

Fig. 8.1 Block diagram representation of a closed-loop system

The closed-loop transfer function is given by:

$$\frac{V_o}{V_{\text{ref}}} = \frac{A}{1 + \beta A} \tag{8.1}$$

The open loop transfer function is defined as:

$$L = \beta A \tag{8.2}$$

Some considerations are worth being considered. Firstly, if the open loop gain is sufficiently high, Eq. (8.1) becomes:

$$\frac{V_o}{V_{\text{ref}}} \approx \frac{1}{\beta}\bigg|_{A \gg \beta} \tag{8.3}$$

In this case, the behavior of the feedback system depends only on the feedback network.

Usually, the denominator of Eq. (8.1) depends both on the system and feedback network transfer functions. Some frequencies could nullify the value of $1 + \beta A$; for these values, the Eq. (8.1) would not be defined. In practice, this lead to a system instability. This situation must be avoided; to this aim, some techniques as compensation networks, or pole placement are adopted. They will be explained in the following.

8.1.2 Stability Analysis

The loss of stability for a system, defined by Eq. (8.1), can be described by the following conditions:

$$\begin{cases} |\beta A| = 1 \\ \text{phase}(\beta) + \text{phase}(A) = 180^\circ \end{cases} \tag{8.4}$$

In this case, the denominator of Eq. (8.1) has a zero that lies in the imaginary axis. From a physical point of view, if the conditions expressed by Eqs. (8.4) occur, the input signal of the feedback system goes through the loop formed by the block described by transfer function A, the feedback network β, then it is rotated of 180° by the minus sign and it is superimposed to the original signal with the same amplitude and phase. A new signal with an amplitude twice of the original is obtained and so on, until a nonlinearity occurs. In such a condition, oscillations, high amplitudes, and the loss of the small signal hypothesis make the system uncontrollable.

Moreover, Eqs. (8.4) express a limit condition that can be reached during normal operation due to several factors as noise, ageing, etc. For these reasons, it is preferable to design the feedback system so to work quite far from Eq. (8.4). This is known as *relative stability* of the system. Two parameters are introduced to characterize the relative stability of the system; they represent how much the system is far from the conditions expressed by Eqs. (8.4).

The first parameter is the *gain margin*, it is defined as the gain, calculated on the open-loop transfer function L, to drive the system into instability. In other words, it is the amount of gain needed to obtain a unitary gain of L when its phase is equal to 180°. Usually the gain is expressed in dB, as a consequence the gain margin is the quantity to be added to the gain of L to obtain 0 dB, when its phase is 180°.

The second parameter is the *phase margin*, it defined as the opposite of the phase shift necessary to make the phase of L equal to $-180°$, when the gain is 0 dB.

To ensure stability, usually a gain margin of 6 dB and a phase margin of about 45° have to be assured; when this is not provided directly from the open loop transfer function, an additional network must be designed.

Both the gain and phase margins can be deduced by the Bode plot of the open-loop transfer function.

As an example of a stable system, the open-loop transfer function defined by Eq. (8.5) can be considered. The feedback network transfer function is supposed unitary for simplicity.

$$L_1(s) = \frac{0.4}{s(s^2+s+1)} \tag{8.5}$$

Its bode plot can be drawn by the following MATLAB® script:

```
*********************************************
%transfer function parameters

a=1;
b=1;
gain=0.4;


s=tf('s');

L=gain/(s*(s^2+a*s+b));

MARGIN (L);

hold
******************************************
```

where the use of the command MARGIN allows both gain and phase margins to be calculated (Fig. 8.2).

It can be noted that, when the gain curve reaches the value of 0 dB (unitary gain condition), the phase is about 61°; when the phase is equal to 180° the gain is about –7.9 dB. These conditions characterize the system defined by Eq. (8.5) as stable.

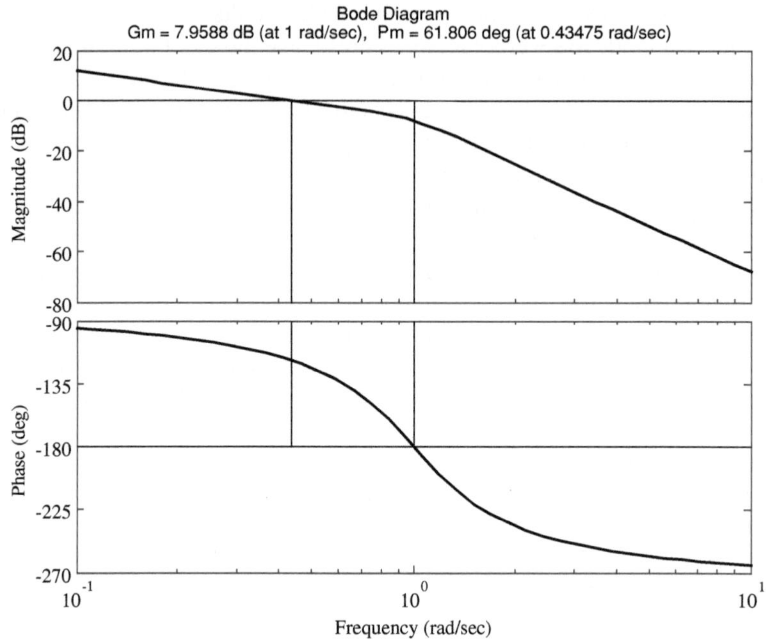

Fig. 8.2 Bode plot of the transfer function defined by Eq. (8.5)

The pole position of the transfer function described by Eq. (8.5) can be obtained by POLE function in MATLAB®

```
*********************************************
>> pole(L)

ans =

         0
  -0.5000 + 0.8660i
  -0.5000 - 0.8660i

*********************************************
```

The system has one pole at the origin and a couple of complex conjugate poles with negative real part.

The system defined by the open-loop transfer function in Eq. (8.6) is now considered. The Eq. (8.6) has a pole at the origin and two real negative poles equal to $p_1 = -9.899$ and $p_2 = -0.1010$. By observing its bode diagram, it can be noted that the system, nevertheless the presence of two real and negative poles, when operated in a closed loop with a unitary feedback ($\beta = 1$ in Fig. 8.1), is unstable. As a matter of fact, the gain margin is negative and the phase margin approaches 0°, it corresponds to superimpose, inside the feedback loop, signals with the same phase and equal amplitude (Fig. 8.3).

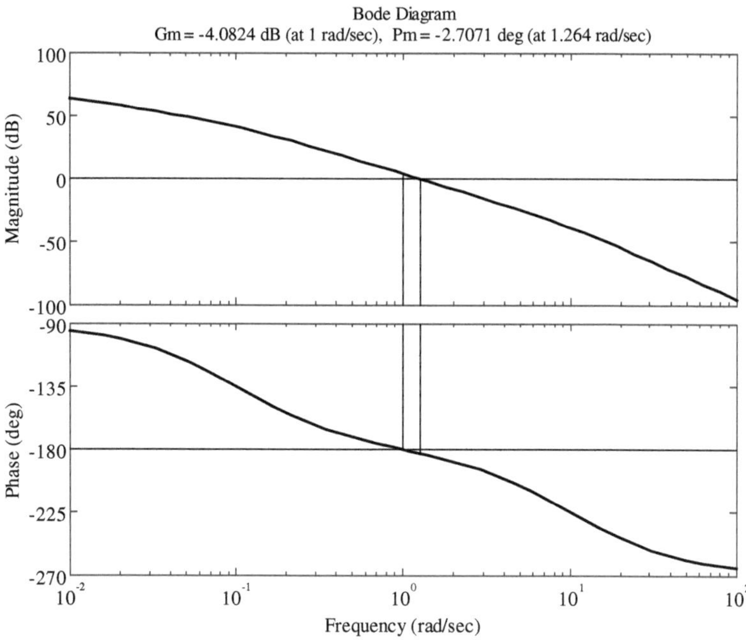

Fig. 8.3 Bode plot of the transfer function defined by Eq. (8.6)

$$L_2(s) = \frac{16}{s(s^2 + 10s + 1)} \tag{8.6}$$

Finally, the open loop transfer function in Eq. (8.7), characterized by a pole at the origin and two complex conjugate poles $p_{1,2} = -0.0500 \pm j0.9987$, represents a critical system in which the mere calculus of gain and phase margin would give a stable system; on the other hand, by the bode plot it is evident that a resonance occurs and the related peak reaches the unitary gain when the phase is 180°. In this condition, the system can become unstable (Fig. 8.4).

$$L_3(s) = \frac{0.08}{s(s^2 + 0.1s + 1)} \tag{8.7}$$

From what explained above, the bode diagram of the open-loop transfer function allows the stability to be studied without the calculus of the closed-loop transfer function. This is usually more complicated; for example, for the system defined by Eq. (8.6), the closed-loop transfer function is given by

$$C_2(s) = \frac{L_2(s)}{1 + \beta L_2(s)} = \frac{\frac{16}{s(s^2+10s+1)}}{1 + \beta \frac{16}{s(s^2+10s+1)}} \tag{8.8}$$

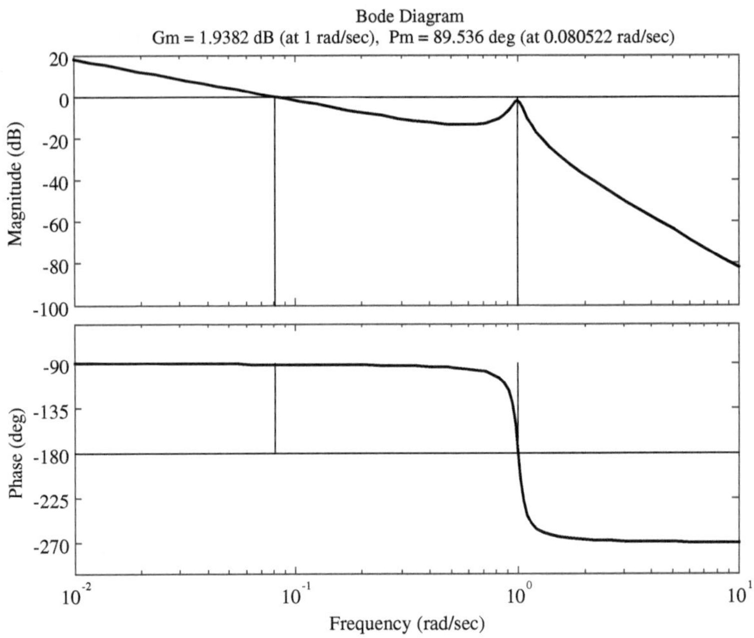

Fig. 8.4 Bode plot of the transfer function defined by Eq. (8.7)

The pole values of $G_2(s)$, for $\beta = 1$ can be calculated by the command RLOCUS in MATLAB®, where L is the open-loop transfer function.

```
**********************************************
>> R=RLOCUS(L,1)

R =

 -10.0587
   0.0294 + 1.2609i
   0.0294 - 1.2609i
**********************************************
```

It can be noted that the feedback system has a couple of complex conjugate poles with positive real part.

In general, the β value can be considered as a parameter, in this case the poles of the closed-loop transfer function describe a locus known as *root locus*. It can be plotted by using the command RLOCUS (H) in MATLAB®.

From what is shown above, a system having an open-loop transfer function with all negative real part poles can become unstable in a feedback connection or its gain and phase margin can be unsatisfactory. In this case, the open-loop transfer function needs a correction obtained by a properly designed compensation networks.

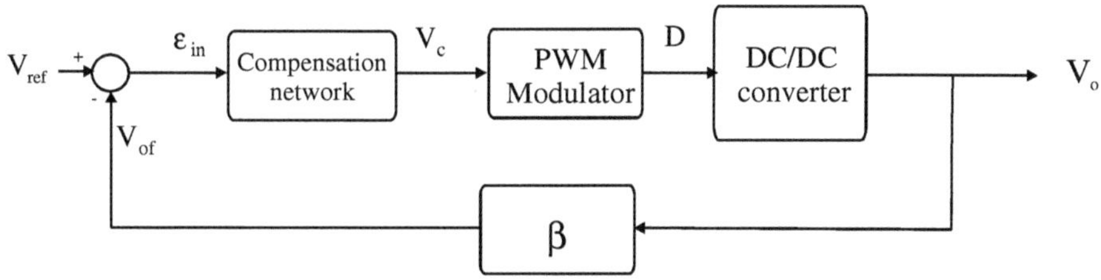

Fig. 8.5 Feedback structure of a DC/DC converter

8.2 Feedback Structure of a DC/DC Converter

A DC/DC converter in a feedback structure can be represented as in Fig. 8.5.

An error signal ε_{in} is obtained by subtracting the output voltage V_o, processed by the feedback network, from the desired voltage signal V_{ref}. A compensation network, having ε_{in} as input, provides the control voltage V_c that is applied to the modulator from which the duty cycle is obtained.

The DC/DC converter transfer function (for the case of buck, boost and buck-boost topologies) has been calculated in Chap. 7, here the other parts are deduced.

8.2.1 Feedback Network Transfer Function

The feedback networks can be formed by a voltage transducer or simply by a voltage divider (see Sect. 7.2.1); their aim is to manage a reduced signal compared to the output value. In the case of a voltage divider, the transfer function is given by:

$$V_{\text{of}} = \frac{R_2}{R_1 + R_2} V_o \tag{8.9}$$

8.2.2 Pulse Width Modulator Transfer Function

The pulse width modulation (PWM) provides the duty cycle D obtained, for example, by comparing a sawtooth signal with the control voltage V_c. This operating mode is called voltage mode control.

The duty cycle ranges from zero to unitary value. The sawtooth frequency corresponds to the DC/DC converter switching frequency. When the control voltage equals the maximum value of the sawtooth, the duty cycle approaches 1. The modulator gain is given by:

$$G_{\mathrm{PWM}} = \begin{cases} 1 & V_c \geq V_{\mathrm{sw}} \\ \frac{1}{V_{\mathrm{sw}}} & 0 > V_c > V_{\mathrm{sw}} \\ 0 & V_c < 0 \end{cases} \tag{8.10}$$

Therefore, any duty cycle value can be obtained by the product of the modulator gain G_{PWM} and the control voltage V_c.

The first and the last conditions in Eq. (8.10) must be avoided to maintain the linear operating conditions.

In current mode control, the duty cycle is obtained by comparing the inductor current with a reference one. This operating mode is not treated here.

8.2.3 Compensation Networks

The aim of a compensation network is to assure the stability of the feedback system by a suitable gain and phase margin. Moreover, a correct design allows the closed-loop system poles to give the desired transient time-domain response.

A great part of compensating networks are based on a linear combination of the error signal of its integral and of its derivative. Among them, the proportional-integrative (PI) network and the proportional-integrative-derivative (PID) network are commonly used.

The design of a compensation network comes from a trade-off among several constraints. The three main system characteristics are:

1. Steady-state performance: they are usually expressed in terms of the maximum error for a given input (step or ramp) and are improved by the presence of integrators.
2. Transient performance: the system is required to satisfy the constraint in terms of rise time, settling time, bandwidth, frequency oscillation, and damping in time and frequency domain, respectively. The increase of the gain or the presence of a derivative action improve the dynamic performance, on the contrary the presence of an integrator has the opposite effect.
3. Stability: by increasing the gain or by adding integrators the stability margins are diminished, on the contrary a derivative action increase the damping of the system, supporting stability.

Since all constraint cannot be contemporarily satisfied, first the stability must be guaranteed, then the others parameters are considered according to the main interest of the designer.

The main components of a compensation network perform proportional (P), derivative (D), or integrator (I) action. It is possible to employ a linear combination of them.

The use of a mere proportional action reduces the steady-state error as its gain is increased but it can lead the system to instability; to reduce this error an integrator action can be introduced so that, when the error is different from zero, the

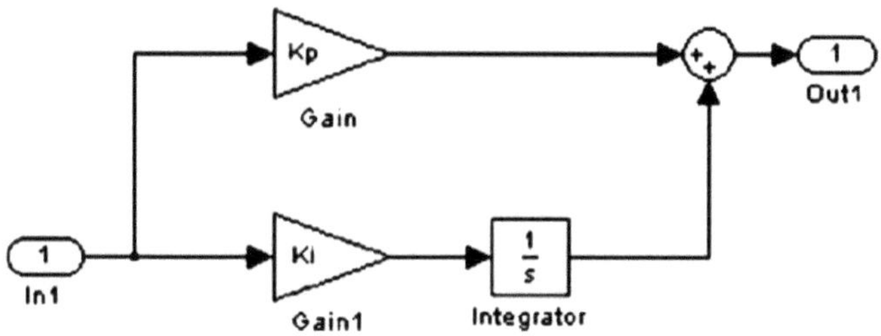

Fig. 8.6 Structure of PI compensating network

integrator increase its contribution. The overall effect of a proportional-integral controller (PI) is to maintain a high gain only at low frequencies.

By using a proportional-derivative-integral (PID) controller, the designer can use three parameters to satisfy the system requirements. The derivative effect reduces the rise time and increases the overshoot of the system response; however, high frequency noise is amplified.

8.2.3.1 PI Compensation

The PI compensation is obtained by adding the input signal, multiplied for a gain K_p, to an other signal obtained by the integration of the input signal multiplied for a gain K_i. A simple block diagram of a PI implemented in Simulink® implementation is given in Fig. 8.6.

The corresponding transfer function has a pole at the origin and a zero whose value is given by $z = K_i/K_p$.

$$\frac{V_c}{\varepsilon_{\text{in}}} = \left(K_p + \frac{K_i}{s}\right) = K_p \frac{s+z}{s} \tag{8.11}$$

It can be noted that Eq. (8.11) exhibits a constant gain for high frequencies. On the other hand, in the case of a DC/DC converter, it is mandatory that, near to the switching frequency, the open-loop gain is sufficiently low to reduce the ripple at the switching frequency. In this case, and additional low pass filter is added to the PI as shown in Fig. 8.7. The obtained transfer function is given by Eq. (8.12).

$$\frac{V_c}{\varepsilon_{\text{in}}} = K_p \frac{s+z}{s(s+p)} \tag{8.12}$$

Figure 8.8 shows the bode diagram of both a PI compensator with $K_p = 3.16$ (corresponding to 10 dB), $z = 1000$ rad/s, and the same compensator with an additional high frequency pole $p = 20000$ rad/s. To compensate the attenuation due to this pole, in this last case, the value of K_p is equal to 63200.

It should be noted that both the networks show a decrease of the gain due to the pole at the origin and a flat zone with 10 dB gain in the region between the zero and the pole. The additional high frequency pole causes a further decrease of the

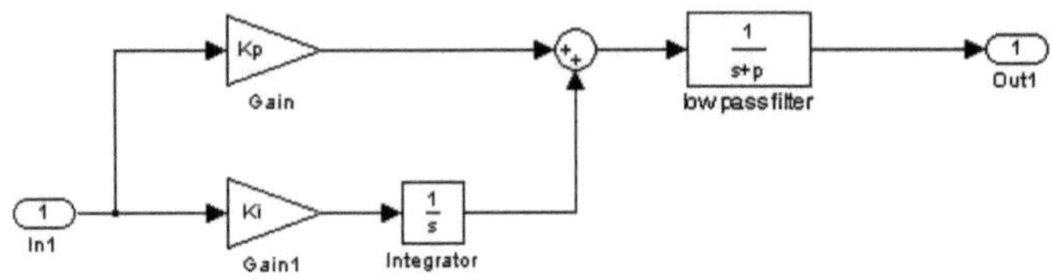

Fig. 8.7 Structure of PI compensating network with additional low pass filter

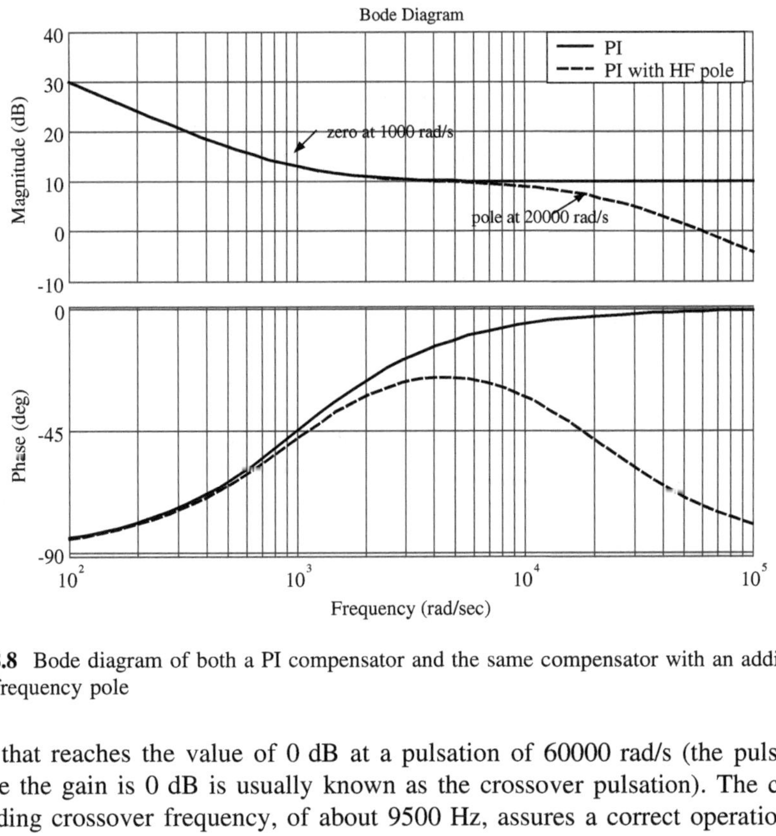

Fig. 8.8 Bode diagram of both a PI compensator and the same compensator with an additional high frequency pole

gain that reaches the value of 0 dB at a pulsation of 60000 rad/s (the pulsation where the gain is 0 dB is usually known as the crossover pulsation). The corresponding crossover frequency, of about 9500 Hz, assures a correct operation for switching frequencies greater than about 40kHz. As a matter of fact, it is necessary to have a switching frequency 4–5 times greater than the crossover frequency to avoid its circulation in the control feedback loop.

Finally, a PI compensator can be realized by a network with capacitors, resistors, and operational amplifiers. In this case, the additional pole presence cannot be avoided because a system physically realizable must have the number of poles greater than the number of zeros. However, if the open-loop transfer function has a satisfactory crossover frequency, the additional pole can be designed so to

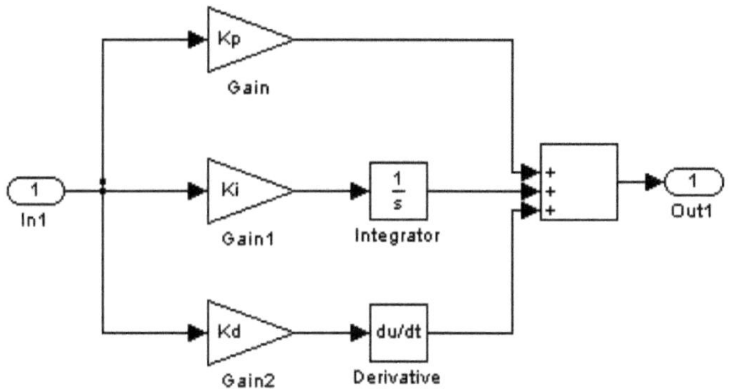

Fig. 8.9 Structure of PID compensating network

have a frequency much greater than the switching frequency. In this case, its presence does not affect the system behavior.

8.2.3.2 PID Compensation

The PID compensation is obtained by adding three signals: the first one is the input signal multiplied for a gain K_p, the second is the integral of the input signal multiplied for a gain K_i , and the last is the derivative of the input signal multiplied for a gain K_d.

A simple block diagram of a PID implemented in Simulink® is given in Fig. 8.9. The corresponding transfer function has a pole at the origin and two zeros. The corresponding transfer function is:

$$\frac{V_c}{\varepsilon_{\text{in}}} = \left(K_p + \frac{K_i}{s} + sK_d\right) = \frac{K_d s^2 + K_p s + K_i}{s} = K_d \frac{(s+z_1)(s+z_2)}{s} \tag{8.13}$$

To obtain a suitable crossover frequency, two high frequency poles can be added. In this case, Eq. (8.13) becomes:

$$\frac{V_c}{\varepsilon_{\text{in}}} = K_d \frac{(s+z_1)(s+z_2)}{s(s+p_1)(s+p_2)} \tag{8.14}$$

Figure 8.10 shows the bode diagram of both a PID compensator with $K_d = 3.16$, $z_1 = 100$ rad/s, $z_t = 1000$ rad/s and the same compensator with two additional high frequency poles $p_1 = 5000$ rad/s and $p_2 = 20000$ rad/s and a gain $K_d = 3.16\ 10^8$.

It should be noted that both the networks show a decrease of the gain due to the pole at the origin and a flat zone in the region between the two zeros. After the second zero, the gain increases with a slope of 20 dB/decade, in the case of a PID with additional poles, the gain trend becomes again flat for the presence of the first pole and finally it decreases with a slope of 20 dB/decade due to the second pole.

8.3 Complete State Feedback (Pole Placement Technique)

As it is known, the poles position is related to the transient response of the system, then, by the knowledge of the time-domain characteristic response, it is possible to determine the system poles which have caused that response.

The complete state feedback technique is based on the sensing of all the state variables of the system that are multiplied for a suitable gain; it allows the poles to be assigned. This last characteristic is of particular importance for system emulation. As a matter of fact, a correct system emulation requires the matching of transient response of the emulator with that of the system to be emulated. In the case of PV emulation, this matching can be achieved by imposing the closed-loop poles to the DC/DC converter used for the source emulation.

A continuous-time linear system is described by the following state space representation (see also (Eq. 7.93)):

$$\dot{\boldsymbol{x}} = \boldsymbol{A}\boldsymbol{x} + \boldsymbol{B}\boldsymbol{u} \tag{8.15}$$

where the input vector $\boldsymbol{u}$, that coincides with the control variable, can be expressed as the product of a matrix $\boldsymbol{F}$ multiplied for the state vector $\boldsymbol{x}$, so that:

$$\boldsymbol{u} = -\boldsymbol{F}\boldsymbol{x} \tag{8.16}$$

By substituting Eq. (8.16) in Eq. (8.15), it becomes:

$$\dot{\boldsymbol{x}} = (\boldsymbol{A} - \boldsymbol{B}\boldsymbol{F})\boldsymbol{x} \tag{8.17}$$

The closed-loop poles are the solutions of the characteristic equation:

$$\det[s\boldsymbol{I} - \boldsymbol{A} - \boldsymbol{B}\boldsymbol{F}] = 0 \tag{8.18}$$

It should be noted that the choice of the $\boldsymbol{F}$ matrix allows the closed-loop poles to be arbitrarily assigned.

For a DC/DC converter, the matrices deduced by the state-space averaging can be used. Equation (7.107), here rewritten concisely, can be considered.

$$\dot{\tilde{\boldsymbol{x}}} = \boldsymbol{A}\,\tilde{\boldsymbol{x}} + \boldsymbol{B}\,\tilde{\boldsymbol{u}} + \boldsymbol{E}\tilde{d} \tag{8.19}$$

Supposing that the perturbations act only on the duty cycle, Eq. (8.19) becomes:

$$\dot{\tilde{\boldsymbol{x}}} = \boldsymbol{A}\,\tilde{\boldsymbol{x}} + \boldsymbol{E}\tilde{d} \tag{8.20}$$

By the second of Eq. (8.10) and considering that the control voltage V_c equals the peak of sawtooth, used to obtain PWM, multiplied by the duty cycle, it follows that:

$$\tilde{d} = \left(\frac{D}{V_c}\right)\tilde{v}_c \tag{8.21}$$

The control voltage $\tilde{v}_c$ can be expressed as the product of a matrix $\boldsymbol{\Phi}$ multiplied for the state vector $\tilde{\boldsymbol{x}}$, so that:

$$\tilde{v}_c = -\boldsymbol{\phi}\tilde{\boldsymbol{x}} \tag{8.16a}$$

Now by substituting Eq. (8.21) in Eq. (8.20) and using Eq. (8.16a),

$$\dot{\tilde{\boldsymbol{x}}} = \left(\boldsymbol{A} - \boldsymbol{E}\frac{D}{V_c}\boldsymbol{\Phi}\right)\tilde{\boldsymbol{x}} \tag{8.22}$$

In this case, $\boldsymbol{\Phi}$ is a row vector with two coefficients that have to be determined to assign the poles. These coefficients can be calculated by the MATLAB® command:

```
**********************************************
>> F=PLACE(A, B, P);
**********************************************
```

where A and B are the matrices of the state space representation and P is the vector of the poles to be assigned.

For the buck converter with closed-loop matrix defined by Eq. (8.22), the coefficients of $\boldsymbol{\Phi}$ are calculated by the MATLAB® command:

```
**********************************************
>> F=PLACE(A,E*D/Vc, P);
**********************************************
```

Other two useful MATLAB® commands are:

```
**********************************************
% the vector poles contains the eigenvalues of the A matrix
>> poles = eig(A)
**********************************************
```

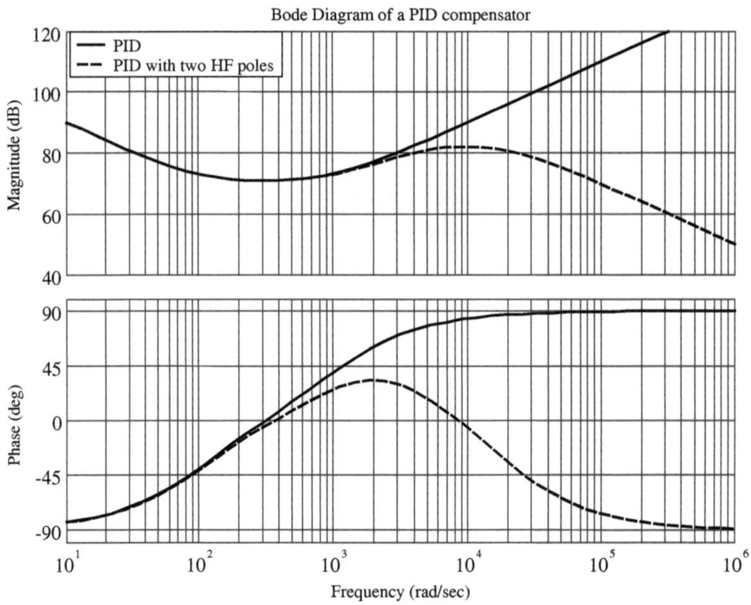

Fig. 8.10 Bode diagram of both a PID compensator and of the same compensator with two additional high frequency poles

and

```
*********************************************
% sys command defines the state space model
% step command calculates the step response of the system
>> sys = (A, B, C, 0)
>> step(sys)
*********************************************
```

In practise, after the definition of the system poles vector $\boldsymbol{P}$, the coefficient of $\boldsymbol{\Phi}$ are calculated, then the step response can be plotted for the feedback system to verify if the curve matches with that of the source to be emulated.

8.4 Enhanced Pole Placement for Buck Converter

Figure 8.11 shows a buck converter similar to that analyzed in Sect. 7.4 with an additional current generator in parallel to the resistive load. It models the current supplied to an inverter connected to its output. The variable w is the mean voltage supplied by the generator V_s.

The buck converter can be considered as supplied by a generator:

$$w = DV_s \tag{8.23}$$

that represents the control variable.

The circuit is described by equations similar to Eqs. (7.38) and (7.39) slightly modified for the presence of the generator I_{o1}.

$$\begin{cases} w = L\frac{di_L}{dt} + r_L i_L + V_o \\ V_o = v_c + r_c i_c \\ i_L = i_c + I_o + I_{o1} \end{cases} \tag{8.24}$$

$$V_o = R\left(i_L - C\frac{dv_c}{dt} - I_{o1}\right) \tag{8.25}$$

The Eqs. (8.24) and (8.25) can be solved for the time derivative of the state variables, i_L and v_c.

$$\begin{cases} \frac{di_L}{dt} = -\frac{Rr_c + Rr_L + r_c r_L}{L(R+r_c)} i_L - \frac{R}{L(R+r_c)} v_c + \frac{1}{L} w + \frac{Rr_c}{L(R+r_c)} I_{o1} \\ \frac{dv_c}{dt} = \frac{R}{C(R+r_c)} i_L - \frac{1}{C(R+r_c)} v_c - \frac{R}{C(R+r_c)} I_{o1} \end{cases} \tag{8.26}$$

Considering that $R \gg r_c$, the following state space representation is obtained:

$$\begin{bmatrix} \dot{i}_L \\ \dot{v}_C \end{bmatrix} = \begin{bmatrix} -\frac{r_c + r_L}{L} & -\frac{1}{L} \\ \frac{1}{C} & -\frac{1}{CR} \end{bmatrix} \begin{bmatrix} i_L \\ v_c \end{bmatrix} + \begin{bmatrix} \frac{1}{L} & \frac{r_c}{L} \\ 0 & -\frac{1}{C} \end{bmatrix} \begin{bmatrix} w \\ i_{o1} \end{bmatrix} \tag{8.27}$$

where:

$$\begin{cases} \boldsymbol{A} = \begin{bmatrix} -\frac{r_c + r_L}{L} & -\frac{1}{L} \\ \frac{1}{C} & -\frac{1}{CR} \end{bmatrix} \\ \boldsymbol{B} = \begin{bmatrix} \frac{1}{L} & \frac{r_c}{L} \\ 0 & -\frac{1}{C} \end{bmatrix} \end{cases} \tag{8.27a}$$

It should be noted that the introduction of the variable w allows the state-space averaging to be directly determined and the matrices $\boldsymbol{A}$ and $\boldsymbol{B}$ to be obtained with constant coefficients.

A new additional state variable is introduced, i.e., the integral of the error on the output voltage (approximated as the capacitor voltage).

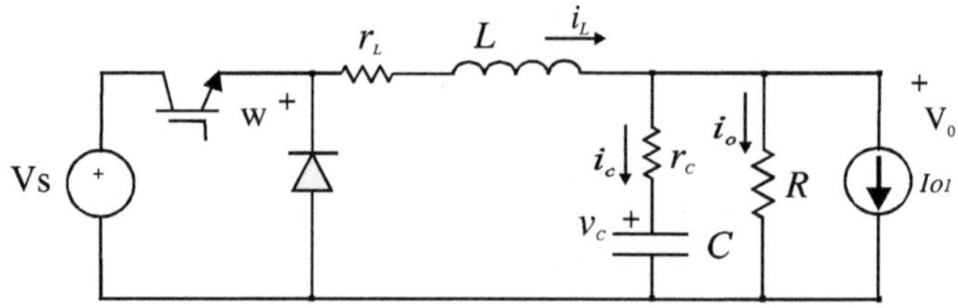

Fig. 8.11 Buck converter with additional current generator as load

$$x_1 = \int (v_C - V_{\text{ref}}) \mathrm{d}t \tag{8.28}$$

The other two state variables are the capacitor voltage and one to be chosen between the inductor current and the capacitor current.

It will be demonstrated, hereinafter, that the use of the capacitor current as a state variable is more advantageous.

Assuming that the state variables are the integral of the error on output voltage, the capacitor voltage and the capacitor current, the circuit of Fig. 8.11 is described by the equations:

$$\begin{cases} \dfrac{di_c}{dt} = -\dfrac{1}{RC} i_c - \dfrac{v_C}{L} + \dfrac{w}{L} - \dfrac{dI_{o1}}{dt} \\ \dfrac{dv_C}{dt} = \dfrac{i_c}{C} \\ \dfrac{dx_1}{dt} = v_c - V_{ref} \end{cases} \tag{8.29}$$

The first of Eq. (8.29) is obtained by the first of Eqs. (8.24).

The control variable given in Eq. (8.23) can be expressed as the product of a matrix multiplied for the state vector, as explained in Eq. (8.16):

$$w = -[g_1 \quad g_2 \quad g_3] \begin{bmatrix} i_c \\ v_c \\ x_1 \end{bmatrix} = -g_1 i_c - g_2 v_c - g_3 x_1 \tag{8.30}$$

where g_1, g_2, and g_3 are three gains to be determined to impose the position of the closed-loop poles.

By transforming Eqs. (8.29) and (8.30) in the Laplace domain:

$$\begin{cases} W(s) = -g_1 I_c(s) - g_2 V_c(s) - g_3 X_1(s) \\ sI_c(s) = -\frac{1}{RC} I_c(s) - \frac{1}{L} V_c(s) + \frac{1}{L} W(s) - sI_{o1}(s) \\ sCV_c(s) = I_c(s) \\ sX_1(s) = V_c(s) - V_{\text{ref}}(s) \end{cases} \tag{8.31}$$

After some manipulations, and eliminating $W(s)$, $X_1(s)$, and $I_c(s)$ in Eq. (8.31), a third order polynomial in s is obtained:

$$-s^2\frac{I_{o1}(s)}{CV_c(s)}+\frac{g_3}{LC}\frac{V_{ref}(s)}{V_c(s)}=s^3+s^2\left[\frac{1}{RC}+\frac{g_1}{L}\right]+s\left[\frac{1}{LC}+\frac{g_2}{LC}\right]+\frac{g_3}{LC} \tag{8.32}$$

By setting $I_{01}=0$ in Eq. (8.32), the closed-loop transfer function is obtained:

$$\frac{V_c(s)}{V_{ref}(s)}=\frac{g_3}{LC}\frac{1}{s^3+s^2\left[\frac{1}{RC}+\frac{g_1}{L}\right]+s\left[\frac{1}{LC}+\frac{g_2}{LC}\right]+\frac{g_3}{LC}}=\frac{g_3}{LC}\frac{1}{p(s)} \tag{8.33}$$

By setting $V_{ref}=0$ in Eq. (8.32), the output impedance is obtained:

$$Z_o(s)=\frac{V_c(s)}{I_{o1}(s)}=-\frac{s^2}{C}\frac{1}{p(s)} \tag{8.34}$$

where $p(s)$ is the characteristic polynomial of the system:

$$p(s)=s^3+s^2\left[\frac{1}{RC}+\frac{g_1}{L}\right]+s\left[\frac{1}{LC}+\frac{g_2}{LC}\right]+\frac{g_3}{LC} \tag{8.35}$$

Finally, it is possible to obtain the relationship between the closed-loop poles of the system and the gains g_1, g_2, and g_3, by equating Eq. (8.35) with the general expression of a three poles polynomial.

$$(s+p_1)(s+p_2)(s+p_3)=s^3+s^2\left[\frac{1}{RC}+\frac{g_1}{L}\right]+s\left[\frac{1}{LC}+\frac{g_2}{LC}\right]+\frac{g_3}{LC} \tag{8.36}$$

The obtained gains are given by:

$$\begin{cases} g_1=L\left(p_1+p_2+p_3-\frac{1}{RC}\right)\\ g_2=LC(p_1p_2+p_3p_2+p_1p_3)-1\\ g_3=LC(p_1p_2p_3)\end{cases} \tag{8.37}$$

The same calculation can be performed by assuming that the state variables are the integral of the error on output voltage, the capacitor voltage, and the inductor current. In this case, the circuit of Fig. 8.9 is described by the equations:

$$\begin{cases} W(s)=-g_1I_L(s)-g_2V_c(s)-g_3X_1(s)\\ I_L(s)=I_c(s)+\dfrac{V_c(s)}{RC}+I_{o1}(s)\\ sI_c(s)=-\frac{1}{RC}I_c(s)-\frac{1}{L}V_c(s)+\frac{1}{L}W(s)-sI_{o1}(s)\\ sCV_c(s)=I_c(s)\\ sX_1(s)=V_c(s)-V_{ref}(s)\end{cases} \tag{8.38}$$

The open-loop transfer function is:

$$\frac{V_c(s)}{V_{ref}(s)}=\frac{g_3}{LC}\frac{1}{s^3+s^2\left[\frac{1}{RC}+\frac{g_1}{L}\right]+s\left[\frac{1}{LC}+\frac{g_1}{LCR}+\frac{g_2}{LC}\right]+\frac{g_3}{LC}}=\frac{g_3}{LC}\frac{1}{p_1(s)} \tag{8.39}$$

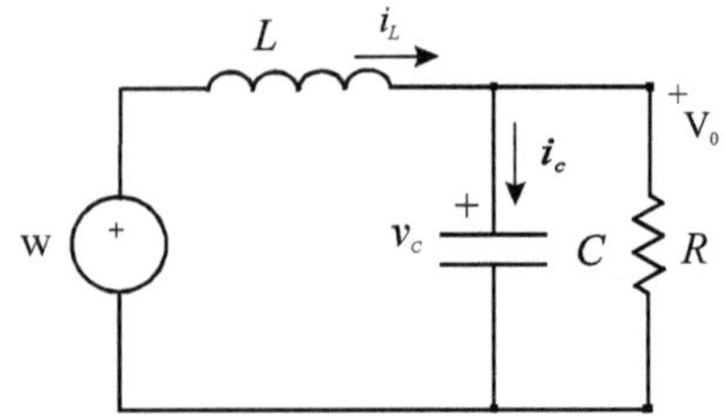

Fig. 8.12 Simplified scheme of buck converter

The output impedance is given by:

$$Z_o(s) = \frac{V_c(s)}{I_{o1}(s)} = -\frac{s(sL + g_1)}{LC}\frac{1}{p_1(s)} \tag{8.40}$$

The relationships between the poles and the gains are:

$$\begin{cases} g_1 = L\left(p_1 + p_2 + p_3 - \frac{1}{RC}\right) \\ g_2 = LC(p_1p_2 + p_3p_2 + p_1p_3) - 1 + \frac{L}{R}\left(\frac{1}{RC} - p_1 - p_2 - p_3\right) \\ g_3 = LC(p_1p_2p_3) \end{cases} \tag{8.41}$$

By comparing Eqs. (8.34) and (8.37) with Eqs. (8.40) and (8.41), respectively, it can be noted that, by choosing the inductor current as state variable, a higher output impedance is obtained and both g_1 and g_2 depend on the load value.

Moreover, since the steady-state mean capacitor current is zero, a smaller volume and cheaper coil sensor, compared with a current sensor placed on the inductance (whose peak-to-peak current is higher and mean value is different from zero) can be used. Finally, the capacitor current can be estimated on the basis of the load voltage using the same sensor required for the measurement of the output voltage. For these considerations the use of the capacitor current as a state variable is preferred.

It should be noted that the possibility to impose the poles allows a fast dynamical response to be obtained. This characteristic avoids interference between the control of the DC/DC converter (used to emulate the PV source) and the load supplied by the source, usually an inverter.

8.4.1 Simulink® Implementation

With reference to the simplified circuit of the buck converter drawn in Fig. 8.12, the state variables can be expressed as in Eq. (8.42).

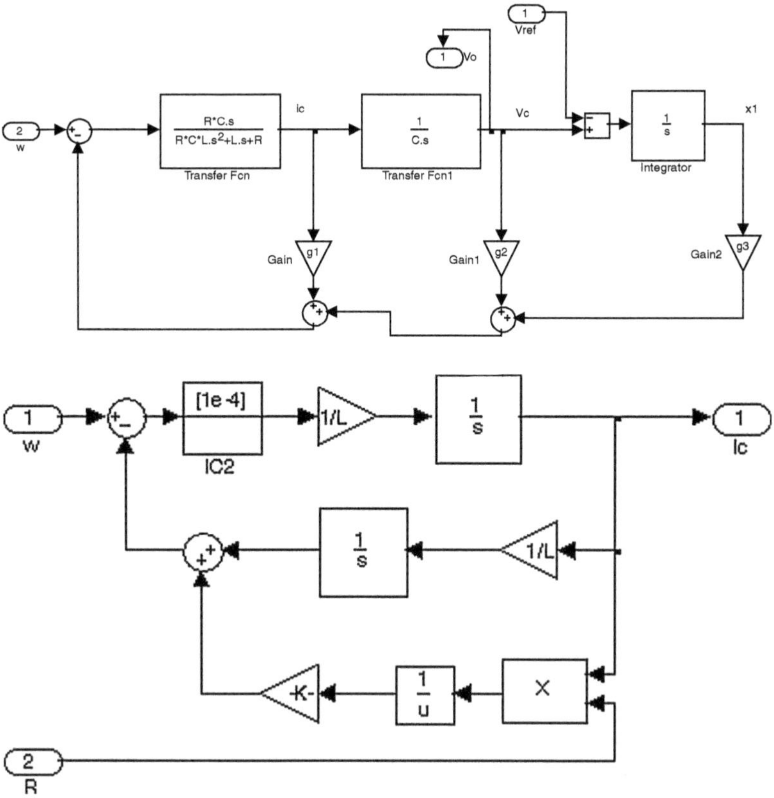

Fig. 8.13 Simulink® implementation of buck converter with pole placement control technique (*top*); Simulink® implementation of the transfer function i_c versus. w allowing load variation (*bottom*)

$$\begin{cases} i_c = \dfrac{sRC}{s^2RLC + sL + R}w \\ v_c = \dfrac{1}{sC}i_c \\ x_1 = \dfrac{1}{s}(v_c - V_{\text{ref}}) \end{cases} \tag{8.42}$$

This formulation can be implemented in Simulink® as shown in Fig. 8.13, top scheme, where the gains are calculated by Eq. (8.37). It should be noted that the use of the block "transfer function" does not allow the parameters to be modified.

The transfer function i_c versus w can be rearranged as in Fig. 8.13, bottom scheme, to simulate load variations.

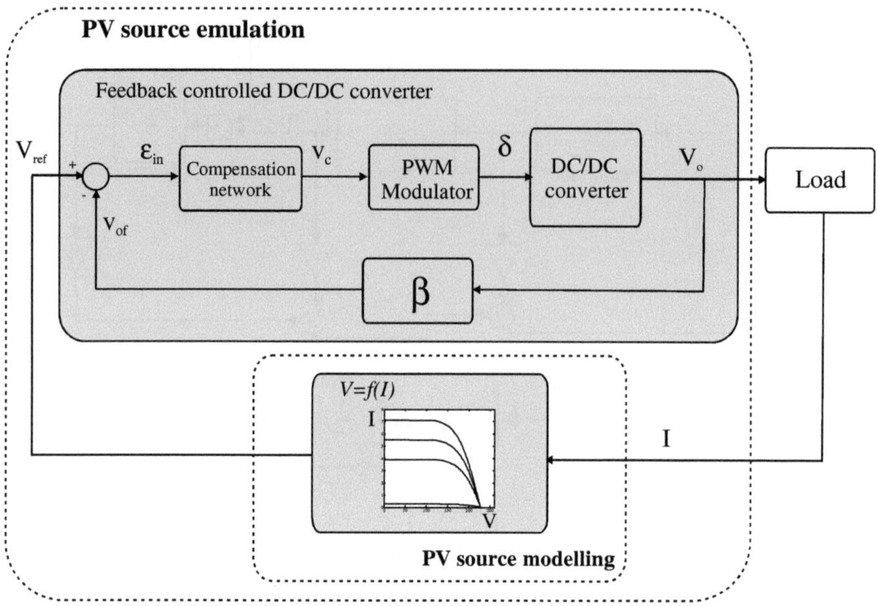

Fig. 8.14 General block diagram of a PV source emulator

8.5 DC/DC Converter-Based Emulation of a PV Source

The block diagram shown in Fig. 8.14 summarizes the concept of a PV source emulation, where the feedback structure of a DC/DC converter and the $V = f(I)$ block are highlighted.

The feedback controlled DC/DC converter is that described in the previous section. It realizes an output equal to that delivered by the I–V relationship. Its output voltage is applied to the electrical load and the obtained current is used as an input for the $V = f(I)$ block.

Figure 8.14 synthesises the content of the two parts of this book as well.

In particular, the block indicated as "PV source modelling" is the result of the I–V characteristic representation, obtained with one of the techniques described in part I.

On the other hand, the block indicated as "Feedback controlled DC/DC converter" is the controlled PV emulator's power stage, whose selection and design derives from the concepts explained in part II.

With specific reference to the DC/DC converter, the issues related to its design constraints and the appropriate selection of its best topology and control, are given in the following.

8.5.1 DC/DC Power Converter Design Constraints

8.5.1.1 Current and Voltage Output

The maximum output current and voltage deliverable by the PV emulator have to be defined on the basis of the operating conditions of the considered PV generator.

It should be observed that the maximum output current is the short circuit current at the highest solar irradiance, while the maximum output voltage is the maximum open circuit voltage.

Once these design constraints are fixed, the maximum allowable load variation must range from the infinity impedance (open circuit condition) to an impedance as much as low to permit approaching the above described short circuit condition.

It is worth noting that the mere short circuit condition, corresponding to a null load is not compatible with the converter operation as deduced by the transfer functions given in Sect. 7.13.

Finally, the electrical load must be able to correctly dissipate the power delivered by the PV emulator.

8.5.1.2 Dynamic Response and Arbitrary Load Problem

Two different issues have to be handled when the PV emulator is required to correctly reproduce the dynamic behavior of a PV generator. The former is related to the need of following the response to rapidly changing environmental conditions or applied load. The latter is known as the arbitrary load problem and it is tied to the use of the PV emulator in a power conversion chain where it is connected to a power converter, intended as an electrical load.

With reference to the first issue, the response of the PV emulator should exhibit a time constant faster than the dynamic response of the model. This is achieved by suitably setting the emulator bandwidth and by setting a switching frequency much higher than the cut-off frequency of the emulator bandwidth.

It should be noted that these constraints are not cumbersome for reproducing varying environmental conditions, since the corresponding time constants are generally high. Anyway, rapid load variations in a stand alone configuration, represent a more critical situation due to the related smaller time constants in terms of voltage response of the PV emulator. For this reason, the assessment tests on the PV emulator model, carried out in part I, are performed using load step variations.

If the minimum rise time of the output voltage is equal to τ_r, the power amplifier bandwidth must be set greater than $1/(\pi\ \tau_r)$.

With reference to the arbitrary load problem, it should be considered that the primary benefit of the PV emulators consists on their use in testing the performance of PV inverters and MPPT algorithms. Therefore, they should be as flexible as possible to properly operate independently from the loading converter.

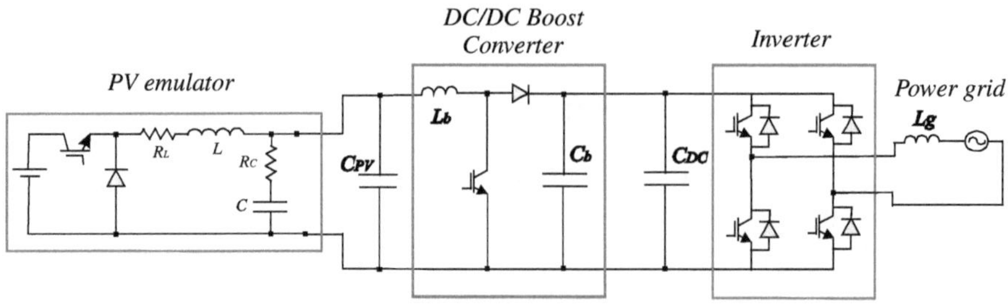

Fig. 8.15 Conversion chain for the grid connection of a PV source emulator

In detail, the loading converter can exhibit different values of the capacitance seen at the emulator terminals and requires a high frequency current (whose fundamental harmonic is equal to the switching frequency).

These parameters are not a priori known.

The output impedance of the PV emulator gives information on the range of frequencies in which it behaves like an ideal voltage source.

A low impedance value is desirable in order to correctly reproduce the current demand coming from the loading converter.

As an example, Fig. 8.15 shows a possible application of a PV emulator, used to test a complete conversion chain for the grid connection of a PV source. Here, the PV emulator replaces a real PV generator, permitting to accurately investigate the whole PV system behavior, including the influence of weather conditions, partial shading and dynamics, without the use of a real outdoor installation, thus reducing the time and the cost of the experiments.

The DC/DC boost converter performs the boosting of the voltage coming from PV source and it can be integrated with the MPPT.

Finally, the single-phase grid-connected inverter, controlled as an active rectifier, is able to maintain the DC link voltage and to properly manage the active and reactive power supplied to the grid.

8.5.2 Selection of the Best Topology

On the basis of the analysis developed in Chap. 7, the considered DC/DC converter topologies, i.e., the buck converter, the boost converter, and the buck-boost converter, are all, in principle, possible candidates to be used for the power amplifier stage of a PV emulator.

Anyway, the following practical considerations suggest the employment, when possible, of the buck topology.

- The DC voltage level for supplying the PV emulator can be easily obtained by the grid voltage through a simple bridge rectifier both in the case of single-phase grid and three-phase grid. In this last case, the obtained DC voltage of about

Table 8.1 Main features under standard test conditions (*stc*) of the PV plant for the emulator design

Parameter	Symbol	Value
Number of assemblies	N_P	2
Number of modules per assembly	N_S	6
Peak power	P_{PV}	1450 W
Open circuit voltage	V_{oc}	228.6 V
Short-circuit current	I_{sc}	9.2 A
Maximum power voltage	V_{MP}	186 V
Maximum power current	I_{MP}	8 A

500 V is suitable to be reduced, via the buck converter, to the rated voltage of a common PV plant of a power up to tens of kW.

- The use of a buck topology permits a simpler control implementation, because the corresponding transfer functions are not dependent on the operating point, for a given electrical load, as illustrated in Table 7.2.

With reference to the first aspect, it is worth observing that the boost converter does not permit to obtain an output voltage less than the supply voltage; this means that operating conditions close to the short circuit are not reproducible.

Furthermore, high output voltages imply high input currents. This last disadvantage is also present in the buck-boost scheme, which, however, allows low voltages to be reproduced.

Regarding to the second aspect, it should be noted that both the boost and the buck-boost converters, show a nonlinear dependence on the duty cycle and the load value as well, thus making difficult a fixed parameter control strategy.

8.6 Example of a PV Source Emulator Design

This section is dedicated to the design and practical setup of a PV emulator, devised by the authors, which is used to reproduce, in a laboratory frame, the electrical behavior of a real PV plant, previously modeled according to a single diode four parameter scheme. The PV generator's I–V relationship is expressed by Eqs. (3.13) and (3.14).

The considered PV plant has been described in detail in Sect. 4.4.2.3, where the parameter identification for the corresponding model has been performed, as well. In particular, the adopted PV model uses a regression law to relate temperature and solar irradiance, as explained in Sect. 4.6

Here, the PV plant specific features are summarized in Table 8.1, for the sake of convenience.

Figure 8.16 shows the PV plant configuration.

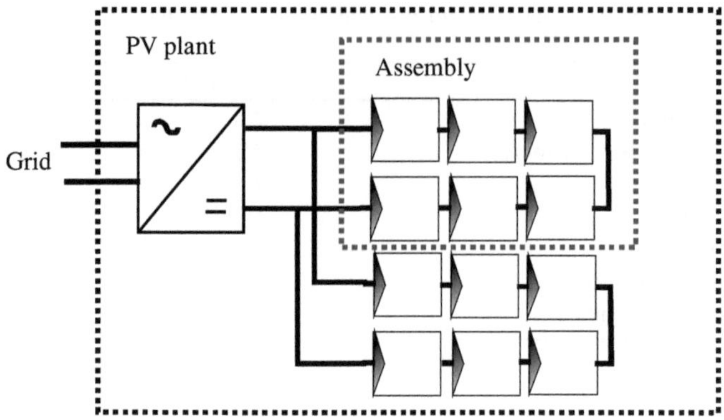

Fig. 8.16 PV plant configuration

8.6.1 Power Stage Design

A switching DC/DC buck converter topology is chosen. The DC/DC converter is operated at a constant switching frequency and input DC voltage, while the regulation of the output voltage is obtained by varying the converter duty cycle (D).

The scheme of the DC/DC buck converter circuit is that illustrated in Fig. 7.13. In such a scheme, the resistive components r_L and r_C represent the parasitic resistance of the output filter inductor and capacitor, respectively.

The design of the converter output LC filter is carried out on the basis of the following conditions:

- continuous-current conduction operation of the converter;
- ripple on the output voltage not exceeding few percent.

The above-mentioned conditions can be obtained from Eqs. (7.23) and (7.19), respectively. Rearranging these equations, the inductance and the capacitance values are calculated on the basis of the design constraints, according to:

$$L \geq \frac{V_s T_s}{2 I_{\mathrm{Lt}}} D(1 - D) \tag{8.43}$$

$$C \geq \frac{T_s^2}{8L} \frac{V_o}{\Delta V_o} (1 - D) \tag{8.44}$$

where I_{Lt} is the limit current between continuous and discontinuous conduction mode, T_s is the switching period, and V_o is the output voltage of the DC/DC converter.

The rated power of the prototype is about 3 kW.

The switching frequency of the DC/DC buck converter, is chosen equal to 10 kHz.

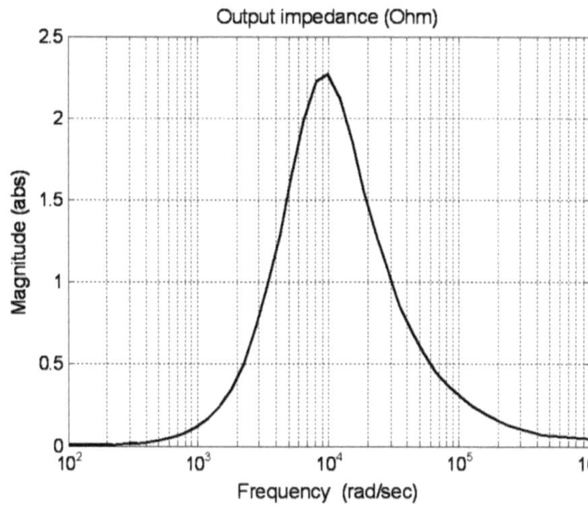

Fig. 8.17 DC/DC converter output impedance versus frequency [From Di Piazza and Vitale (2010)]. Used with kind permission from Elsevier

The maximum voltage output of the PV emulator is fixed to 235 V, in consideration of the electrical features of the PV source. Therefore, an input voltage $V_s = 350$ V is chosen.

According to Eqs. (8.43) and (8.44), the following values of capacitance and inductance of the low pass output filter are adopted, respectively:

- $C = 33$ μF;
- $L = 3.7$ mH.

This last value gives $I_{\mathrm{Lt}} = 1.18$ A for $T_s = 10^{-4}$ s.

8.6.2 Pole Placement Voltage Controller

As previously highlighted, a good dynamic behavior is required in order to take into account the loading effect of the power converter supplied by the PV emulator.

The possibility to impose the closed-loop poles, by using the pole placement technique, described in Sect. 8.4, permits to obtain a fast dynamical response.

The output impedance of the DC/DC converter gives information on the range of frequencies in which the PV emulator behaves like an ideal voltage source.

As previously said, a low impedance value is desirable in order to correctly reproduce the current demand coming from the loading converter.

The closed-loop poles are chosen as a pair of complex and conjugate and a real one. The corresponding k_i coefficients are chosen equal to $k_{1,2} = 2\pi(500 \pm j500)$ and $k_3 = 2\pi 500$. This choice assures a dynamical response with a rise time in the order of milliseconds.

The equivalent output capacitance corresponds to about 10 μF that is smaller than a typical expected load capacitance.

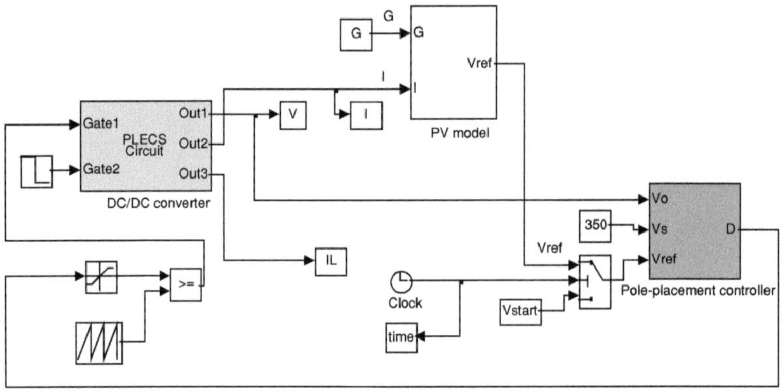

Fig. 8.18 Implemented model of the PV emulator

The obtained output impedance versus frequency curve, calculated by Eqs. (8.34) and (8.35), is shown in Fig. 8.17. It is possible to observe that it exhibits a maximum equal to 2.25 Ω at 1.6 kHz (10^4 rad/s).

8.7 PLECS-Based Simulation of PV Source Emulator

In order to evaluate the performance of the PV emulator, developed according to the previously described design constraints, a simulation analysis is carried out. It exploits the association of Matlab/Simulink® environment and PLECS® (Piecewise Linear Electrical Circuit Simulation for Simulink) toolbox.

As specified in Chap. 5, PLECS® allows an actual plant, for example a power electronic circuit, to be implemented, as subsystem, while the related control is developed using standard Simulink® blocks. Its main advantage is the very short simulation time.

As for the considered application, the DC/DC buck converter has been built using PLECS® libraries, while the control algorithm has been implemented in Simulink® and directly interfaced to the circuit-based simulation model.

Figure 8.18 shows the whole implemented model, including the PV model, the DC/DC converter, and the pole placement controller.

It can be noted that, at the beginning of simulation, a constant value of reference voltage is given to the controller in an open loop mode to avoid handling high voltage error signals. This constant voltage is indicated as V_{start}; after a short transient of 0.02 s the closed-loop operation is enabled.

The PV model implementation is done using the form $V = f(I)$, to obtain the voltage reference for the DC/DC converter.

In order to correctly reproduce partial shading conditions, the whole PV model is represented as the composition of the single module models. In particular, each module is considered with its own temperature and solar irradiance.

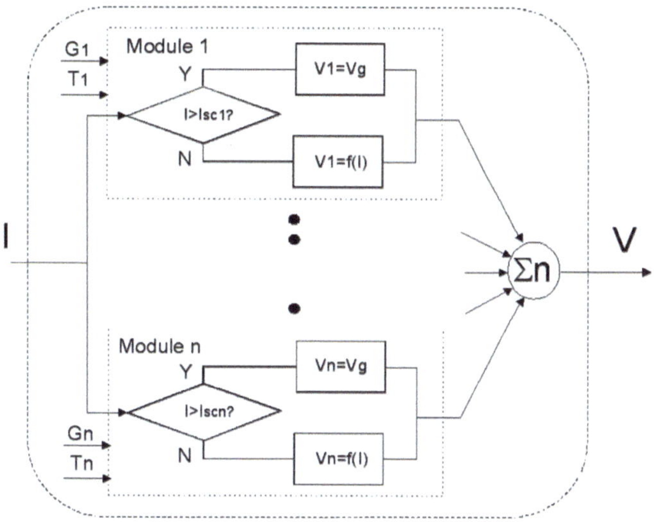

Fig. 8.19 Implemented algorithm for obtaining the I–V characteristics of a PV assembly. (From Di Piazza and Vitale (2010)). Used with kind permission from Elsevier

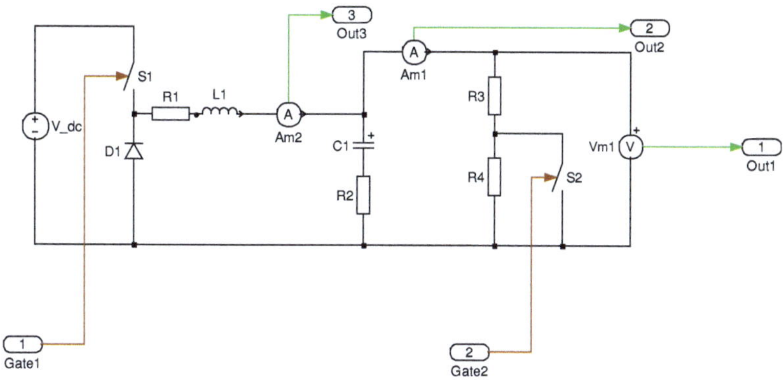

Fig. 8.20 Circuit scheme of the DC/DC converter implemented in PLECS®

In this way, the PV plant behavior is implemented by summing the voltages of each module and the currents of parallel connected assemblies. Moreover, with reference to a single assembly, as shown in Fig. 8.16, if the current is greater than the short-circuit current of a module, the module output voltage is set to the threshold voltage of the bypass diode (V_g), otherwise the voltage is obtained by the model. This approach allows the PV assembly to be accurately analyzed under partial shading conditions.

The above described rules are summed up on the flow chart drawn in Fig. 8.19.

Figure 8.20 illustrates the PLECS® model of the DC/DC buck converter; it encompasses the output filter and its parasitic parameters. In particular, the load is

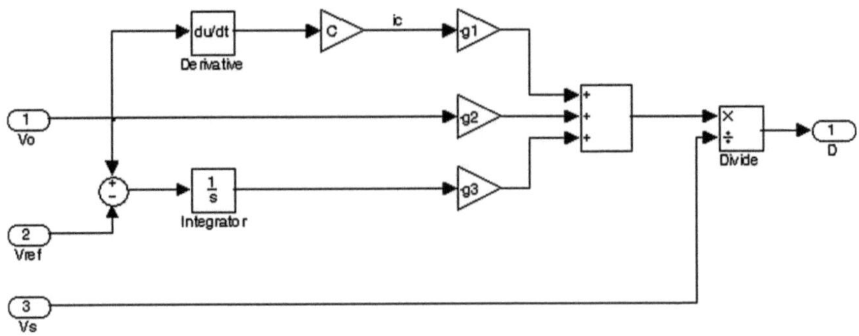

Fig. 8.21 Detail of the pole placement controller block

Fig. 8.22 Representation of the assembly under partial shading condition

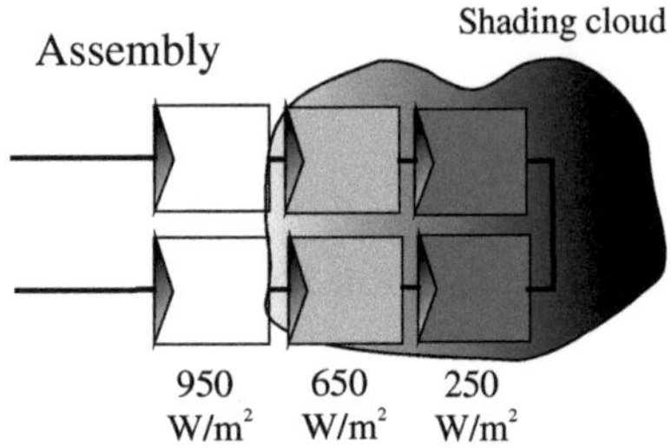

realized by two series connected resistors where a switch, parallel connected to one of them, is used to impose a sudden variation of the load.

The PLECS® model has two inputs and three outputs. The inputs are the switching command for the IGBT, modeled as an ideal switch, and the command for the switch used to obtain the load step transition. The outputs give the inductor current, used to assess the continuous conduction mode, the output voltage, and the output current of the PV emulator.

The scheme shown in Fig. 8.18 is conceived for testing the PV emulator behavior with step load transitions and constant solar irradiance.

Other testing conditions are obviously possible, for example, step transitions of irradiance or load/irradiance variations according to a chosen profile.

The pole placement controller block is shown in detail in Fig. 8.21.

It should be observed that the values of the three gains g_1, g_2, and g_3 are calculated according to Eq. (8.37). This calculation is done by a MATLAB® script, run before operating the PV emulator for tests.

As an example, the static I–V characteristics of a PV assembly are determined both under uniform solar irradiance of 950 W/m^2 and under the partial shading condition represented in Fig. 8.22. This condition is given by two modules with an irradiance of 950 W/m^2, two modules with an irradiance of 650 W/m^2, and the last two with an irradiance of 250 W/m^2. The obtained I–V curves are shown in Fig. 8.23.

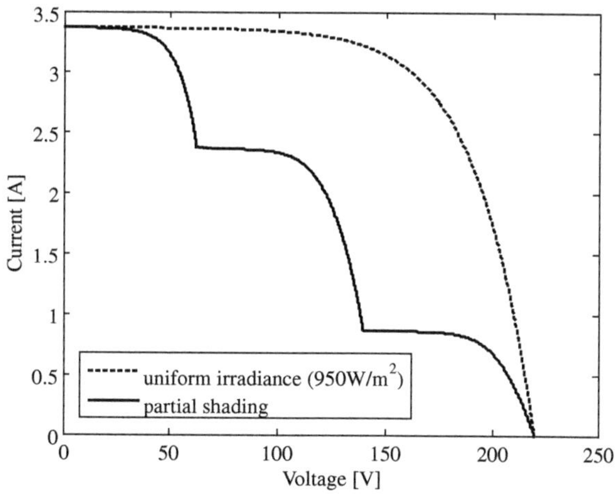

Fig. 8.23 I–V characteristics with uniform irradiance and under partial shading condition, reproduced by the PV model

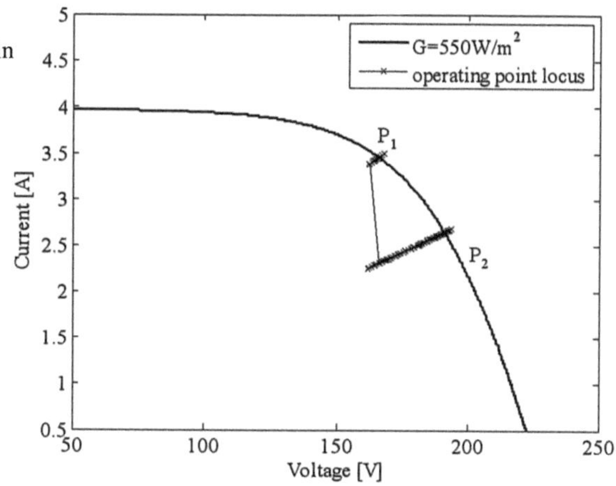

Fig. 8.24 Operating point locus in Test #1

The following tests are carried out:

- Test #1: step load transition with constant solar irradiance;
- Test #2: irradiance step transition with constant load resistance.

In Test #1 solar irradiance is set equal to 550 W/m^2, while the load is switched between an initial value of 50 Ω, nearly corresponding to the MPP, to a final value of 75 Ω. Due to this transition, the operating point goes from P_1 to P_2 on the I–V curve corresponding to $G = 550$ W/m^2, as illustrated in Fig. 8.24. Here, it can be noted that the sudden variation of the load resistance produces an initial quasi-vertical trajectory in which the operating point jumps from the straight line corresponding to the load resistance of 50 Ω to the straight line corresponding to the final resistance load of 75 Ω. Then it continues toward the final point remaining on the same line. The slight deviations from points P_1 and P_2 exhibited in the operating point locus are due to the switching operation of the emulator.

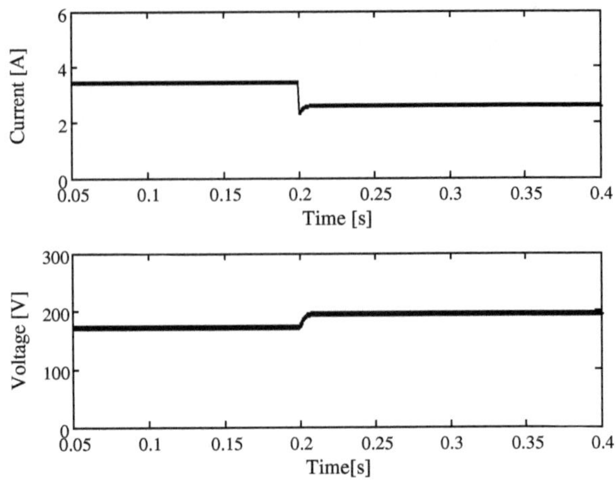

Fig. 8.25 Current and voltage transition, corresponding to Test #1, versus time

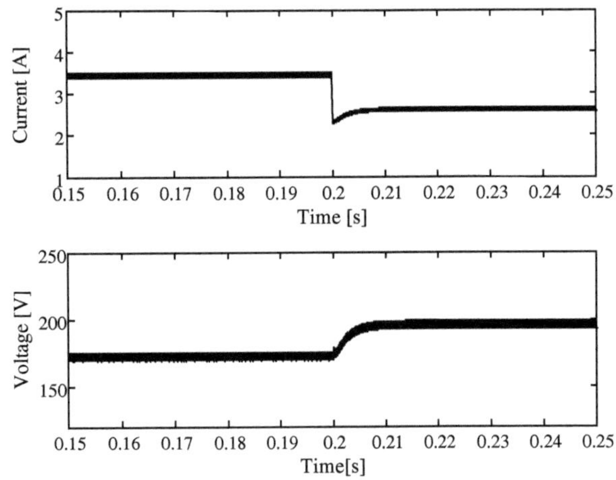

Fig. 8.26 Zoom of the current and voltage transitions, corresponding to Test #1, versus time

Figure 8.25 shows the voltage and current transitions, corresponding to Test #1, versus time. A zoom of these transitions is drawn in Fig. 8.26, where it can be noted the sudden current variation due to the step load variation and the current and voltage response whose dynamics is dominated by the choice of the emulator poles. As indicated in Sect. 8.6.2, these poles are fixed to $p_{1,2} = -2\pi(500 \pm j500)$ and $p_3 = -2\pi 500$, on the basis of the PV plant to be emulated.

As previously explained, the suitable selection of the emulator poles, thanks to the pole placement technique, allows the desired dynamic behavior to be reproduced.

In order to put in evidence the effect of the poles position on the emulator dynamics, Test #1 is repeated with three different settings in which the real pole assumes the following values:

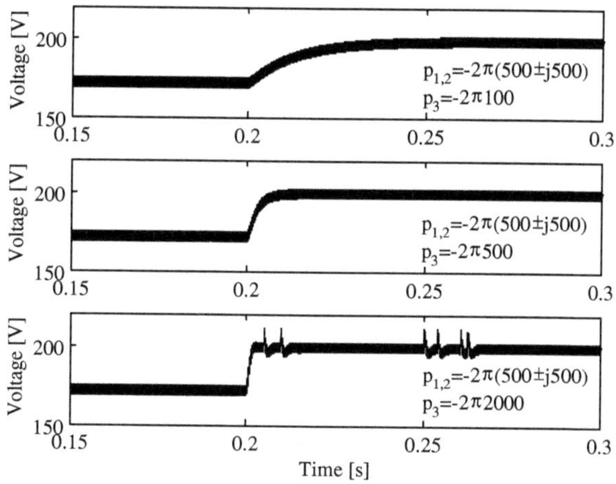

Fig. 8.27 Voltage transitions, with different poles configurations (Test #1)

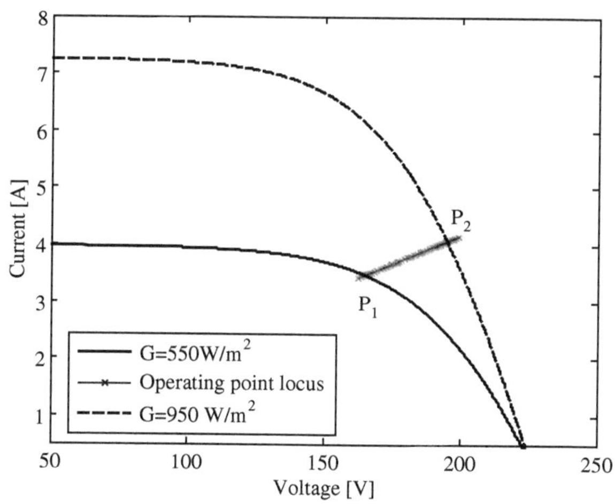

Fig. 8.28 Operating point locus in Test #2

- $p_3 = -2\pi 100$ rad/s
- $p_3 = -2\pi 500$ rad/s
- $p_3 = -2\pi 2000$ rad/s

while the complex conjugate pair of poles remains unchanged.

The voltage transition has been observed in all cases, whose corresponding time domain waveforms are given in Fig. 8.27.

It is possible to observe that the first choice implies an over-damped response with a slower dynamics.

The second choice leads to an over-damped behavior and a faster dynamics.

Finally, with the third choice, additional oscillations appear.

In Test #2 the load resistance is kept constant and equal to 50 Ω, while a step variation of the solar irradiance from 550 to 950 W/m^2 is imposed.

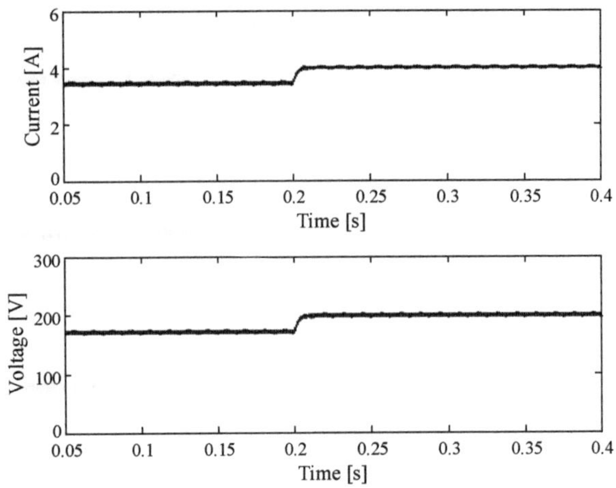

Fig. 8.29 Current and voltage transition, corresponding to Test #2, versus time

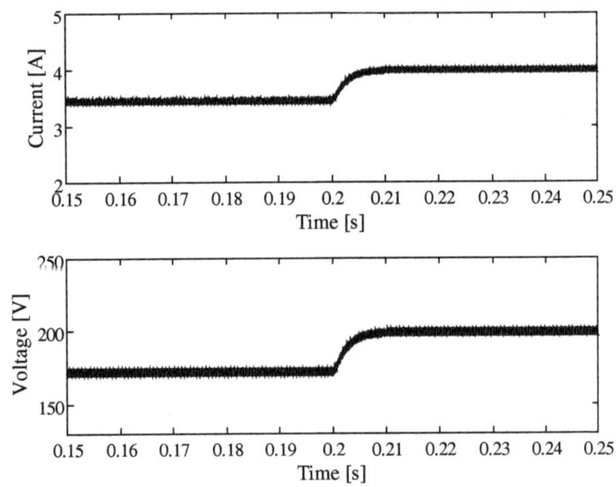

Fig. 8.30 Zoom of the current and voltage transitions, corresponding to Test #2, versus time

In this case, the trajectory of the operating point goes from the I–V curve corresponding to 550 W/m^2 to the I–V curve corresponding to 950 W/m^2, remaining on the straight line imposed by the load resistance.

This trajectory is shown in Fig. 8.28, where the slight deviations from points P_1 and P_2 exhibited in the operating point locus are due to the switching operation of the emulator.

The voltage and current transitions in Test #2, versus time, are plotted in Fig. 8.29.

A zoom of these transitions is drawn in Fig. 8.30, where, differently from Test #1, it can be noted both the current and voltage variations follow practically the same variation trend with a rise time of about 10 ms.

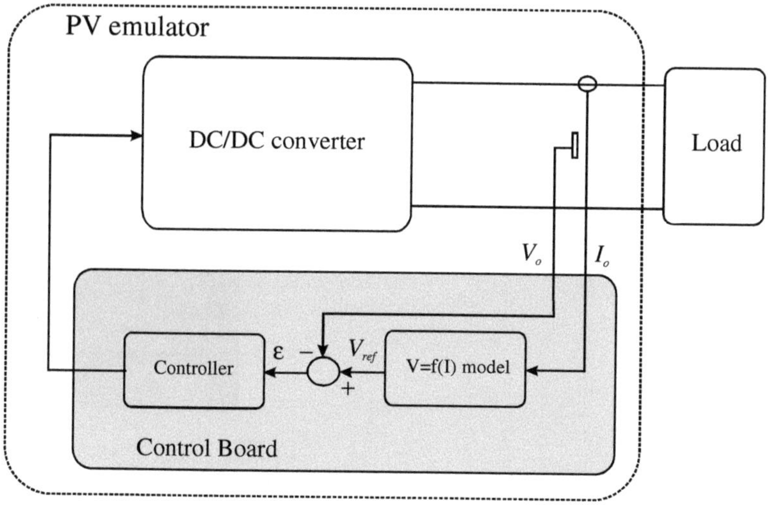

Fig. 8.31 Principle block diagram of the PV emulator

8.8 Experimental Implementation of the PV Source Emulator

The principle block diagram of the PV emulator is represented in Fig. 8.31.

It encompasses the DC/DC buck converter and a control board, whose inputs are the converter output voltage and current and whose output is the control signal of the buck converter.

The control board implements both the PV model and the control algorithm.

On the basis of the output current, the PV model calculates the reference voltage which is compared with the actual output voltage of the DC/DC converter.

The error signal is processed by the controller that outputs the command for the switching device.

In the following subsections, the experimental set-up of the whole PV emulator equipment is described in detail.

8.8.1 DC/DC Buck Converter

The DC/DC buck converter is supplied by the TDK-Lambda GEN600-5.5 DC power supply because, with this solution, the maximum deliverable current can be electronically limited and the PV emulator is galvanically isolated from the power grid.

As an alternative, an isolation transformer and a bridge rectifier can be used to obtain the supply voltage for the PV emulator.

The DC/DC converter employs the SKM50GB123D power IGBT whose rated current is equal to 40 A at 80° C. This high current value allows to manage powers up to 10 kW.

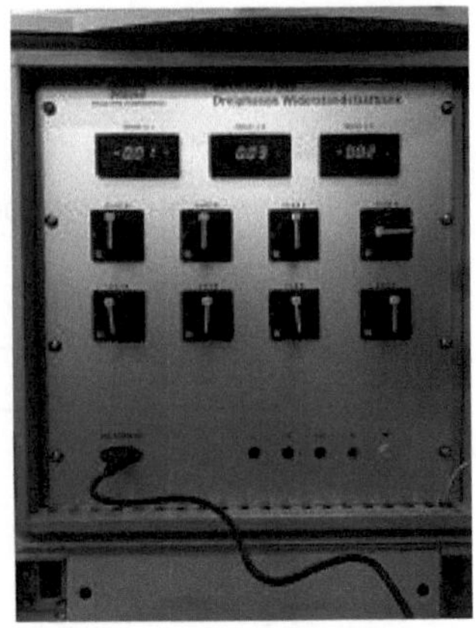

Fig. 8.32 REO load 302 three-phase resistive load bank

The used driver circuit is the hybrid dual MOSFET driver SKHI22AR.

A snubber circuit is realized by using a 0.22 μF polypropylene capacitor, i.e., the MKPC4BS, suitable for high-frequency applications.

As previously indicated, the output filter of the buck converter is composed of a 3.7 mH inductor and a 33 μF capacitor. The corresponding parasitic components, experimentally evaluated, are, respectively, $r_L = 0.4\ \Omega$ and $r_c = 2.06\ \Omega$.

The output voltage is sensed by a LEM LV25-P transducer, while the output current is measured by a closed-loop current sensor, i.e., the Honeywell CSNP661. Both these sensors exhibit a wide bandwidth and their range can be varied by a suitable setting of an external precision resistor.

As for the current transducer, its maximum range is equal to 50 A, coherently with the chosen IGBT. For the emulation of the considered plant this value has been lowered to 10 A by winding five turns around the sensor core.

Finally, the resistive load is realized by a REO load 302 three-phase resistive load bank.

A view of the resistive load is given in Fig. 8.32.

8.8.2 Control Board

The PV emulator control is experimentally implemented using the DSP-2 board developed at the Institute of Robotica of Faculty of Electrical Engineering and Computer Science in Maribor, Slovenia.

The DSP-2 board is a high performance, floating-point digital signal processor-based inverter controller. This board, in combination with DSP-2 library, can be successfully used for several industrial applications.

In general, the DSP-2 board allows control algorithms, set up in Matlab/Simulink® environment, to be implemented and verified in suitable development systems, at a significantly low cost.

Fig. 8.33 Picture of the DSP-2 board

The board is based on the Texas Instruments TMS320C32 DSP and the FPGA XCS40-4PQ240C, member of Xilinx Spartan family. It is equipped with 4 × 12 bit simultaneous A/D converters with serial output and a two channels 12 bit D/A converter with serial input and unipolar output from 0 to 4 V.

DSP-2 library for Simulink® enables rapid transition from the simulation in Simulink® to the real time operation on the DSP-2 board.

DSP-2 library in combination with DSP-2 Terminal interface enables online changing of the Simulink® block parameters, by the parameter inspector window, while the code is executed.

In Fig. 8.33 a picture of the DSP-2 board is shown.

In general, for the implementation of the PV emulator control, it is necessary to use four inputs, i.e., current, voltage, solar irradiance, and temperature. In particular, current and voltage are those sampled at the emulator output, while solar irradiance and temperature are set by the operator or given by external sensors. In the specific case, since a PV model using a regression law to relate temperature and solar irradiance is adopted, only the solar irradiance values are given as input via software to the control board, according to the test to be performed.

The output of the board is the duty cycle generated by PWM block, taken from the DSP-2 library. When the output of the PWM is set to "1", the power device is active.

In Fig. 8.34 the block diagram of the emulator control, including the PV model, the pole placement-based controller and the PWM modulator, is sketched.

The current and voltage are acquired by two analog inputs indicated as DSP-2 AI.

The solar irradiance G is set by software either if constant or if following a given profile.

The block indicated as DSP-2 PWM is a three-phase pulse width modulator suitably handled to give the switching command for the IGBT (only one input is used). Finally, the blocks indicated as DSP-2 TT allow the corresponding variables to be displayed in the DSP-2 Terminal interface.

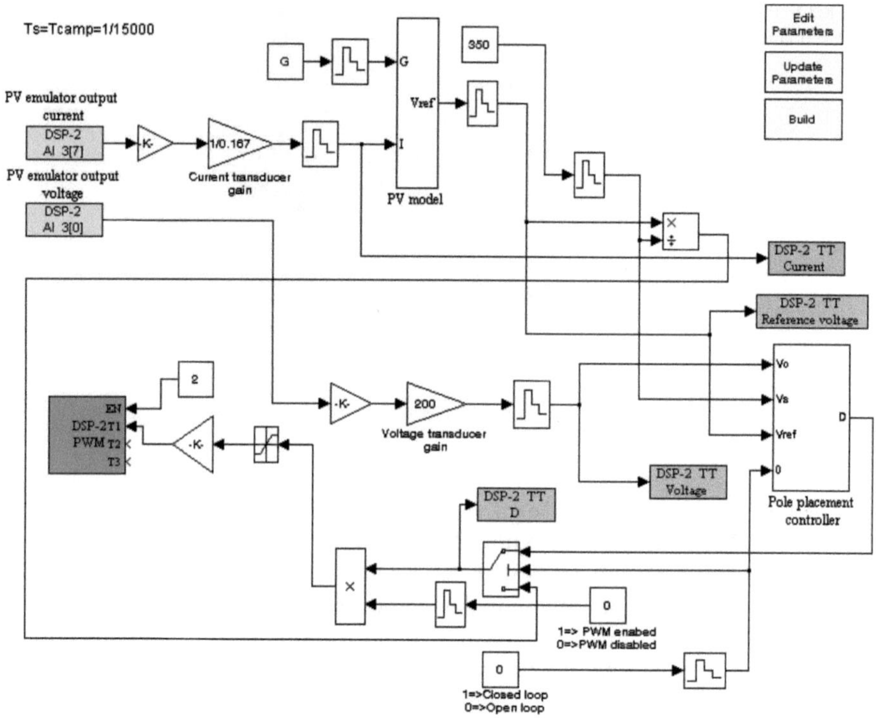

Fig. 8.34 Block diagram of the PV emulator control

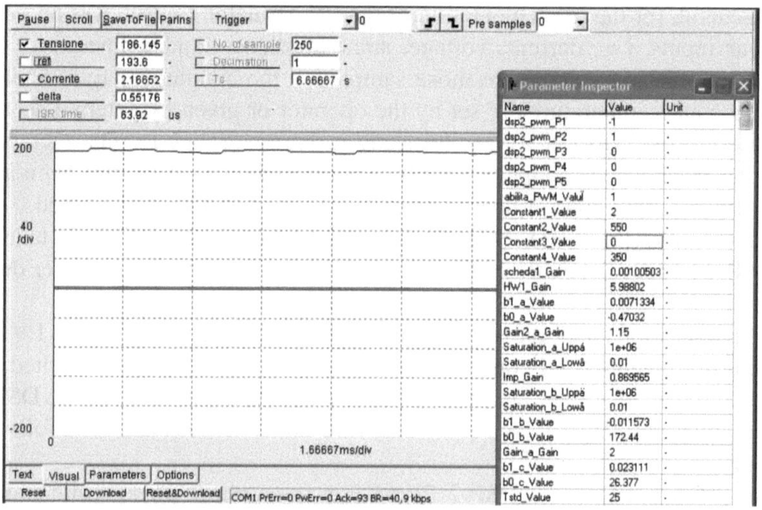

Fig. 8.35 DSP-2 Terminal interface

In Fig. 8.35 Terminal interface, with superimposed the parameter inspector window is shown, as an example.

8.8.3 DC/DC Boost Converter for the MPPT

The optimal exploitation of a PV source is obtained by maximizing the delivered power. This means that the optimal resistance load must be set to V_{MP}/I_{MP}. Since the MPP is variable with environmental parameters, the load value should be continuously changed to follow this optimal operating condition.

A possible way to realize a variable load is to use a further power converter connected to the output of the PV source.

If a DC/DC boost converter is used for this purpose, on the basis of (7.46) and (7.47), its input resistance is given by:

$$R_i = \frac{V_s}{I_s} = \frac{V_o}{I_o}(1-D)^2 = R(1-D)^2 \tag{8.45}$$

It can be noted that the boost input impedance can be regarded as a variable load resistance controlled by the duty cycle D for the PV source. In this case, the task of the MPPT algorithm is to properly impose the value of the duty cycle, at any one instant.

In our case, a boost converter has been devised to test the PV emulator behavior with an electronic load, which corresponds to most of the actual practical configuration of PV plants.

Once the PV emulator behavior is assessed, the whole system composed of the PV emulator and the boost converter is useful to test MPPT algorithm performance. In addition, if an inverter is included in the conversion chain, as illustrated in the example of Fig. 8.15, the inverter control algorithm can be tested, as well.

It is important to observe that, compared to a purely resistive load, an electronic load requires a high frequency current, in addition to the DC component, due to the switching operation. Hence, the high-frequency output impedance of the PV emulator should match the one of the real PV plant to achieve a realistic profile of delivered current and voltage.

The design of the boost converter is carried out imposing the conditions of continuous-current conduction operation and ripple on the output voltage lower than few percent.

The following values for the inductance and capacitance are fixed to: $L_b = 2$ mH and $C_b = 23.5$ μF.

In the case of a grid connection, by an active rectifier, the DC link voltage is mantained at a constant value, then it is possible to control the input voltage of the boost converter, and consequently the output voltage of the PV emulator, so to track the MPP.

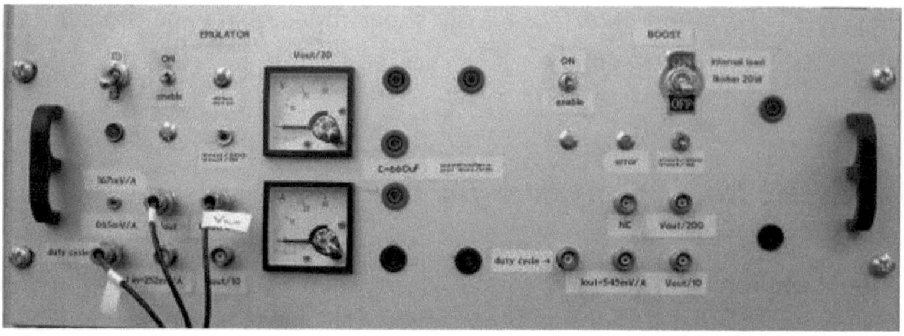

Fig. 8.36 Front panel of the PV emulator

Fig. 8.37 Rack containing the PV emulator

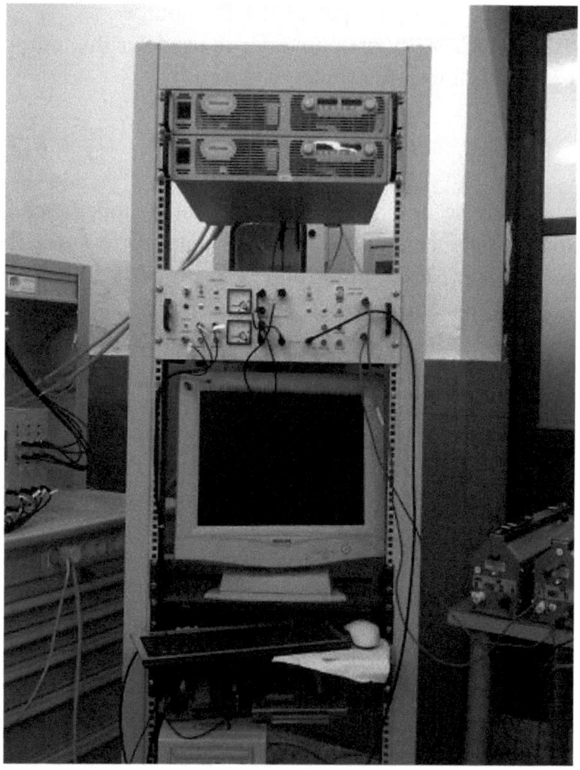

The power switch is a SKM50GB123D power IGBT, switched at 10 kHz, as well as the sampling frequency of the control system.

A decoupling capacitance $C_{PV} = 650\ \mu F$ is placed between the PV emulator and the boost converter in order to prevent oscillations toward the emulated PV source.

The boost circuit is allocated in the same case containing the power stage of the PV emulator. As shown in Fig. 8.36, the boost signal and power connectors are positioned in the right part of the front panel of the whole equipment.

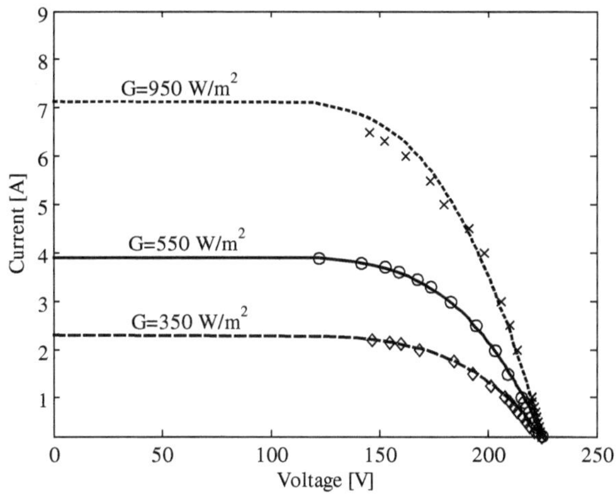

Fig. 8.38 Theoretical I–V curves with superimposed experimental points obtained by the PV emulator

8.9 Experimental Results

Figure 8.37 shows a picture of the experimental rack containing the PV emulator. At the top of the rack the DC power supply can be observed.

The operation of the PV emulator has been tested by laboratory measurements. The used measurement system is composed by:

- a digital oscilloscope with a bandwidth of 1 GHz and a maximum sampling frequency of 2.5 GHz, 20 GS/s, i.e., the TEKTRONIX TDS7254B;
- a 100 MHz high voltage differential probe, i.e., the TEKTRONIX P5205;
- a current measurement system, including an amplifier, i.e. the Tektronix TCPA 300;
- a current probe, i.e., the Tektronix A6303.

First, the static I–V characteristics of the PV plant (described in Table 8.1 and Fig. 8.16) are determined under different uniform solar irradiance values. In particular, the irradiance levels of 350, 550, and 950 W/m^2 are considered. The static experimental points on the I–V characteristics are obtained by imposing a constant solar irradiance value on the DSP board and by suitable values of the resistive load.

In Fig. 8.38 the I–V characteristics of the PV generator, deduced by the model with superimposed experimental points, obtained by the PV emulator, are shown. It is possible to observe that the PV emulator reproduces appropriately the theoretical operation of the PV generator. A slight deviation of the experimental points from the I–V curve is observed at higher irradiance.

Another test has been performed to show the ability of the PV emulator to reproduce the theoretical behavior of the PV generator even under partial shading conditions. In particular, the assembly drawn in Fig. 8.39 has been emulated, according to the method explained in Sect. 8.7.

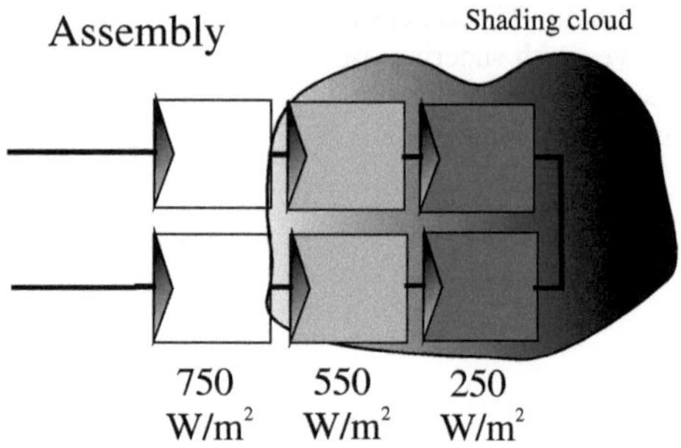

Fig. 8.39 Representation of the assembly under partial shading condition, reproduced by the PV simulator

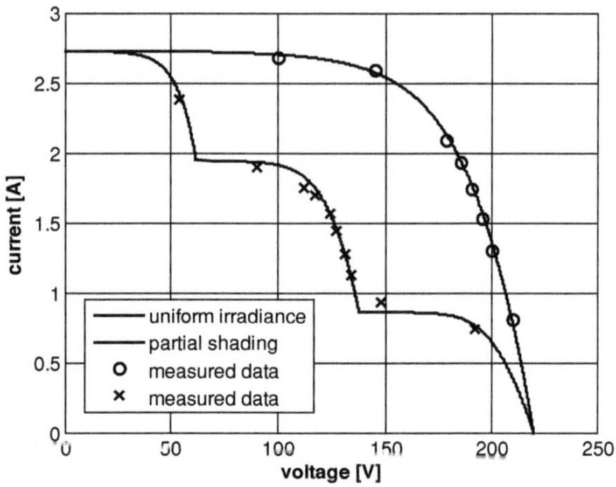

Fig. 8.40 I–V characteristics with uniform irradiance and under partial shading condition with experimental points superimposed [From Di Piazza and Vitale (2010)]. Used with kind permission from Elsevier

Figure 8.40 illustrates the I–V curves obtained by the model with the static experimental points superimposed. A good matching can be noted.

The transition between the two points belonging to the same I–V curve has been realized using the PV emulator. In particular, Test #1 described in Sect. 8.7 has been performed. In this test, solar irradiance is set equal to 550 W/m^2, while the load is switched between an initial value of 50 Ω, nearly corresponding to the MPP, to a final value of 75 Ω. This load step variation is obtained by an abrupt commutation of the resistance load value using the resistive load bank.

The corresponding experimental current and voltage time domain waveforms are shown in Fig. 8.41. The obtained experimental results are in good agreement with those obtained by simulation with PLECS®. This is evident comparing Fig. 8.41 with Fig. 8.26. It should be noted that, to put in evidence the voltage variation, a scale of 20 V/div has been used in the oscilloscope and a DC offset has been set; for this reason, the reference level is not visible.

The transition between the two points belonging to different I–V curves has been realized using the PV emulator. In particular, Test #2 described in Sect. 8.7 has been performed. In this test, the load resistance is kept constant and equal to

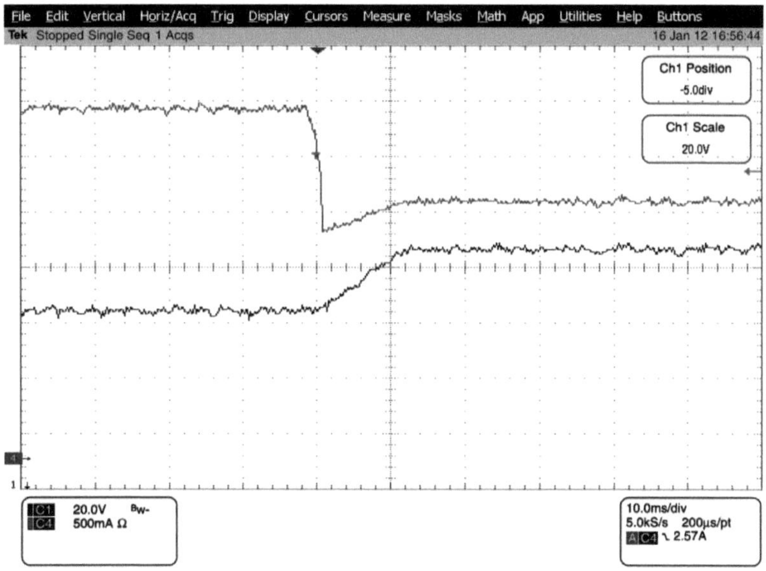

Fig. 8.41 Current (top trace) and voltage (bottom trace) transition, corresponding to Test #1 obtained by the PV emulator

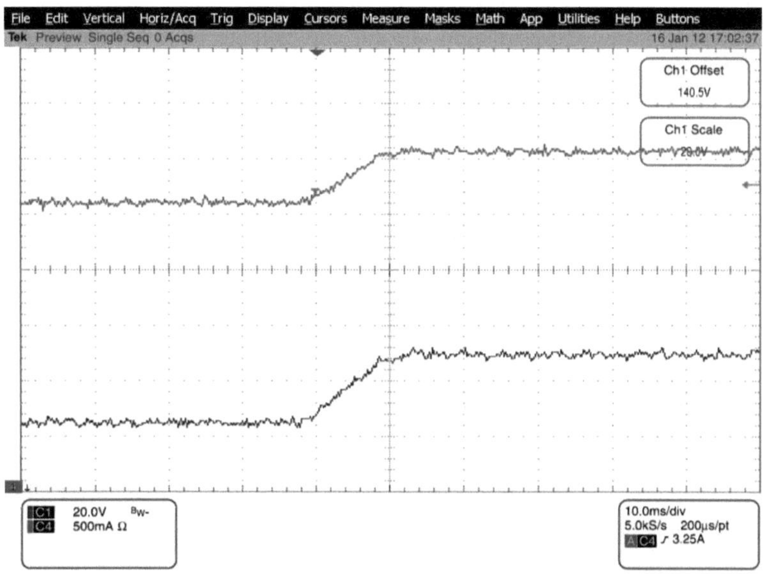

Fig. 8.42 Current (top trace) and voltage (bottom trace) transition, corresponding to Test #2 obtained by the PV emulator

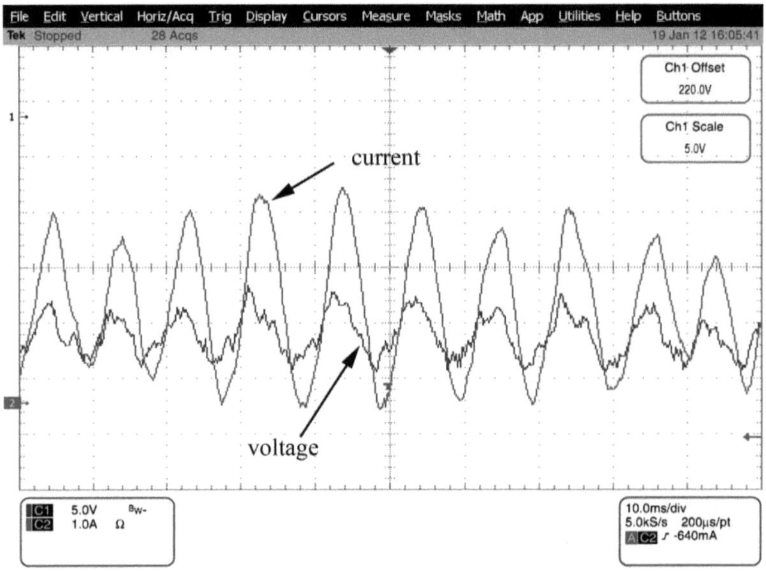

Fig. 8.43 Current and voltage AC components supplied by the PV emulator in grid-connected configuration

50 Ω, while a step variation of the solar irradiance from 550 to 950 W/m^2 is imposed via software by the DSP-2 Terminal interface.

The corresponding experimental current and voltage time domain waveforms are shown in Fig. 8.42. Even in this case, the obtained experimental results are in good agreement with those obtained by simulation with PLECS®. This is confirmed by the comparison of Fig. 8.42 with Fig. 8.30, where the same rise time is noticeable.

A further assessment on the PV emulator operation is done to demonstrate that it is able to reproduce the typical fluctuations due to the power injected into the grid by a single-phase inverter. Such a power is expressed as:

$$p(t) = P_0[1 + \sin(2\omega t)] \tag{8.46}$$

where ω is the grid frequency in radians per second and P_0 is the average power delivered at the fundamental frequency of the grid.

In order to perform this test, the PV emulator is connected to a grid inverter by a DC/DC boost converter, according to the scheme shown in Fig. 8.15. In particular, the boost converter is that described in Sect. 8.8.3; the grid inverter is a commercial device, i.e., the Sunny Boy 1100.

Figure 8.43 shows both the current and voltage AC components supplied by the PV emulator. It can be noted that the frequency of the two waveforms is equal to 100 Hz, i.e., twice the fundamental of the grid frequency ($f = 50$ Hz). In order to

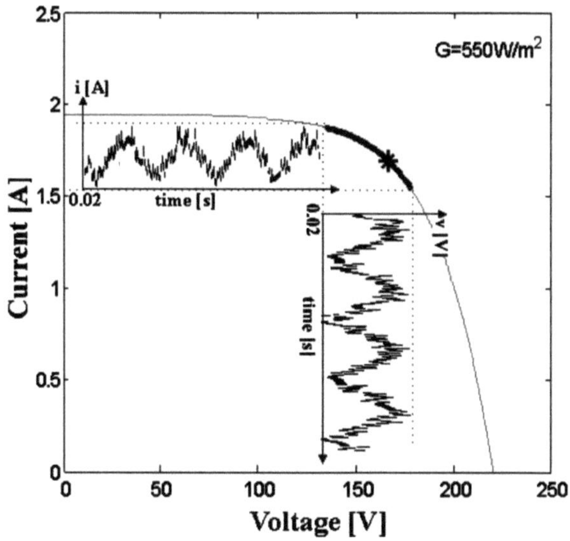

Fig. 8.44 Current and voltage fluctuations superimposed to a static I-V characteristic (From Di Piazza and Vitale (2010)). Used with kind permission from Elsevier

show the two waveforms superimposed, a suitable DC offset has been set in the oscilloscope. The voltage fluctuation is lessened by the presence of a DC link capacitor at the PV emulator output as it occurs in real operating conditions.

Finally, Fig. 8.44 shows the current and voltage fluctuations in grid-connected configuration, superimposed to a static I–V characteristic. It is shown in particular, the effect of the grid connection on the operating point, whose position varies with pulsation 2ω around the MPP. In this test, to highlight the voltage fluctuation, a lower DC link capacitance is used.

8.10 Conclusions

This Chapter presents an analysis of the fundamentals of feedback control for DC/DC converters and an example of emulator design realized by a buck converter.

The choice of the buck scheme is justified on the basis of the possibility to implement an appropriate control strategy for the emulation purpose.

Both the PV model and the power converter, used for the emulator set up, are first simulated in Matlab-PLECS® environment, then the practical implementation of the control algorithm on a DSP board and the overall PV emulator equipment are described.

Experimental results show that the developed PV emulator is able to reproduce correctly the electrical behavior of a real PV source under any environmental situation, including partial shading and rapidly changing conditions.

The pole placement control technique allows to achieve a dynamics and an output impedance that makes the emulator suitable to be properly used in association with any converter connected at its output. All these features are obtained by a quite cheap equipment.

The whole equipment can be used to test MPPT algorithm performance and PV inverters both in stand alone and grid-connected operation.

Bibliography

Bolzern P, Scattolini R, Schiavoni N (2008) Fondamenti di controlli automatici, McGraw Hill, 3rd edn

D'Azzo JJ, Houpis C (1995) Linear control system analysis and design. Mc Graw Hill, NY, USA

Di Piazza MC, Vitale G (2010) Photovoltaic field emulation including dynamic and partial shadow conditions. Appl Energy 3(87):814–823

Hercog D, Curkovic M, Edelbaher G, Urlep E (2003) Programming of the DSP2 board with the Matlab/Simulink. In: Proceedings IEEE international conference industrial technology, Vol.2 pp 709–713

Ioannidis G, Kandiais A, Manias SN (1998) Novel control design for the buck converter. IEEE Proc Electron Power Appl 145(1):39–47

Li Q, Zhoun K (2010) Introduction to feedback control. Prentice hall, Upper Saddle River, NJ, USA

Ollila J (1995) A medium power PV-array simulator with a robust control strategy. In: Proceedings IEEE power electron specific conference pp 40–45

MIX
Papier aus verantwortungsvollen Quellen
Paper from responsible sources
FSC® C105338

If you have any concerns about our products,
you can contact us on
ProductSafety@springernature.com

In case Publisher is established outside the EU,
the EU authorized representative is:
Springer Nature Customer Service Center GmbH
Europaplatz 3, 69115 Heidelberg, Germany

Printed by Libri Plureos GmbH
in Hamburg, Germany